Advances in Turfgrass Science

Managing Turfgrass Pests

by

Thomas L. Watschke
Peter H. Dernoeden
and
David J. Shetlar

Library of Congress Cataloging-in-Publication Data

Watschke, Thomas L.
Managing turfgrass pests / Thomas L. Watschke, Peter H. Dernoeden, David J. Shetlar.
p. cm. – (Advances in turfgrass science)
Includes bibliographical references (p.) and index.
1. Turfgrasses – Weed control. 2. Turfgrasses – Diseases and pests – Control. 3. Turf management. I. Dernoeden, Peter H. II. Shetlar, David J. III. Title. IV. Series.
SB608.T87W37 1994
635.9′6429 – dc20 94-10164
ISBN 0-87371-999-9

Direct all inquiries to CRC Press, Inc., 2000 Corporate Blvd., N.W., Boca Raton, Florida 33431.

Lewis Publishers is an imprint of CRC Press

International Standard Book Number 0-87371-999-9

Library of Congress Card Number 94-10164

Printed in the United States of America
1 2 3 4 5 6 7 8 9 0

Printed on acid-free paper

Cover Photo Credits:

Front cover, top: Prairie Dunes Country Club. Photo courtesy of P. Stan George.

Front cover, bottom; and back cover, top, and bottom two: Photos courtesy of Thomas L. Watschke.

Back cover, second: Photo courtesy of David J. Shetlar.

Thomas L. Watschke received his BS from Iowa State University in 1967, and his MS (1969) and PhD (1971) from Virginia Polytechnic Institute and State University. He is now Professor of Turfgrass Science at The Pennsylvania State University, where he has been on the faculty for 23 years.

Dr. Watschke teaches four courses in turfgrass management, and coordinates the undergraduate degree program in Turfgrass Science. The principal areas of his research activity are turfgrass weed control, growth regulation of turfgrasses, effects of fertilizers on turfgrass growth and metabolism, turfgrass physiology and microclimate, and roadside vegetation management.

Dr. Watschke has been honored nationally by the Golf Course Superintendents Association of America and the American Sod Producers Association, and has been accorded Fellow status by both the Crop Science Society of America and the American Society of Agronomy. He is recognized throughout the world for his research accomplishments in weed science, plant growth regulation, and water quality. He has made presentations in France, Australia, Scotland, and several locations in Canada, and is active in the International Turfgrass Society.

Peter H. Dernoeden received his BS (1970) and MS (1976) degrees in Horticulture (Turfgrass Science Option) from Colorado State University. He earned his PhD in Plant Pathology in 1980 at the University of Rhode Island, and served in the U.S. Army Field Artillery from 1970 to 1973.

Dr. Dernoeden joined the University of Maryland in 1980, where he is a Professor of Agronomy. His appointment includes research and extension components, and he teaches a course in pest management strategies for turfgrasses. He has a very active integrated pest management research program, which has focused on the impact of moving, irrigation, and fertility on disease severity and weed encroachment in turf. Other research interests have been devoted to disease etiology; nontarget effects of fungicides; chemical control of diseases and weeds; and the management of zoysiagrass and fine leaf fescue species for low maintenance turf sites.

David J. Shetlar obtained his BS in Zoology from the University of Oklahoma in 1969 and studied a little-known group of insects, the Zoraptera, for his MS in Zoology, completed there in 1969. He then entered a PhD program in Entomology at The Pennsylvania State University. After completing his work on the systematics (taxonomy) of the Nearctic Ascalaphidae, a group of insects related to ant-lions, he joined Penn State faculty to teach beginning entomology and perform research on nursery, greenhouse, and Christmas tree pests.

In 1984, Dave joined the staff of Chemlawn Services R&D in Columbus, Ohio as a research scientist working on turfgrass insect management. For the next ten years, he developed improved methods for evaluating turf insect control products and traveled extensively throughout the United States to work on regional insect problems. In 1990, he joined the faculty at The Ohio State University as the Landscape Entomologist.

Dr. Shetlar is a member of the Entomological Society of America and serves as a technical advisory to the Ohio Turfgrass Foundation. He has authored and coauthored trade magazine and journal articles, book chapters, scientific publications, extension fact sheets and bulletins, and a book. His current research emphasis is in development of alternative materials and methods for management of turfgrass insects.

Series Preface

Turfgrass science is comparatively young compared to most areas of applied plant science. The first revolutionary breakthrough was the invention of the reel mower in 1830. Then in 1906 the first book, entitled *Lawns and How to Make Them*, was published by author Leonard Barron. This was followed by a number of books published in the next six decades that were based on the art of turfgrass culture, developed by trial-and-error methods for the most part. A small pool of basic knowledge of turfgrass science was developed through research in the 1920s and then was greatly expanded after 1950. The publication of *Turfgrass: Science and Culture* in 1973 documented a new era in the science of turfgrass culture. The generation of research information and technical developments in turfgrass science has continued to accelerate even more rapidly since then.

Now there is a need for even more specific science-based books that encompass a practitioner orientation for the individual technical areas within the turfgrass science and culture fields. Accordingly, this ADVANCES IN TURFGRASS SCIENCE series has been developed as a coordinated, well-planned series of books encompassing the major areas of turfgrass science, culture, IPM, management, construction, and equipment.

This book, *Managing Turfgrass Pests*, by Drs. Thomas L. Watschke, Peter H. Dernoeden, and David J. Shetlar, represents the current state of applied knowledge concerning the integrated management of pests of turfgrasses. Specific IPM strategies for the major turfgrass weeds, diseases, and insects are discussed, in addition to the overall concepts of integrated pest management. This book provides the basis for strategies in minimizing the development of turfgrass pests through practices that ensure a healthy, competitive turfgrass community with good recuperative potential.

—Dr. James B. Beard
Series Editor

Preface

Turfgrass pests frequently cause a decline in turfgrass quality (both aesthetically and functionally). The decline in quality can be severe enough that the turfgrass stand becomes sufficiently damaged to require costly renovation. The degree of turf deterioration is a function of the grass species, type of pest, and the environmental conditions before, during, and following the pest involvement.

Often pests occur in combination and/or as a result of one another. For example, a disease may weaken a turfgrass plant, making it more susceptible to damage from an otherwise subthreshold level of a particular insect. The resulting stand thinning due to the insect activity could allow weed species to germinate and invade the damaged area. Crabgrass levels are frequently proportional to the amount of foliar blighting associated with common springtime diseases. Such a scenario would be aggravated by the existence of adverse environmental conditions, which is often the case.

Inevitably, pests become part of any ecosystem and managing turfgrasses to optimize their competitiveness is the essence of any pest management program. The presence of pests in a turfgrass stand is usually an excellent indicator that management effectiveness has broken down somewhere. An overall examination of the interactive components of the management system is required before an effective pest management strategy can be implemented. Rather than rely on a particular pesticide for control, successful pest management must include the manipulation and fine tuning of cultural practices so they better address the question of why the pest is present in the first place.

Most turfgrass pest control literature relies heavily on appropriate pesticide use for dealing with pests. While pesticides are an important tool in devising sound pest management programs, they should only be used as a supplement to well coordinated cultural programs. The concept of integrated pest management (IPM) has been practiced by turfgrass managers for decades, and this book will emphasize the philosophy of minimizing pests through well defined and organized cultural practices.

These practices may include changes in mowing height and/or frequency, proper irrigation procedures, renovation to convert to resistant species and/or cultivars, adjustments to nitrogen source as well as fertilizer rate and/or timing of application, improved surface and/or subsurface drainage, soil modification, soil pH adjustment, compaction and traffic control, and others. Without well conceived pest management and cultural programs, the conditions that lead to pest occurrence may not be altered and the pest problem will persist. Without properly adjusting cultural practices, any pesticide solution to a pest problem will not be as efficacious as it could be, and is unlikely to be permanently successful. Unfortunately, when a particular pesticide fails to adequately control a pest, turf managers often are inclined to select an alternative pesticide rather than assessing potential cultural flaws in their management system.

Pesticides are indispensable tools for many pest management situations, particularly when no alternative exists or the pest situation is so bad that a "pesticidal rescue" is required. Most pesticides provide very effective pest control, and their use has evolved in some situations to the point where they are the first line of defense in a pest control program. However, pesticide effectiveness is markedly enhanced when all possible cultural solutions have been imposed. Pesticides often temporarily reduce pest problems, but sometimes the benefits are merely cosmetic. The underlying mismanagement problems that have allowed the pest to become a serious problem in the first place must be resolved. Hence, mismanagement is often the indirect cause of the presence of a pest, and only after corrective changes in management occur will the effectiveness of any pesticide be maximized.

This book has been written in a manner that emphasizes cultural systems. It contains specific recommendations for each pest with respect to integrating all available management tactics. While pesticides are considered a necessary component of managing turfgrass pests, they are referenced in this book in the context of being part of the overall strategy. Proper management systems do not always preclude the need for pesticides, but do enhance their effectiveness and reduce the amount of their use.

To increase the global usefulness of this book, references to pest solutions are made using turfgrass species (warm or cool season) rather than any specific geographic location in the world. However, some pests unique only to a specific region are discussed. It is hoped that any reader will be able to identify with the management recommendations proposed based upon the turfgrass species being managed. This book is intended to assist turfgrass managers in achieving successful pest control through integrating management strategies.

Contents

1: TURFGRASS WEEDS AND THEIR MANAGEMENT

(*Thomas L. Watschke*)

Introduction 1
Managing Summer Annual Grasses 4
Crabgrasses 5
Goosegrass 6
Foxtail 8
Barnyardgrass 9
Fall Panicum 10
Dallisgrass 11
Sandbur 12
Managing Winter Annual Grasses 12
Poa annua 13
Managing Perennial Grasses and Sedges 15
Creeping Bentgrass 15
Tall Fescue 16
Orchardgrass 17
Timothy 18
Smooth Brome 19
Quackgrass 20
Nimblewill 21
Bermudagrass 22
Zoysiagrass 23
Nutsedge 24
Rescuegrass 25
Managing Summer Annual Broadleaf Weeds 26
Oxalis 26
Knotweed 27
Spotted Spurge 28
Purslane 29
Prostrate Pigweed 30
Lambsquarters 31
Puncturevine 32
Ragweed 33

Carpetweed.... 33
Kochia.... 34
Florida Pusley.... 35
Sow Thistle.... 36
Brass Buttons.... 37
Managing Winter Annual Broadleaf Weeds.... 37
Common Chickweed.... 38
Henbit.... 39
Shepherdspurse.... 40
Corn Speedwell.... 41
Bedstraw.... 42
Dog Fennel.... 43
Peppergrass.... 44
Managing Biennials.... 44
Yellow Rocket.... 45
Wild Carrot.... 46
Black Medic.... 47
Managing Perennial Broadleaf Weeds.... 47
Wild Garlic.... 48
Dandelion.... 49
White Clover.... 50
Plantain.... 51
Buckhorn.... 52
Mouse-Ear Chickweed.... 53
Mallow.... 54
Ground Ivy.... 55
Sheep Sorrel.... 56
Canada Thistle.... 57
Chicory.... 58
Curly Dock.... 59
Bull Thistle.... 59
Healall.... 60
Ox-Eye Daisy.... 61
Hawkweed.... 62
Thyme-Leaf Speedwell.... 63
Creeping Speedwell.... 64
Violet.... 65
Stitchwort.... 66
Bindweed.... 67
Bellflower.... 68
Cinquefoil.... 68
Yarrow.... 69
Moss.... 70
Betony.... 71
Beggarweed.... 72

Bur Clover 73
English Daisy 73
Oxalis (Creeping) 74
Weed Management IPM 75
References 76

2: TURFGRASS DISEASES AND THEIR MANAGEMENT

(*Peter H. Dernoeden*)

Introduction 87
Monitoring Disease and Establishing Thresholds 88
Relationship of Environmental Conditions and Cultural Practices to Disease 90
Biological Control of Turfgrass Diseases 95
Collecting and Sending Diseased Samples 98
Winter Diseases 102
Pink Snow Mold or Fusarium Patch 102
Gray Snow Mold or Typhula Blight 104
Yellow Patch or Cool Temperature Brown Patch 105
Low Temperature Pythium or Snow Blight 106
Diseases Initiated in Fall or Spring That May Persist into Summer 106
Red Thread and Pink Patch 107
Helminthosporium Leaf Spot and Melting-Out 108
Take-All Patch 110
Necrotic Ring Spot 113
Spring Dead Spot 114
Large Patch of Zoysiagrass 115
Pythium Root Rot 116
Dollar Spot 117
Stripe Smut and Flag Smut 118
Powdery Mildew 119
Anthracnose 120
Ascochyta and Leptosphaerulina Leaf Blights 121
Yellow Tuft or Downy Mildew 122
Diseases Initiated During Summer That May Persist into Autumn 123
Bermudagrass Decline 123
Brown Patch or Rhizoctonia Blight 124
Pythium Blight 126
Summer Patch 128
Fusarium Blight 130
Southern Blight 130
Helminthosporium Leaf Spot, Melting-Out and

Red Leaf Spot 131
Curvularia Blight 131
Nigrospora Blight 132
Copper Spot 133
Fairy Rings 133
Superficial Fairy Ring 137
Localized Dry Spot 138
Yellow Ring 139
Gray Leaf Spot 139
Cercospora Leaf Spot 140
White Blight 140
Slime Mold 141
Rust 141
Seedling Diseases and Damping-Off 142
Virus Diseases 144
St. Augustine Decline and Centipede Mosaic 144
Bacterial Diseases 145
Bacterial Wilt 145
Plant Parasitic Nematodes 146
Algae and Black-Layer 150
Algae 150
Black-Layer 151
Factors Associated with Fungicide Use 156
Fungicide Use in Lawn Care 158
Fungicide Use on Golf Courses 159
Types of Fungicides 161
Fungicide Application 162
References 170

3: TURFGRASS INSECT AND MITE MANAGEMENT
(*David J. Shetlar*)

Goal of Insect and Mite Management in Turf 171
Pest Identification 173
Pest Life Cycles 177
Turf, A Unique Habitat 180
Monitoring 181
Selecting Appropriate Controls 186
Leaf and Stem Infesting Insect and Mite Pests 206
Bermudagrass Mite 206
Clover Mite 208
Banks Grass Mite 210
Winter Grain Mite 212
Greenbug 214

Sod Webworms (Lawn Moth)—Introduction 217
Cutworms and Armyworms—Introduction.................. 227
Other Turf Infesting Caterpillars 238
Stem and Thatch Infesting Insect and Mite Pests 242
Chinch Bug and Hairy Chinch Bug 242
Southern Chinch Bug 246
Twolined Spittlebug 248
Rhodesgrass Mealybug....................................... 250
Bermudagrass Scale... 252
Billbugs—Introduction 254
Annual Bluegrass Weevil 262
Cranberry Girdler 264
Burrowing Sod Webworms 265
European Crane Fly or Leather Jacket 267
Frit Fly 269
Thatch and Root Infesting Insect and Mite Pests 270
Mole Crickets—Introduction................................. 271
Ground Pearls 279
White Grubs—Introduction.................................. 280
Black Turfgrass Ataenius..................................... 286
Asiatic Garden Beetle... 288
European Chafer... 289
Green June Beetle 291
Japanese Beetle 293
Northern Masked Chafer..................................... 295
Southern Masked Chafer..................................... 297
Oriental Beetle.. 299
May and June Beetles, *Phyllophaga* 300
Nuisance Invertebrate, Insect, and Mite Pests 302
Earthworms 303
Slugs and Snails.. 305
Spiders and Tarantulas.. 306
Chiggers 307
Ticks 308
Sowbugs and Pillbugs... 311
Centipedes 312
Millipedes.. 313
Earwigs... 314
Bigeyed Bugs 316
Leafhoppers... 317
Ground Beetles 319
Rove Beetles.. 321
Fleas 322
March Flies (Bibionids) 324
Ants—General.. 325

Fire Ants 328
Cicada Killer 330
Nuisance Vertebrate Pests 332
Common Grackle 333
Starling 334
Redwinged Blackbird 335
Moles 336
Pocket Gophers 338
Skunks and Civet Cats 339
Raccoon 341
Ninebanded Texas Armadillo 341
References 342

Index 345

Advances in Turfgrass Science

Managing Turfgrass Pests

CHAPTER 1

Turfgrass Weeds and Their Management

Thomas L. Watschke

INTRODUCTION

Much has been written about turfgrass weed control over the years. Weeds are defined as plants out of place. Therefore, some species in turf can be both weeds and desired plants, depending on the situation. Early information emphasized cultural controls that combined best management practices and mechanical techniques to provide the desired species with a competitive advantage. As the chemical age developed, turfgrass weed control took on another dimension—the use of herbicides. Initially, chemicals used in turfgrass weed control were nonselective, mostly acidic compounds that did little to advance the state of the art of chemical weed control. However, after the introduction of 2,4-D in 1944, chemical control quickly became the dominating force in turfgrass weed control programs. Over the past 40 years, many herbicides have been commercialized for use in turfgrass. Chemical weed control utilizes materials that are used for preemergence, postemergence, or total vegetation control. There are contact, systemic, selective, nonselective, and various combinations of such chemicals now available for use. At the end of this chapter, a table containing a listing of herbicides registered in the United States for use on turfgrasses can be found. Proper herbicide selection and use is a function of accurate weed identification and complete familiarity with the herbicide label.

Recommendations for herbicide use can be very specific with respect to location and grass species; therefore, it is vitally important that individual turfgrass managers consult all possible sources of expertise in their area before choosing and applying a herbicide.

Alternatives to herbicides are not as prevalent as alternatives to insecticides and fungicides. Some fungi and bacteria have been investigated for their potential herbicidal effects on certain weeds. Limited success has been achieved; however, interest remains high and research programs continue to investigate the potential of various substances. Recently, a substance identified in corn meal gluten has been shown to control some annual grasses when applied preemergent. However, the amount of nitrogen that accompanies the rate of application needed for commercially acceptable control may be considered excessive. Researchers in turfgrass weed science will continue to develop weed control substances that can be alternatives to herbicides as we know them today.

In the final analysis, successful weed control programs begin with cultural practices that favor the competitive nature of the desired turfgrass species over all others. The existence of weeds indicates a lack of turfgrass health and vigor. Consequently, one or more cultural practices and/or environmental influences have reduced the competitiveness of the desired turfgrass.

Certain weeds serve as excellent indicators of correctable conditions that have favored their development. Examples are as follows:

Weeds as Indicators

- Low pH (sheep sorrel)
- Compaction (goosegrass)
- Low N (legumes)
- Poor soil (quackgrass)
- Poor drainage (sedges)
- Excess surface moisture (algae)
- High pH (plantain)

Therefore, once a weed has been accurately identified, the proper course of action is to determine why a void existed in the turf that allowed the weed to encroach in the first place. The causes for voids generally can be arranged into two groups. The first group could be considered as "natural causes" such as:

Natural Reasons for Voids

- Environmental stresses
- Disruptive use of the area (ballmarks, divots, wear)
- Floods, hail, lightning
- Animals
- Diseases, insects, nematodes

The second group could be considered as management causes for voids that include:

Management Reasons for Voids

- Improper mowing height/frequency
- Improper fertilization (rate/timing)
- Improper irrigation (too little or too much)
- Lack of core cultivation (or poor timing)
- Lack of drainage program
- Lack of soil test information
- Lack of thatch management
- Lack of traffic control

Consequently, the steps for successful weed management are as follows:

Steps in Weed Control Strategy

- Identify weed
- Discern how introduced
- Impose cultural solution
- Use appropriate herbicide, if needed

For chemical control, weeds can be divided into four functional categories: annual grasses, broadleaf weeds, sedges, and perennial grasses. A typical herbicide approach to annual grass control is the use of preemergence herbicides. When preemergence control is not successful, postemergence controls can be used. Postemergence herbicides are generally not as consistent or effective as preemergence materials. Although some annual broadleaf weeds can be controlled preemergence, broadleaf weeds most often are controlled postemergence. Postemergence control of broadleaf weeds can be accomplished selectively, but control of perennial grassy weeds cannot be accomplished as readily using herbicides. Selective control of perennial grasses is the result of spot treatment applications of nonselective herbicides. Selectivity is the result of application placement rather than chemical specificity.

Although the positive effects of herbicides on turf-weed competition are well documented, there is some evidence that herbicide use can cause potentially detrimental side effects. Some preemergence herbicides are known to injure some species more than others. Dinitroanaline compounds have been shown to reduce the re-rooting of some turfs. Injury as a result of preemergence herbicide use has not been reported to have caused increased encroachment by the target weed.

Due to the effectiveness of existing herbicides and combinations thereof, most weeds in turfgrass can be controlled. However, those species that have no known chemical control are becoming more serious problems on sites that have had the more easily controlled weeds removed. Some speedwell species, for example, are more prevalent now than twenty years ago. Future research will be required to develop cultural and chemical strategies for these weeds. Therefore, herbicide use has partially been a selecting force in the population dynamics of turfgrass swards.

MANAGING SUMMER ANNUAL GRASSES

Summer annual grasses are tropical plants that persist and are competitive with turf during frost-free periods. They depend upon seed production as the sole means of propagation and survival. Examples of summer annuals are smooth crabgrass, large crabgrass, barnyardgrass, dallisgrass, fall panicum, green and yellow foxtail, goosegrass, sandbur, and rescuegrass. In general, summer annual grasses begin to germinate when the soil temperature in the vicinity of the seed has been 55°F (13°C) for four or five consecutive days. A soil thermometer is an inexpensive and valuable management tool for monitoring soil temperature. It is important to monitor soil temperature near the surface (it may require angular insertion of sensing unit) where most of the previous year's seed production was deposited. The majority of any given summer annual grass population is mostly the result of the germination of seed produced the previous season. Of the summer annual grasses listed, goosegrass requires higher soil temperatures for germination than smooth crabgrass.

Smooth crabgrass, goosegrass, dallisgrass, sandbur, and rescuegrass are more commonly problems in established turf while the foxtails and barnyardgrass are more often a problem during turf establishment. Foxtails, barnyardgrass, and dallisgrass produce their seed on culms that grow taller than most turf is mowed. Consequently, their reproductive capability in established turf is quite limited.

Chemical control of summer annual grasses with the exception of dallisgrass is best achieved by using preemergence herbicides. Proper timing is imperative for successful control. Application of preemergence herbicides should precede seed germination by several days. Rainfall and/or irrigation are required to distribute the active ingredient of the herbicide into the upper soil profile. A chemical barrier is created that is toxic to the germinating summer annual grassy weeds. This barrier will last from 6 to 12 weeks or longer, depending upon the chemical. Chemical residual is a function of biodegradability, which is influenced by weather (particularly rainfall) and microbial activity. Preemergence herbicide effectiveness can be influenced by physical factors as well. Any soil disturbance, from earthworm activity to core cultivation, will disrupt the chemical barrier. For a high degree of control, the integrity of the chemical barrier should be maintained.

Controlling crabgrass in cool season grasses can usually be accomplished with a single application of preemergence herbicide. In climatological zones located near the high temperature limit of cool season grass growth, two applications of preemergence herbicide are often required. In instances where an herbicide has provided excellent control of at least 95%, half rates of the same herbicide can be used in subsequent years and commercially acceptable control can be attained.

Postemergence control of summer annual grasses that have invaded cool-season grasses can be accomplished using organic arsenicals, fenoxaprop, and dithiopyr. Postemergence control, however, is not as effective as preemergence control. Organic arsenicals require repeated application, sometimes three times, to gain control. Summer annual grasses have a lower tolerance of arsenic than cool-season turfgrasses. The first application causes a distinct yellowing of the weedy grasses, but does not result in control. Repeated applications (10 to 14 days apart) are necessary, but increase the risk of injury to the desired species. The desired grasses also are susceptible to the toxic effects of arsenic, but at a higher internal concentration. Postemergence control of annual grassy weeds in bermudagrass can be achieved by combining metribuzin with organic arsenical. In St. Augustinegrass, asulam is required for postemergence control of crabgrass and goosegrass.

The following descriptions and control strategies are provided for the summer annual grasses.

CRABGRASSES

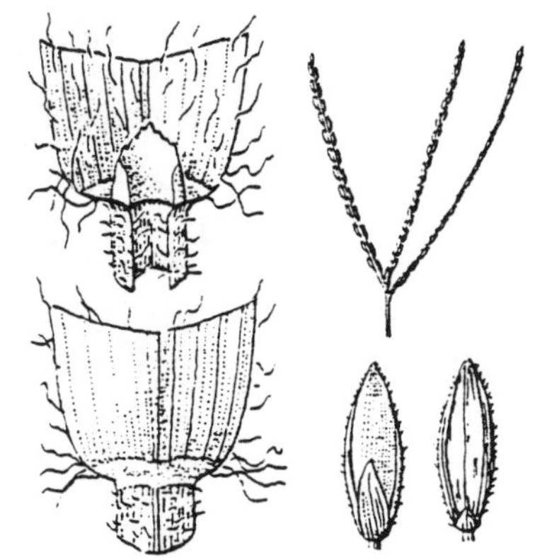

Digitaria sanguinalis L. (Scop)
(Large or hairy crabgrass)

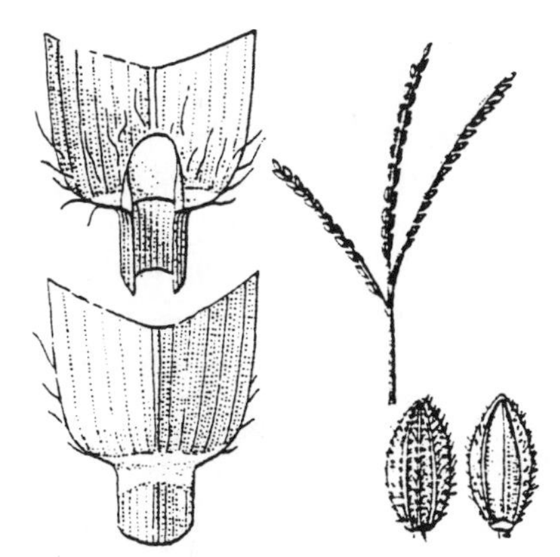

Digitaria ischaemum Schreb.
(Smooth crabgrass)

Description

Smooth and large crabgrass are warm-season annuals reproducing by seed. They are light green and have a prostrate growth habit. Smooth crabgrass is smaller and less hairy than large crabgrass. Crabgrass leaf blades are short, pointed, hairy to sparsely hairy, and rolled in the bud. The leaf sheaths are split, compressed and sparsely hairy. Auricles are present. The collar is broad with hairs along the margin. Ligule is large, membranous, and toothed for large crabgrass and is smooth for smooth crabgrass. Seedhead consists of 3 to 9 spikes atop the main stem. Both species can root at culm nodes.

Cultural Strategies

Crabgrass species are relatively easy to manage through sound cultural practices, use of leaf spot resistant species for cool season grasses, and the proper use of preemergence herbicides when needed.

Strategy I: Avoid establishing turf at the time crabgrass normally is germinating (soil temperature near the surface of 55°F [13°C]). Avoid verticutting or core cultivation at the time crabgrass is germinating or after application of a preemergence herbicide. Irrigate to avoid wilting of turf.

Strategy II: Raise mowing height slightly at the time crabgrass seed is germinating. Lower height of cut slightly and collect clippings during time that crabgrass is setting seed.

Strategy III: Reduce available nitrogen when crabgrass is most competitive (tropical-type weather) via timing and low application rates of soluble nitrogen sources and/or proper selection of slow release nitrogen sources.

Strategy IV: Use a preemergence herbicide when soil temperature under turf areas has been 55°F (13°C) for four to five consecutive days (at one-half inch [13 mm] soil depth). If seed germination has occurred, use a postemergence herbicide according to label and local recommendations or combinations of pre- and postemergence products.

GOOSEGRASS

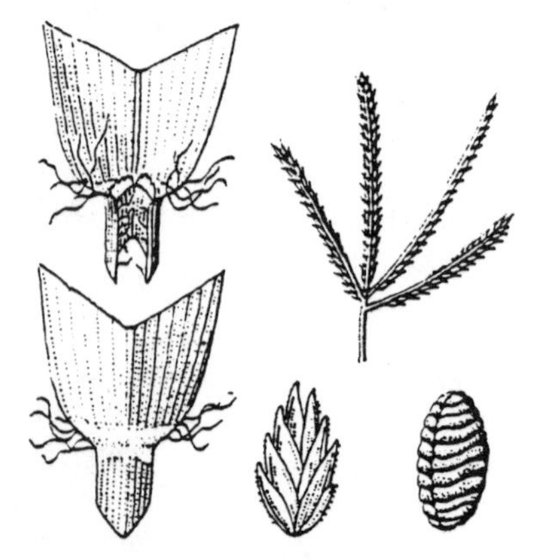

Eleusine indica (L) Gaertn.

Description

Goosegrass is a coarse-textured summer annual, with leaves folded in the bud, that generally germinates later than smooth crabgrass. Soil temperatures required for germination are 60 to 65°F (15 to 18°C) for 12 to 15 consecutive days in the upper one-half inch (13 mm) of the soil. Goosegrass has a cluster of tillers arising from the central part of the plant. These tillers typically grow prostrate giving the plant a rosette-like appearance. The leaf sheath is strongly flattened and usually whitish to silvery in color, which has given rise to the common name silver crabgrass in some locations. However, goosegrass as the common name is more frequently used. The most distinguishing feature of goosegrass is the seedhead. Spikes, 3 to 10 in number, radiate from the end of the seed stalk in a finger-like appearance.

Cultural Strategies

Goosegrass is particularly competitive during the summer months and in compacted soil conditions. Proper cultural practices and the use of preemergence herbicides can provide reasonably good management of goosegrass. Where few plants are competing, the most economical and practical cultural solution is to physically remove plants with a knife.

Strategy I: See Crabgrass.

Strategy II: See Crabgrass.

Strategy III: Relieving soil compaction through core cultivation or by reducing or redirecting traffic will increase the competitiveness of the desired turfgrass species. Goosegrass grows better in noncompacted soil than in compacted soil, but competition from desired species is considerably greater when soils are not compacted.

Strategy IV: Apply preemergence herbicide approximately 14 to 17 days later than for smooth crabgrass. If lilac is present, petal fall is a fairly reliable indicator for proper timing of application. Check herbicide label for goosegrass control rates, as they frequently differ from those for smooth crabgrass. Do not core cultivate following application. Postemergence control generally is less effective than for smooth crabgrass, but postemergence treatment prior to tillering can be very successful.

FOXTAIL

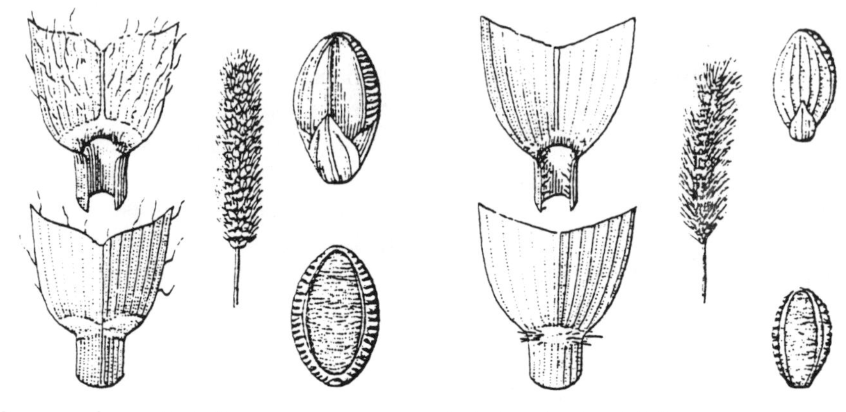

Setaria glauca (L.) Beauv.
(Yellow foxtail)

Setaria viridis (L.) Beauv.
(Green foxtail)

Description

Foxtails are bunch-type summer annuals that are rolled in the bud and do not persist in established turf. Yellow foxtail is encountered more commonly in turf and tends to be more persistent than green foxtail. The primary distinguishing feature between yellow and green foxtail is the hair on the sheath. Yellow foxtail is smooth to sparsely hairy, while green foxtail is hairy along the sheath margin. Seeds are formed in very dense panicles that resemble "foxtails" in mid to late summer. Foxtails germinate when soils warm to above 65°F (18°C).

Cultural Strategies

Foxtails are not severely competitive in established turf, but can seriously compete with desired species when they are being established from seed.

Strategy I: See Crabgrass.

Strategy II: Lower mowing height slightly when foxtails are producing seedheads and collect the clippings. At turf mowing heights of 2 inches (50 mm) or lower, foxtails do not produce large quantities of viable seed.

Strategy III: See Crabgrass.

Strategy IV: Since foxtails are primarily a problem during turf establishment, choice of a preemergence herbicide is restricted to that which does not inhibit seed germination of the desired species. When foxtails have been a problem during the establishment year, a preemergence herbicide should be used in the spring of the following year. Postemergence herbicides are effective on foxtails, but desired species in the juvenile stages often are sensitive to applications of postemergence annual grass herbicides.

BARNYARDGRASS

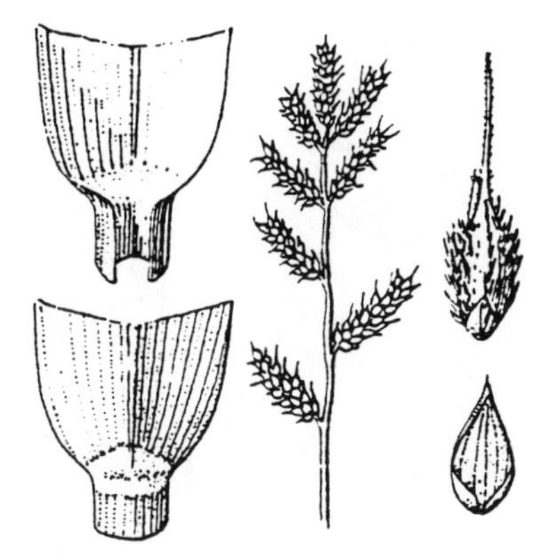

Echinochloa crusgalli (L.) Beauv.

Description

Barnyardgrass is an erect, bunch-type coarse-textured summer annual that typically is shallow rooted and not competitive in established turfs. Barnyardgrass leaves often appear reddish to purple and the seedheads appear reddish. Barnyardgrass seeds often have very long awns that are quite conspicuous when observing the seedhead. The upright growth habit and production of seed at several inches above the soil surface limits the competitiveness of barnyardgrass to the turf establishment phase. Barnyardgrass germinates in mid to late spring once soils have warmed to 60 to 65°F (15 to 18°C).

Cultural Strategies

Barnyardgrass is not competitive in established turf, but can seriously compete when the desired turfgrass species are being established.

Strategy I: See Crabgrass.

Strategy II: See Foxtails.

Strategy III: See Crabgrass.

Strategy IV: See Foxtails.

FALL PANICUM

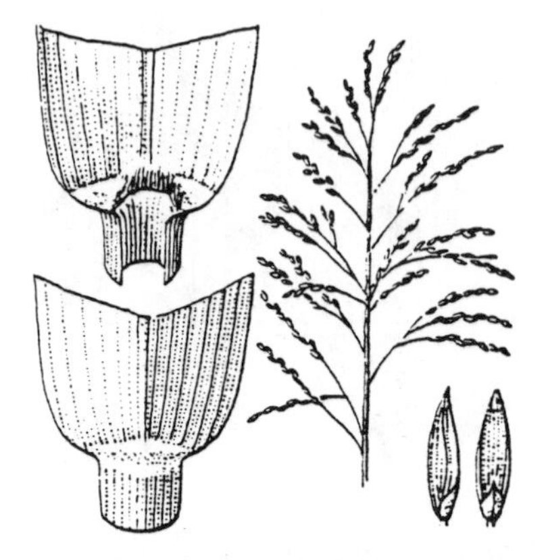

Panicum dichotomiflorum Michx.

Description

Fall panicum is a divergently branching, bunch-type summer annual that is shallow rooted and is not competitive in established turfs. It germinates when soils warm to 60°F (15°C) and can be troublesome during turf establishment. The seed stalk is bent abruptly at joints near the base and the seedheads are panicles that are very open. The leaves of fall panicum are rough to the touch. Production of viable seed frequently does not occur in turf maintained at 2 inches (50 mm) or lower.

Cultural Strategies

Fall panicum does not compete well in established turfs, but can cause problems when desired species are established during a change of season from conditions favoring cool season grasses to that which favors warm season grasses.

<u>Strategy I:</u> See Crabgrass.

<u>Strategy II:</u> See Foxtails.

<u>Strategy III:</u> See Crabgrass.

<u>Strategy IV:</u> See Foxtails.

DALLISGRASS

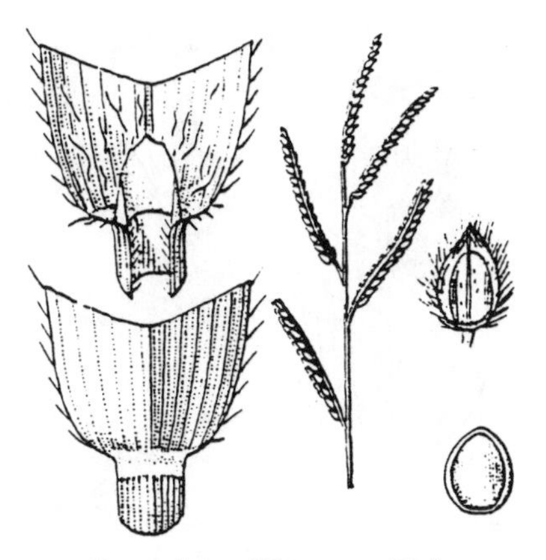

Paspalum dilatatum Poir.

Description

Dallisgrass is a coarse-textured, bunch-type grass that spreads primarily by seed, but may have short rhizomes. Lower leaves may be hairy, while all are noticeably shiny. The seed stalks of dallisgrass are distinctive and contrast dramatically in desired turf, particularly at higher heights of cut. The seedheads are loosely ascending with spikelets distinctly egg-shaped and tapering to a point. The seeds are arranged along the spikelet similar to goosegrass.

Cultural Strategies

Dallisgrass occurs more commonly throughout warmer climate regions. Dallisgrass and hairy crabgrass invade more turf acreage in warmer climates than any other undesirable grasses.

Strategy I: Avoid core cultivation when soils are warming into the range of 60–65°F (15 to 18°C) in mid to late spring when dallisgrass is germinating.

Strategy II: Maintain mowing heights as close to the lowest level tolerated by given turfgrass species during the time dallisgrass is producing seedheads.

Strategy III: Use a preemergence herbicide when appropriate. Postemergence control in St. Augustinegrass and centipedegrass cannot be accomplished with currently available herbicides. Postemergence control in bermudagrass and zoysiagrass can be accomplished with disodium and monoammonium arsenates.

SANDBUR

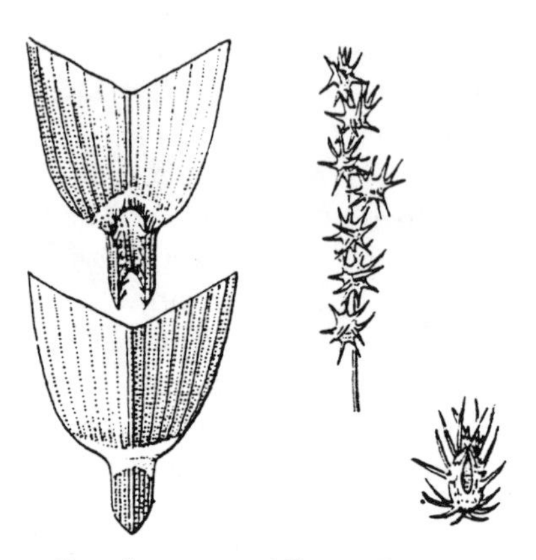

Cenchrus pauciflorus Benth.

Description

Sandbur is a low-growing summer annual grass that produces seed in a spiny bur which is extremely sharp. The burs are formed on short spikes, the seed stalks being decumbent to erect, sometimes with many spreading branches.

Cultural Strategies

Sandbur is most competitive on sandy soils or in conditions of low rainfall (< 50 cm). It does not commonly produce a dense cover. Cultural practices that improve stand density will limit encroachment by sandbur.

Strategy I: Modification of sandy sites with organic amendments to improve water-holding capacity prior to turf establishment will improve competitiveness of the desired turf species.

Strategy II: Maintain proper irrigation management to avoid loss of competitiveness from desired turf species due to moisture stress that may induce dormancy.

Strategy III: Use slow-release sources of nitrogen or frequent light applications of soluble nitrogen sources (foliar feeding) to maintain vigor and competitiveness of desired species.

Strategy IV: Use a preemergence herbicide in early spring or a postemergence herbicide once sandbur is actively growing in the juvenile state.

MANAGING WINTER ANNUAL GRASSES

Winter annual grasses are temperate species that germinate predominately when soil temperatures are below 60°C (15°C). Germinated seedlings survive as juvenile plants during winter months, and reproduce via seed set in late spring. Annual bluegrass is the dominate winter annual grassy weed in the United States. In accordance with the definition of a weed being a plant out of place, annual bluegrass may or may not be one, depending on whether the management personnel considers it out of place.

In many locations where temperatures are usually moderate, annual bluegrass can be competitive throughout the season. As a result, it is not unilaterally considered to be a weed. Indeed, many very high quality turf sites contain a high percentage of annual bluegrass. In warmer climatic zones such as the southern United States, annual bluegrass is almost unilaterally considered to be a weed in dormant warm-season turf species.

Control of annual bluegrass is best achieved through cultural and chemical practices intended to slow its competitiveness, while increasing the competitiveness of desired turf species.

More research efforts have been targeted at annual bluegrass than any other winter annual grass species. Successful control has been more readily attained in the southern United States through the use of herbicides that are not labeled for application on cool-season turfgrasses because of low tolerance.

The following descriptions and control strategies are provided for annual bluegrass.

POA ANNUA

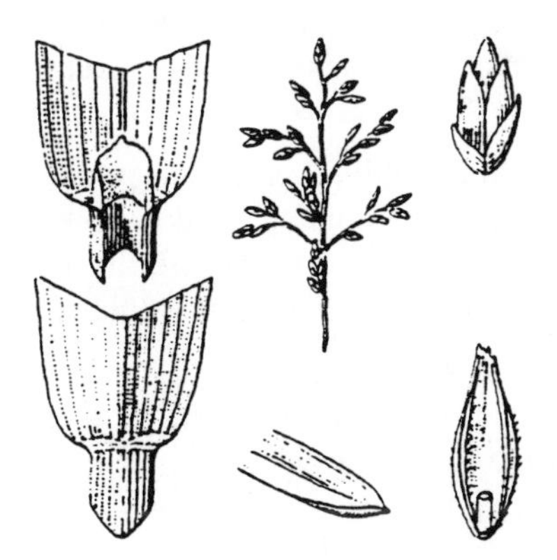

Poa annua var. *annua* L. Timm (annual type)
Poa annua var. *reptans* (Hauskins) Timm (perennial type)

Description

Annual bluegrass is mostly bunch-type in growth habit, although the subspecies of a perennial type frequently has stolons. Stoloniferous plants usually are found in irrigated, closely mowed situations. Annual bluegrass is folded in the bud and can be distinguished from Kentucky bluegrass very readily by comparing ligules. Annual bluegrass has a very conspicuous, membranous ligule, while Kentucky bluegrass has a ligule that is a very short, blunt membrane. Annual bluegrass can be easily distinguished from creeping bentgrass because bentgrass is rolled in the bud and the veins on the blade are prominent. Most of the time, estimates of annual bluegrass are made in late spring when a majority of the plants have seedheads. Annual bluegrass can produce viable seed at all cutting heights, even as low as a golf course putting green. Seedheads are a pyramidal-like panicle.

Cultural Strategies

Annual bluegrass is most competitive under conditions of close mowing, frequent irrigation, moderate to heavy nitrogen fertilization during cool weather, and moderate soil compaction. Cultural practices employed to minimize conditions favoring annual bluegrass must be used in combination with herbicides and/or plant growth regulators.

Strategy I: Correct any drainage problems, raise height of cut when possible, even a millimeter can help, fertilize to favor desired turf species (for creeping bentgrass, nitrogen fertility should be applied during warmer months), and allow turf to wilt during periods of moisture stress. Annual bluegrass does not have a dormancy mechanism for drought, while other cool-season grasses may become dormant and will recover when irrigated. Annual bluegrass that is allowed to go to the permanent wilting point dies due to the inability to go dormant.

Strategy II: Renovate turf stand by establishing turf-type perennial ryegrass as the predominant species, and use ethofumesate according to label instructions as a pre-/post-emergence herbicidal control.

Strategy III: Avoid turf cultivation during periods when soil temperature has consistently cooled causing annual bluegrass seed to germinate. Maintain maximum shoot density during this time of the year, as annual bluegrass seedlings require space and light to become established.

Strategy IV: Use mowing equipment that causes the least amount of turf wear and soil compaction. Collect clippings, especially during the peak seed production period of the year (late spring). Light verticut and collect clippings when seedheads are being produced.

Strategy V: Apply a preemergence herbicide prior to weed seed germination. Two applications may be required during the dormant season in the southern United States. Postemergence control is attained most readily in the southern United States through the use of a herbicide that is not tolerated by cool-season species or by using a nonselective herbicide when bermudagrass is dormant. Applications of growth regulators labeled for conversion of turf from *Poa annua* to creeping bentgrass may be made per label instructions. Successful conversion has been inconsistent and appears to be associated with the length of time that applications have been made. In locations where at least two applications are made for a period of four to five years, noticeable conversion has been observed. However, growth regulator use must be done in conjunction with the other strategies outlined above (except II).

MANAGING PERENNIAL GRASSES AND SEDGES

Perennial grasses are extremely difficult to control because they frequently have morphology and physiology similar to the desired species. Indeed, perennial grassy weeds also are considered as desired species in certain situations. Selective postemergence herbicides are not readily available.

As with annuals, perennial weedy grasses can be either a cool- or warm-season species. Examples of cool-season species are tall fescue, bentgrass, orchardgrass, quackgrass, timothy, and velvetgrass; while examples of warm-season species are bermudagrass, nimblewill, Johnsongrass, and zoysiagrass. Perennial grasses are considered weeds because of their leaf texture, frost sensitivity, means of spread (growth habit), mowability, etc. They cause a lack of uniformity in the sward, thus significantly reducing overall quality.

Perennial grassy weed problems often can be avoided during turf establishment, as seed from such species may occur in mulching materials; therefore, using clean mulch, certified seed sprigs, plugs, or sod, clean topsoil, and not using manure in the seedbed is recommended. Seedbeds containing vegetative propagules of unwanted perennial grasses, usually rhizomes, can be fumigated or treated with nonselective total vegetative herbicides prior to the seeding of desired turf species.

The following descriptions and control strategies are provided for perennial grasses.

CREEPING BENTGRASS

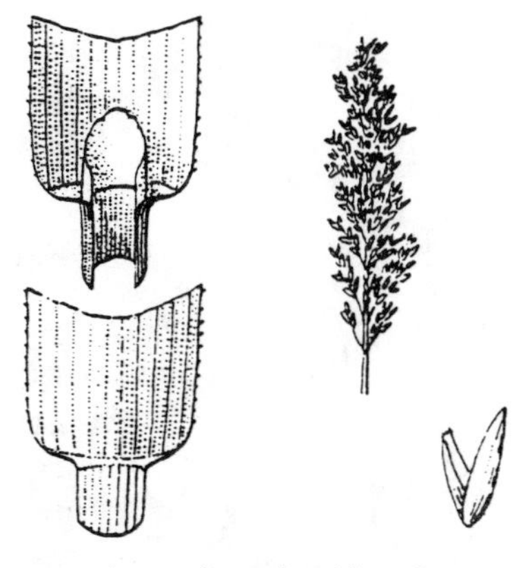

Agrostis stolonifera L.

Description

Creeping bentgrass is rolled in the bud, having leaves tapering to a point with prominent veins. It spreads through prolific stolon production that makes it incompatible with bunch-type species or those spreading by rhizomes. Patches of bentgrass become prevalent in the stand. Creeping bentgrass turf quality deteriorates when mowed above one-half inch (13 mm). Therefore, creeping bentgrass use is restricted to monostands that are intensively maintained.

Cultural Strategies

Creeping bentgrass should only be seeded with other creeping bentgrasses. When creeping bentgrass begins to encroach into a desired turf, physical removal with care taken to remove all stolons is required. Avoid close mowing of other species, particularly Kentucky bluegrass.

Strategy I: Mow desired species as high as feasibly possible. Fertilize moderately with complete fertilizer twice a year, with at least two-thirds of the yearly nitrogen applied in the fall. Lime to a pH of at least 6.7 if needed, as creeping bentgrass is more competitive under acidic soil conditions.

Strategy II: Spot treat with translocatable total vegetative killer herbicide. Treat an area at least twice the size of the visible patch of creeping bentgrass. Stolons spreading outward from the visible patch are underneath the canopy of the desired species and often can escape contact with the herbicide.

Strategy III: Fumigate the site and establish desired turf species from seed or vegetative means.

TALL FESCUE

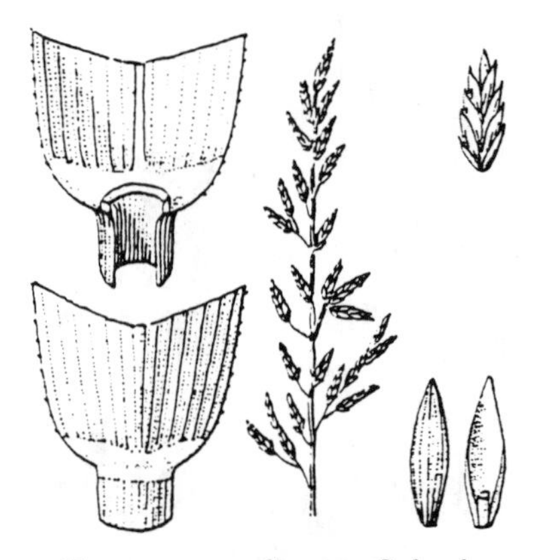

Festuca arundinacea Schreb.

Description

Tall fescue is a bunch-type, coarse textured perennial grass. It is rolled in the bud with prominently veined, long leaves that are shiny underneath. When seeded heavily alone, tall fescue can provide an acceptable turf. New turf-type tall fescues appear to be reasonably compatible with turf-type perennial ryegrasses, but not with rhizomatous-type species.

Cultural Strategies

Tall fescue usually can be eliminated as a problem during turf establishment by using certified seed of desired species and by not mixing tall fescue with any other grass. If an established site becomes infested with tall fescue, a different strategy is required.

Strategy I: Eliminate clumps of tall fescue from the seedbed by physical removal or soil screening. Use seed that is free of contamination by tall fescue. No percentage of tall fescue is allowable in seed mixtures containing fine textured species.

Strategy II: Fumigate site contaminated with tall fescue or use a translocatable nonselective total vegetative control herbicide as a spot treatment. Use topical applications of grass herbicides labeled for tall fescue control; do not apply such herbicides to desired turf.

ORCHARDGRASS

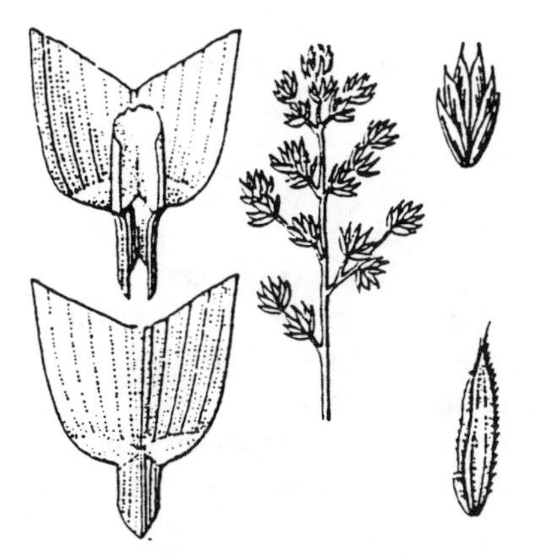

Dactylis glomerata L.

Description

Orchardgrass is a coarse textured, bunch-grass perennial grass. The leaves are light green, tapering to a boat-shaped tip, and are not prominently veined. The leaves are folded in the bud and the ligule is a prominent membrane.

Orchardgrass is not compatible with other species because of leaf texture and very clumpy growth habit.

Cultural Strategies

Orchardgrass is most easily eliminated as a problem by using certified seed for turf establishment and clean straw as a mulching source.

Strategy I: See tall fescue.

Strategy II: See tall fescue.

TIMOTHY

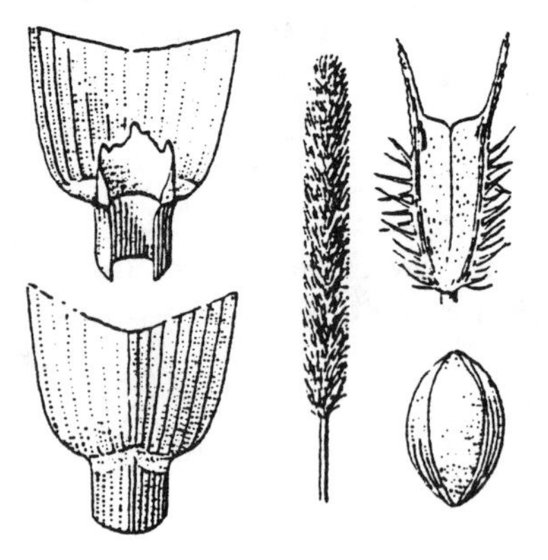

Phleum pratense L.

Description

Timothy is a coarse textured, bunch-type perennial grass. The leaves are rolled in the bud and the sheaths are compressed. The ligule is a large notched membrane. The base of each tiller usually is swollen or bulb-like, and the seedheads are a dense spike 2 to 5 inches (50 to 125 mm) long. Timothy is not compatible with fine textured grasses because of leaf texture and clumpy growth habit. European turf-types are used in some lawn sites and on athletic fields, but not typically in mixtures with other species.

Cultural Strategies

Timothy is most easily eliminated as a problem by using certified seed for turf establishment and clean straw as a mulching source.

Strategy I: See tall fescue.

Strategy II: See tall fescue.

SMOOTH BROME

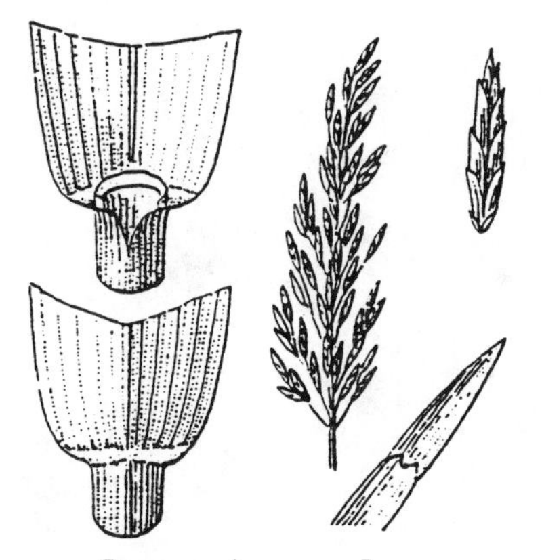

Bromus inermus Leyss.

Description

Smooth bromegrass is a dull-green, rhizomatous, coarse textured perennial grass. Leaves are rolled in the bud, and the sheaths are compressed with hairs near the bottom. The wide leaf blades have a constriction on the upper leaf surface resembling the letter “M” or “W”. Smooth bromegrass is considered a weed because of an open coarse leaf texture which makes it incompatible in mixtures containing fine textured species.

Cultural Strategies

Smooth bromegrass is most competitive under low maintenance conditions. Lowering the mowing height and increasing nitrogen fertility will reduce the competitiveness of smooth bromegrass and increase the density of desired turf species.

Strategy I: See tall fescue.

Strategy II: See tall fescue.

QUACKGRASS

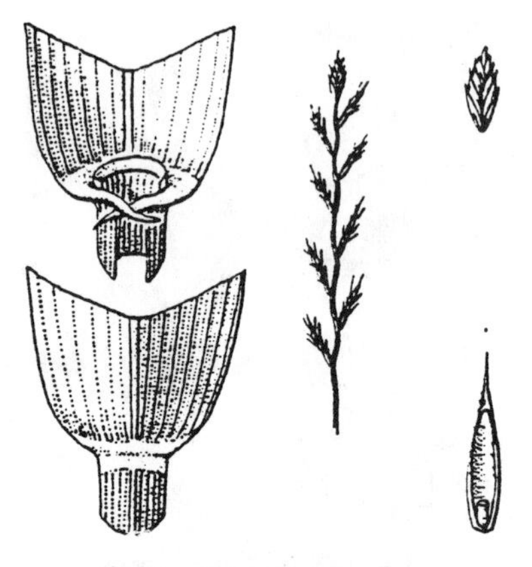

Agropyron repens L.

Description

Quackgrass is a coarse textured, rhizomatous perennial grass. The leaves are rolled in the bud and the auricles are conspicuously clasping. Quackgrass has been shown to have allelopathic effects on other species that increases its competitiveness. The internodal length on quackgrass rhizomes is quite long compared to many other rhizomatous grasses; therefore, quackgrass does not produce a dense stand.

Cultural Strategies

Quackgrass can spread by seed or rhizomes from areas adjacent to turf sites. However, quackgrass is stressed by high temperature and can only encroach areas where the turf is thin. It often is introduced along with balled and burlap nursery stock into landscaped settings.

Strategy I: See tall fescue.

Strategy II: See tall fescue.

Strategy III: Mow at the lowest point of the tolerance range for the desired turf species. Increase nitrogen fertility, particularly during the warmest weather, and be sure the nursery stock and topsoil sources are free from quackgrass contamination.

NIMBLEWILL

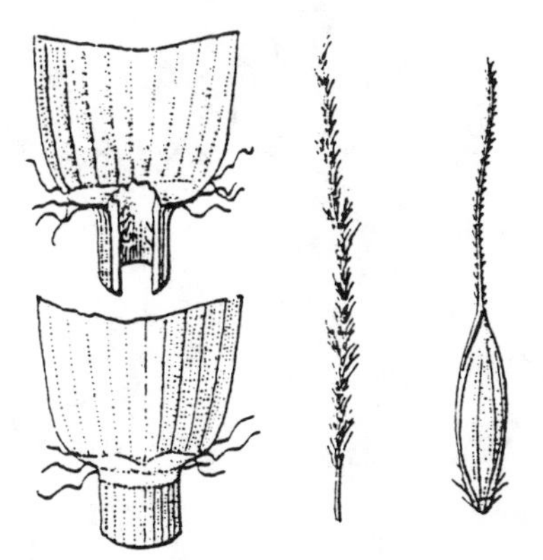

Muhlenbergia scherberi Guel.

Description

Nimblewill is a relatively fine textured warm-season perennial that spreads by seed, stolons, and rhizomes. Frost will cause nimblewill to go off color, leaving brownish patches throughout the turf. Nimblewill is rolled in the bud and the leaves are short, flat and come to a point. The leaves are oriented horizontally, providing the illusion that nimblewill is dense; however, the actual shoot density (plants per unit area) is not particularly good. The growth habit and frost sensitivity make nimblewill undesirable in most other species.

Cultural Strategies

Nimblewill patches are extremely difficult to manage and competitiveness is severe, particularly when turf is underfertilized with nitrogen. Nimblewill does not invade closely mowed sites (< 13 mm).

Strategy I: Increase overall nitrogen fertility level, particularly when frost has forced nimblewill into dormancy.

Strategy II: Apply a translocatable, nonselective herbicide to nimblewill when it is actively growing. Be sure application is to twice the area of observed nimblewill infestation to ensure the entire infestation has been treated.

BERMUDAGRASS

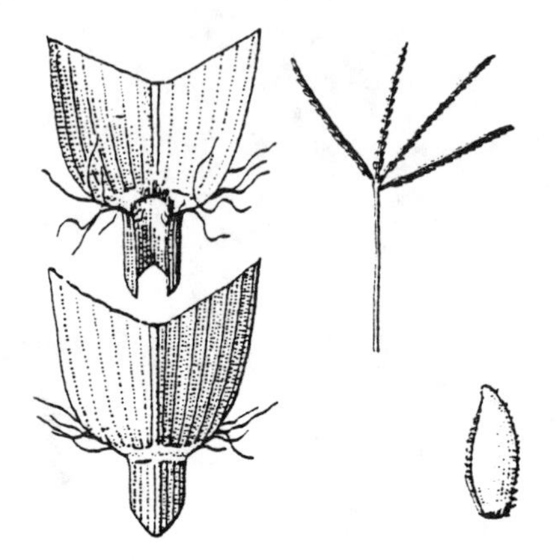

Cynodon dactylon L.

Description

Bermudagrass is a warm-season perennial that spreads by seeds, stolons, and rhizomes. Like nimblewill, bermudagrass is forced into dormancy by chilling temperatures of < 55°F (13°C) and the resulting straw-like color is not compatible when mixed with cool-season turfgrasses. Bermudagrass is folded in the bud and the ligule is a fringe of hairs.

Cultural Strategies

Bermudagrass is difficult to manage once it is established and it is strongly competitive during the summer months.

Strategy I: See nimblewill.

Strategy II: See nimblewill.

Strategy III: Physical barriers (edging, geotextiles, etc.) can impair the spread of bermudagrass into areas where it is not wanted. To reduce bermudagrass encroachment into St. Augustinegrass, maintain a cutting height of 2 to 3 inches (50 to 75mm) so that St. Augustinegrass shades out bermudagrass.

ZOYSIAGRASS

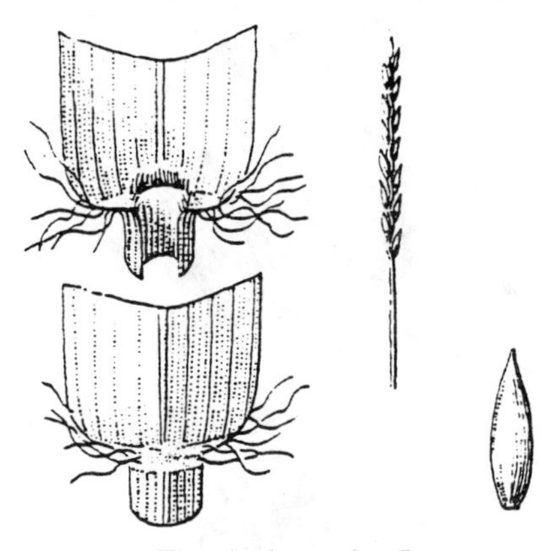

Zoysia japonica L.

Description

Zoysiagrass is a warm-season perennial that spreads by stolons and rhizomes. Like nimblewill and bermudagrass, zoysiagrass is forced into dormancy near freezing temperatures and the resulting straw-like color is not compatible with cool-season grasses. Zoysiagrass is rolled in the bud and slower growing than bermudagrass. Although slow to establish, once it is established, zoysiagrass is extremely competitive.

Cultural Strategies

Zoysiagrass is difficult to eliminate once it is established and is strongly competitive during the warmest months.

Strategy I: See nimblewill.

Strategy II: See nimblewill.

Strategy III: See bermudagrass.

NUTSEDGE

Cyperus esculentus L. (yellow)
Cyperus rotundus L. (purple)

Description

Sedges are grass-like plants that are perennial with three-ranked and triangular stems. Yellow nutsedge, sometimes called nutgrass, is more widely encountered in the United States, with purple nutsedge more commonly found in the southern states. Yellow nutsedge is pale green to yellow in color and has a rapid vertical shoot growth rate in the spring and early summer. Sedges are spread by seed, rhizomes, and nutlets, and are good indicators of poor drainage. Purple nutsedge can be distinguished from yellow nutsedge by its brownish to purple seedhead. Yellow nutsedge has a straw-colored seedhead.

Cultural Strategies

Sedges are somewhat difficult to eliminate from turf, but are less competitive in dense turf situations. Keeping turf from stress during the times that nutsedge is most vigorous will prevent significant encroachment.

Strategy I: Improve drainage, both surface and subsurface. Avoid topsoil from poorly drained sites that likely has nutsedge propagules.

Strategy II: Lower height of cut (on species that will tolerate < 20 mm) and increase mowing frequency during rapid nutsedge growth.

Strategy III: When nutsedge has invaded cool season turf, fertilize with nitrogen in the fall after frosts have become frequent. Nutsedge vigor is reduced by frost more than most cool-season turfgrasses.

Strategy IV: Use a postemergence herbicide recommended for your area during periods of rapid nutsedge growth.

RESCUEGRASS

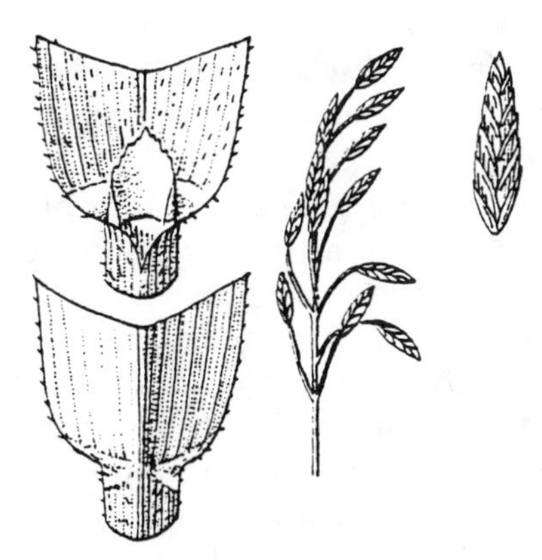

Bromus catharticus Vahl.

Description

This basal tillering perennial grass is predominately a weed problem in areas of warm season grass adaptation. Originally used as a forage, it has subsequently invaded warm-season turfgrasses. The leaves are wide, flat, and long with a prominent membranous ligule. Rescuegrass is mainly a problem during cooler weather when competition from warm-season turf species is minimal. Rescuegrass does not compete as well during hot weather.

Cultural Strategies

Commonly a problem when warm-season turf species are vegetatively established, rescuegrass can be physically removed when the infestation is spotty and not severe.

Strategy I: Maintain shoot density through proper cultural practices, as rescuegrass requires space for seed germination and good seedling development.

Strategy II: Use a triazine herbicide when rescuegrass has invaded St. Augustinegrass, zoysiagrass, centipedegrass, or carpetgrass.

Strategy III: Spot treat with a nonselective translocated total vegetation killer herbicide or with a plant growth regulator.

MANAGING SUMMER ANNUAL BROADLEAF WEEDS

Summer annual broadleaf weeds germinate when soils first thaw and throughout soil warming to 70°F (21°C), depending on the species. Most species are frost sensitive and therefore germinate only after frost danger has past. Regardless of the seed germination date, summer annual broadleaf weeds are most competitive with turfgrasses during the warmer months when they are in the vegetative stage of growth. As day length shortens, flowering and seed set occur and the seeds ripen and shatter usually before the first killing frost. Many species are killed by frost and, in some years, early frosts can reduce seed set.

For cool season turfs, summer annuals can be avoided by establishing the turf when frosts can occur, such as in the northern United States. In such locations, late spring and summer turf establishment of cool-season grasses often results in poor stands due to the competition from broadleaf summer annuals. Thinning caused by summer diseases and insect problems on established cool season turf sites often results in the invasion of summer annual broadleaf weeds. Preemergence herbicides are effective for controlling some broadleaf summer annuals, but timing of application is critical. Since the effective preemergence herbicides commonly are used for crabgrass, a second application for later germinating summer annuals may not be appropriate. Consequently, most summer annual broadleaf weeds are controlled using postemergence herbicides. Summer annuals can also be a problem when establishing warm season species. Closer spacing of plugs and sprig rows helps reduce invasion, but most summer annuals are controlled using postemergence herbicides after establishment.

OXALIS

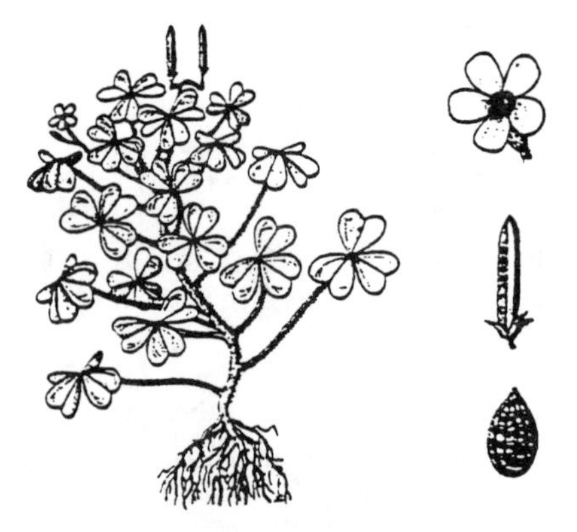

Oxalis stricta L.

Description

Oxalis is a tri-foliate, warm-season annual. The stems branch from the base and are slightly hairy. The leaflets are pale green to yellowish and are somewhat heart-shaped. The flower petals are vivid yellow and the seed forms in five-sided pods that are pointed. At maturity the seed pods burst and can disseminate seeds up to several feet (> 0.6 m). The leaflets contain calcium oxalate which gives them a sour or acidic taste. Rooting may occur at nodes when stems are prostrate, but new plants do not form at the rooting site.

Cultural Strategies

Oxalis is best avoided in cool season grasses by maintaining a dense stand of turf during warm weather when the seed most commonly germinates. Hand removal of a few isolated plants when first noticed can impede encroachment.

Strategy I: Physically remove when sparse in turf. Also, remove from flower beds and landscape plantings, as the adjacent turf areas can be readily infested due to exploding seed pods.

Strategy II: Control insects that may weaken the cool season turf in midsummer. In many cases, turf thinned by grubs and/or chinch bugs is invaded by oxalis.

Strategy III: Use a preemergence herbicide when and where appropriate, but more commonly use a postemergence herbicide following label instructions. Usually only spot treatment is required. Broad area application of an herbicide is usually not necessary.

KNOTWEED

Polygonum aviculare L.

Description

Prostrate knotweed is the earliest germinating summer annual broadleaf weed. Soon after a frozen soil thaws, seed germination can occur. This plant is a low-growing, spreading plant that does not root at nodes, although individual plants may form a thick mat of vegetation up to three or four feet (0.9 or 1.2 m) in diameter. The stems are leafy and at each node there is a paper-like sheath. The leaves are bluish-green when plants are growing in compacted soils, but tend to be bright green in non-compacted soil situations. Newly emerged seedlings appear grass-like. Knotweed is capable of persisting in severely compacted soils due to its ability to function under conditions of low soil oxygen diffusion rates.

Cultural Strategies

Prostrate knotweed is an excellent indicator of soil compaction and, thus, cultural practices designed to avoid, reduce, minimize, or eliminate soil compaction should be employed.

Strategy I: Core cultivation and traffic control to reduce compaction.

Strategy II: Increase nitrogen fertility in the fall after prostrate knotweed has become reproductive and is less competitive with turf.

Strategy III: Use a postemergence herbicide when knotweed is young, upright, and actively growing. Treat before plants begin to grow prostrate.

SPOTTED SPURGE

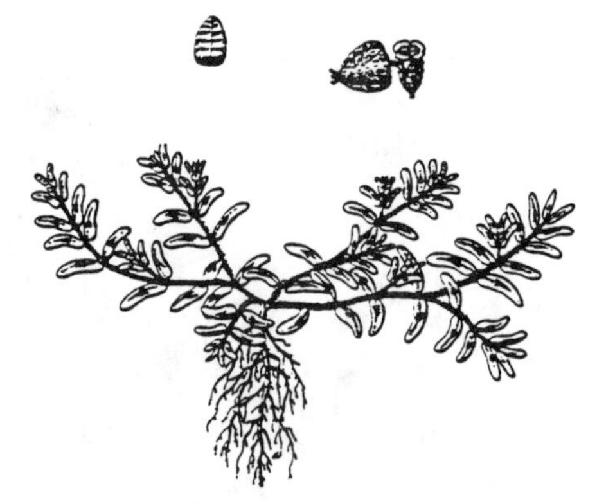

Euphorbia supina Raf.

Description

Prostrate spurge is a low-growing, spreading plant that does not root at the nodes, and resembles prostrate knotweed. Like knotweed, prostrate spurge has stems that radiate from a taproot, and the individual plants can form mats three to four feet (0.9 to 1.2 m) across. The stem of prostrate spurge contains a milky sap that is not found in knotweed. Prostrate spurge germinates much later than knotweed as soil temperatures in the 60 to 65°F (15 to 18°C) range are required. The underside of the leaves is hairy and a purplish spot may be present on the upper surface.

Cultural Strategies

Prostrate spurge is an indicator of soil compaction although it will not tolerate the severity of compaction that knotweed can tolerate. Cool season turf sites subject to moisture stress that are thinned by disease and insect attacks are typically invaded by prostrate spurge in midsummer. Control of disease and insect pests and relief of soil compaction allows turfs to effectively compete with prostrate spurge. Dense, vigorously growing turf is not readily invaded by prostrate spurge.

Strategy I: Relieve soil compaction through core cultivation and traffic control.

Strategy II: See Oxalis.

Strategy III: Irrigate when moisture stress occurs to avoid wilting or dormancy of desired turf species.

Strategy IV: Use a preemergence herbicide when and where appropriate, but more commonly use a postemergence herbicide.

PURSLANE

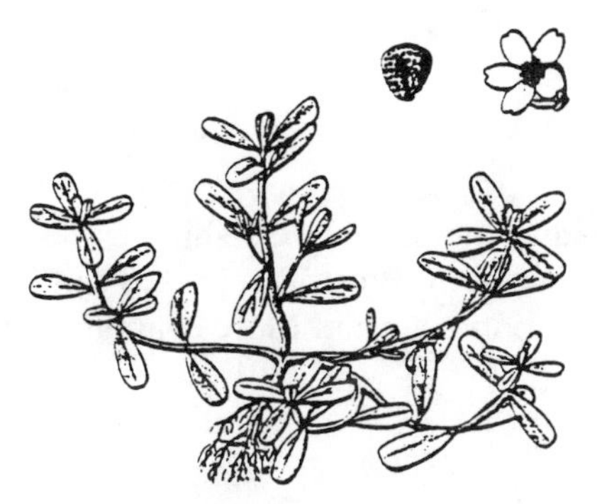

Portulaca oleracea L.

Description

The leaves and stems of this prostrate growing annual are succulent to fleshy and reddish in color. Stems may root, particularly when spreading in very thin turf. The leaves are oblong, clustered, and arranged alternately on the stem.

Cultural Strategies

Avoiding late spring and summer seedings of cool-season turf species is the most effective cultural approach to solving a potential purslane problem.

Strategy I: Establish cool-season turfs in the late summer or fall.

Strategy II: See Oxalis.

Strategy III: Use appropriate postemergence herbicide while purslane is in seedling stage or delay herbicide treatment for 3 or 4 mowings and use a postemergence broadleaf herbicide.

PROSTRATE PIGWEED

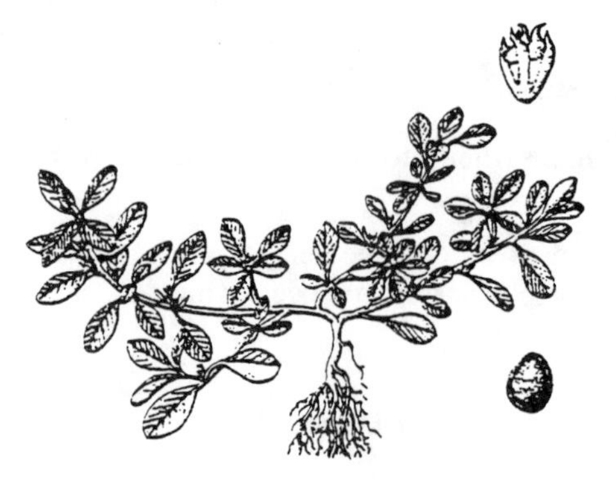

Amaranthus blitoides S.

Description

This prostrate, spreading annual has very shiny leaves. The stems are light green to pinkish and are smooth. Prostrate pigweed can form mats of vegetation up to two feet (0.6 m) in diameter. More common in the western United States.

Cultural Strategies

See Common Purslane.

LAMBSQUARTERS

Chenopodium album L.

Description

This tap-rooted annual has erect-growing stems with ovate alternate leaves. The upper leaves are sometimes sessile. The stems are somewhat square and typically have reddish streaks running along their length. The seedheads are irregular spikes clustered in panicles at the end of branches.

Cultural Strategies

Strategy I: Lower height of cut to at least 30 mm after turfgrass seedlings have become established.

Strategy II: Establish cool-season turfs in the late summer or fall.

Strategy III: Use appropriate postemergence herbicide while purslane is in seedling stage or delay herbicide treatment for 3 or 4 mowings and use a postemergence broadleaf herbicide.

PUNCTUREVINE

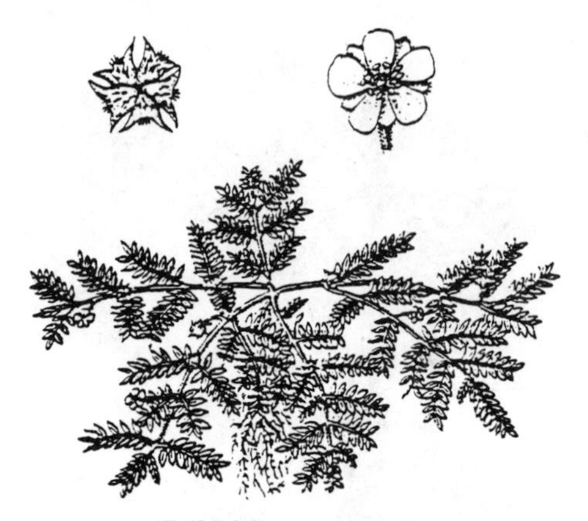

Tribulus terrestris L.

Description

This prostrate, branching annual is slightly hairy. Stems may reach lengths of 5 or 6 feet (1.5 or 1.8 m). Leaves are arranged opposite, pinnately compound, and bright green. Leaflets are narrow, about one-half inch (13 mm) long and hairy. Seed pods have 5 sections, each with sharp spines resembling horns.

Cultural Strategies

Puncturevine is most competitive when soils are compacted and nitrogen fertility is low. Cultural practices to reduce soil compaction should be employed.

Strategies I, II, and III: See Prostrate Knotweed.

RAGWEED

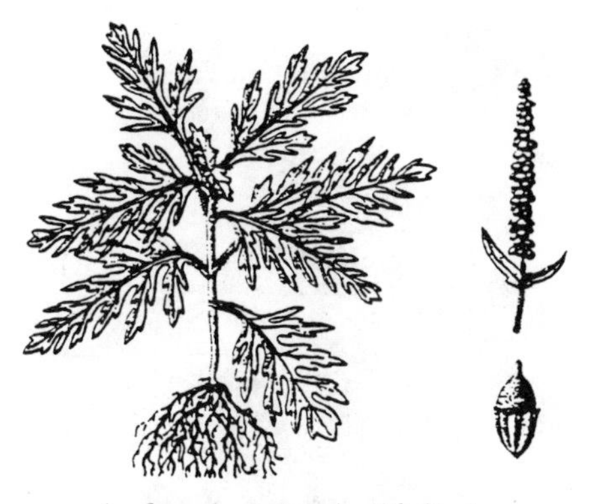

Ambrosia artemissiifolia L.

Description

This annual, erect-growing weed has hairy stems. Leaves are mostly smooth and deeply lobed. It usually is not found in established turfs, but can be competitive in new seedings, minimum maintenance turf areas, or where a turf has been thinned by insect attack or disease incidence.

Cultural Strategies: See Lambsquarters.

CARPETWEED

Mollugo verticillata L.

Description

This late germinating, but quick growing annual is prostrate and forms mats in all directions from the taproot. The stems are green, smooth and branched with 5 to 6 leaves forming in whorls at each node along the stem. This growth habit is quite distinctive and most obvious when thin turf has been invaded. This weed is most often a problem during establishment.

Cultural Strategies: See Common Purslane.

KOCHIA

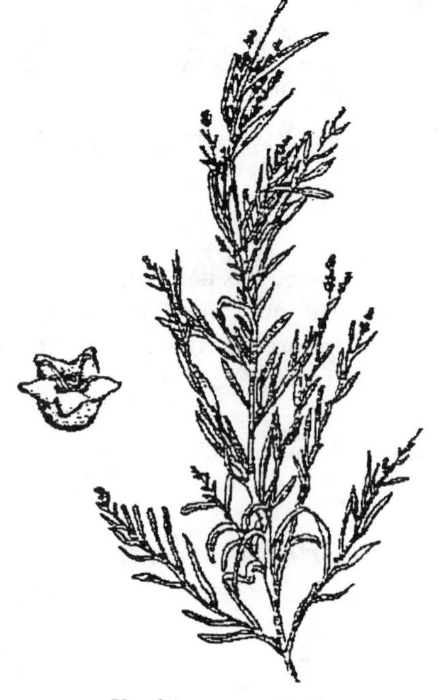

Kochia scoparia L.

Description

A widely adapted annual that survives many environmental conditions. In mowed turf it grows prostrate, but in nonmowed situations can attain a height of up to eight feet (2.7 m). Germination begins in early spring and continues throughout the summer months. The first leaves are very hairy and appear silvery, and the stems vary in color from light green to greenish red.

FLORIDA PUSLEY

Richardia scabra L.

Description

Stems and leaves are hairy. Plants grow prostrate in mowed situations, but tend to be more erect when mowing is infrequent. Leaves are opposite, somewhat oval, and mostly flat. Plants, when growing prostrate, form dense patches. Flowers have 6 parts and appear star-like at the base of upper leaves. The plant most commonly is found in locations where warm season grasses are the predominant turf species.

Cultural Strategies

Although an annual plant, management is primarily through use of postemergence herbicides. Consult local extension recommendations for your area.

Strategy I: Physically remove any small infestations, and keep turf from wilting through proper irrigation management.

Strategy II: In St. Augustinegrass or dichondra, apply a broad spectrum combination of broadleaf herbicides.

Strategy III: Management of Florida pusley in warm-season grasses that are sensitive to phenoxy herbicides can be attained by using a triazine herbicide according to label recommendations.

SOW THISTLE

Sonchus oleraceus L.

Description

This widely found annual forms a rosette from a taproot. The plant is quite leafy and deeply lobed, somewhat similar to shepherdspurse. Although referred to as a thistle, it is not in the thistle family. However, spines do occur along the margins of the leaves. The flower is bright yellow and is similar to dandelion.

Cultural Strategies

This weed is quite unsightly in turf, but can be physically removed, being sure to remove at least two-thirds of the taproot. Sow thistle tends to establish where a turf has been thinned and is not competitive.

Strategy I: Maintain dense stand through proper rate and timing of nitrogen fertility.

Strategy II: On zoysiagrass, centipedegrass, and St. Augustinegrass use a postemergence control according to extension recommendations for your area.

Strategy III: Spot treatment with a translocatable total vegetative herbicide can be effective, but usually causes injury to desired turf species.

BRASS BUTTONS

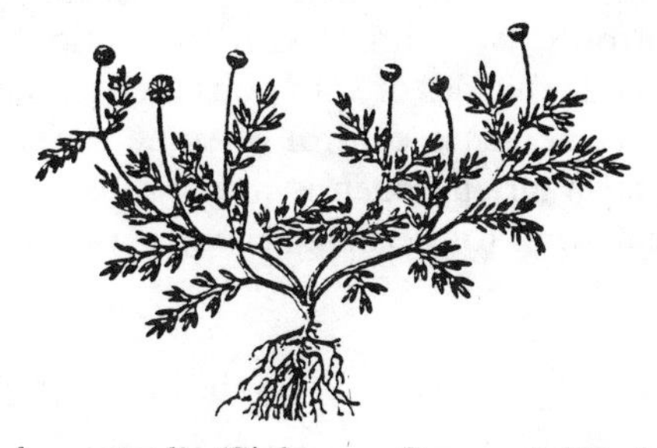

Cotula australis (Sieber ex Spreng.) Hook. f.

Description

This annual is not strongly competitive and usually does not grow taller than three inches (75 mm). Plants tend to grow in clumps, with deeply lobed leaves. In weak or thinned turf, this weed can become a significant problem. Flowering can occur in established turf and the flowers are a distinctly yellowish-green and button-like in appearance.

Cultural Strategies

Brass buttons is a fairly weak species that does not readily invade dense, competitive turfs. Cultural practices, particularly providing adequate nitrogen, will reduce the potential of brass buttons encroachment.

Strategy I: Maintain good disease and insect control programs to prohibit losses of shoot density.

Strategy II: Spot treat with a total vegetative control herbicide.

MANAGING WINTER ANNUAL BROADLEAF WEEDS

The preferred time for the establishment of cool-season turfgrasses is late summer through early fall, in part to avoid competition from summer annual weed species. However, several winter annual weed species can become significant weed problems in the seedbed when cool season turf is established from seed in the fall.

In southern latitudes, winter annuals are significant problems because they germinate and become established into warm-season turfgrasses after the onset of winter dormancy. They also can compete with cool-season grasses that have been overseeded into warm-season grasses for winter turf.

Control of broadleaf winter annuals in seedings of cool-season turfgrasses can be accomplished by using proper establishment methods to ensure good plant competition, a postemergence herbicide application when the weeds are in the seedling stage (if competition is severe), or after the stand has begun to mature after 3 or 4 mowings. In dormant warm-season turf, broadleaf winter annuals can be controlled by treating with conventional broadleaf plus preemergence tank-mixed herbicides or with nonselective plus preemergence tank-mixed herbicides. Preemergence control of some winter annuals also can be achieved; however, such an approach is not feasible in the seedbed. Consult extension recommendations for herbicidal control of broadleaf winter annuals in your area.

COMMON CHICKWEED

Stellaria media (L.) Cyrillo

Description

Common chickweed, found mostly in moist shaded locations, is low-growing when mowed, but erect when infrequently mowed. A single row of hairs extends along one side of a squarish stem. The leaves are teardrop-shaped and opposite on the stem.

Cultural Strategies

Vigorously growing, dense turf will effectively keep common chickweed from encroaching. Chemical treatment may be required in seedbed situations where common chickweed germination is excessive and the desired turf is unable to compete successfully.

Strategy I: Reduce shade by eliminating unnecessary trees, removing understory growth and trimming lower limbs of remaining trees up to eight feet (2.4 m) above ground.

Strategy II: Use a postemergence herbicide in the seedbed that is safe for application to seedling turfgrasses or a preemergence herbicide where appropriate.

Strategy III: Use a combination of postemergence herbicides after seedling cool-season turfgrasses have matured.

Strategy IV: Use a combination of a nonselective herbicide with a preemergence herbicide on dormant warm-season species.

HENBIT

Lamium amplexicaule L.

Description

This cool-season annual grows upright, but some prostrate stems may sprout from the base of the plant. The stems are squarish and reasonably weak. The leaves are hairy and the basal ones have a petiole, while those near the top of the plant are sessile. The upper leaf surface is deeply-veined with small lobing along the margin. Flowers are blue-purple and trumpet shaped.

Cultural Strategies

This winter annual is more competitive than common chickweed, but is not a significant problem in vigorously growing, dense turf. More commonly it is present in seedbeds and dormant warm-season turf.

Strategy I: See Common Chickweed.

Strategy II: See Common Chickweed.

Strategy III: See Common Chickweed.

Strategy IV: See Common Chickweed.

SHEPHERDSPURSE

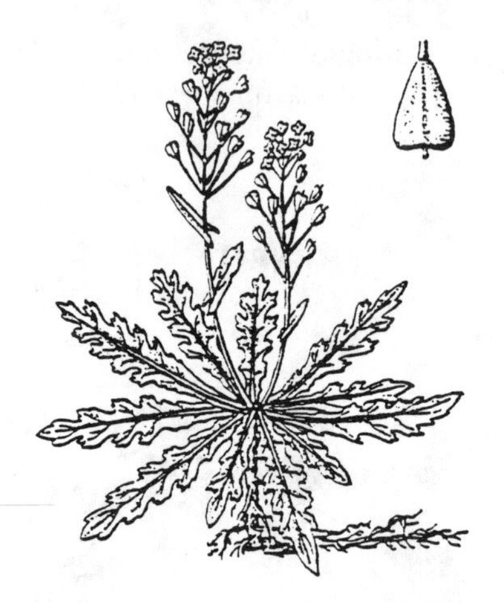

Capsella bursa-pastoris (L.) Medik.

Description

This annual germinates when soil temperatures cool below 60°F (15°C) and forms a rosette of lobed leaves. Considerable variation exists relative to the lobing. Some are deeply lobed all the way to the vein, and others are only slightly lobed. The seed are produced in small heart-shaped capsules that resemble a purse. Small white flowers may be produced in early spring on erect stems.

Cultural Strategies

Shepherdspurse is not competitive in established turf unless there has been a stand loss. Typically, shepherdspurse will germinate in voids that are the result of insect or disease damage which has occurred in mid to late summer.

Strategy I: See Common Chickweed.

Strategy II: See Common Chickweed.

Strategy III: See Common Chickweed.

Strategy IV: See Common Chickweed.

CORN SPEEDWELL

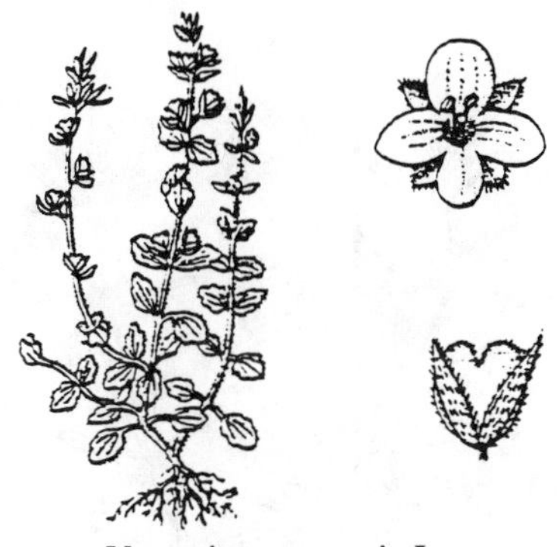

Veronica arvensis L.

Description

Corn speedwell germinates in mid-fall and forms a fairly weak seedling that overwinters in the cool season grass stands but becomes quite conspicuous in dormant warm-season turfgrasses. The entire plant is finely hairy, grows erect, and is most obvious when mowing is infrequent. The upper leaves are smaller than the lower leaves and more pointed.

Cultural Strategies

Corn speedwell does well in thin, open or dormant turf and under these conditions can produce a fairly dense stand. Early fall nitrogen fertilization of cool season grasses, along with control of insects and diseases, help reduce the competitiveness of corn speedwell.

Strategy I: See Common Chickweed.

Strategy II: See Common Chickweed.

Strategy III: See Common Chickweed.

Strategy IV: See Common Chickweed.

BEDSTRAW

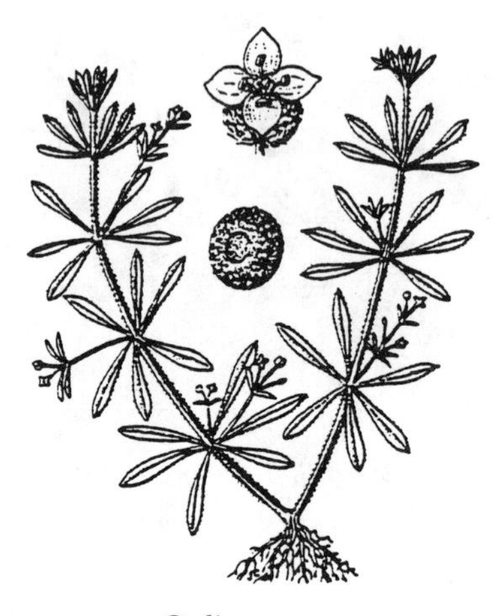

Galium spp.

Description

Bedstraw generally is considered a winter annual in turf situations and several species exist. Bedstraw is most competitive in the shade. The square stems have serrated structures that often cause the stems to stick together. The leaves are arranged in whorls along the stem and stipales often are present.

Cultural Strategies

Elimination or reduction of shade will help competitiveness of desired turf species.

Strategy I: See Common Chickweed.

Strategy II: Apply postemergence herbicide when bedstraw is actively growing. Consult recommendations for your area.

DOG FENNEL

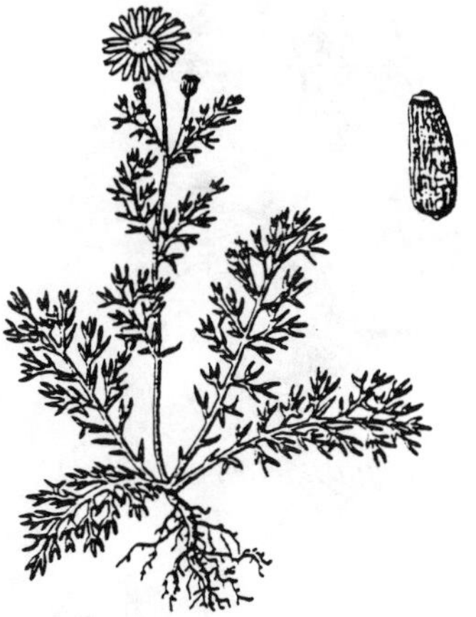

Anthemis cotula L.

Description

This early germinating winter annual forms a low-growing compact rosette. The leaves are small and usually twice pinnately divided, resulting in a fern-like appearance. Crushed leaves give off a very strong, disagreeable odor. Usually dark green, with lower leaves opposite and upper ones alternate. Rarely occurs in established turf, but can germinate in voids and can be competitive in fall seedings.

Cultural Strategies

Physical removal from the seedbed of cool season grasses often is sufficient to eliminate this weed. It is not often dense, even in new seedbeds.

Strategy I: Physically remove from seedbed.

Strategy II: See Common Chickweed.

Strategy III: See Common Chickweed.

Strategy IV: See Common Chickweed.

PEPPERGRASS

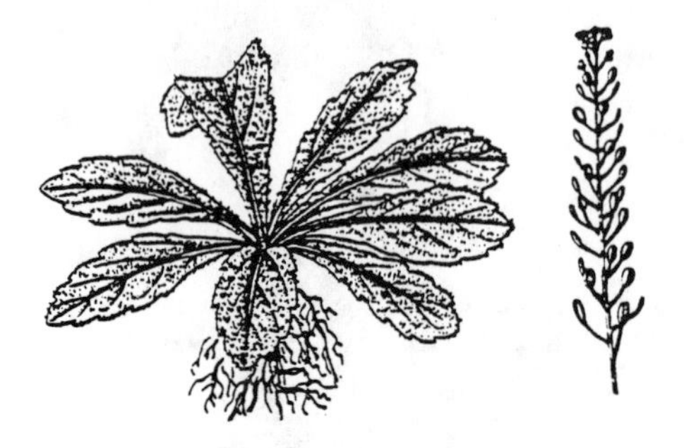

Lepidium virginicum L.

Description

Peppergrass is an annual that germinates as soils cool below 65°F (18°C) and forms a strong rosette. Peppergrass more often is noticed following cold weather, as it resumes growth very early. The apple-green leaves grow rapidly and are slightly lobed and hairy. Peppergrass commonly invades fall seedings of cool season grasses that fail to develop a good stand.

Cultural Strategies

Physical removal when few plants exist usually is adequate. Overseeding of newly established areas that have not produced good turf cover helps reduce pepperweed encroachment.

Strategy I: Mow below 30 mm in spring to discourage development of flower stem. Plants will not flower under close mowing.

Strategy II: Same as Common Chickweed.

Strategy III: Same as Common Chickweed.

MANAGING BIENNIALS

Biennial weeds require two years to complete their life cycle. The first year they grow vegetatively, producing food reserves that are stored in various vegetative plant parts. Consequently, it is during this first year when they are vegetative that biennials are most competitive with turfgrasses. They are leafy and aggressively occupy any open space. In the second year, biennials grow primarily in a reproductive mode using the food reserves stored earlier. In many instances, biennials are less competitive during the second year, as mowing practices continually remove any inflorescence produced by the plant and stored food reserves become depleted. Successful flowering and seed production by biennials usually only occur when mowing is infrequent or the height of cut is higher than normally used in lawn turf. For the most part, biennials are not difficult to manage culturally by maintaining a dense turf, through proper mowing, fertilization practices, and through the use of recommended postemergence herbicides.

YELLOW ROCKET

Barbarea verna (Mill.) Aschers.

Description

This biennial typically germinates when soils cool below 70°F (21°C) and becomes most noticeable in the following spring, as the vigorous early growth exceeds that of most turfgrasses. The leaves are deeply lobed, arranged in a rosette, and appear bright green with the terminal lobe large and rounded. In the seedling stage it is quite similar to shepherdspurse. Under infrequent mowing, yellow rocket will produce a bright yellow flower with four petals.

Cultural Strategies

Maintain mowing height of $<$ 35 mm in early spring to prohibit seed production and consequent reinfestation.

Strategy I: Maintain proper management of disease and insect pests that cause stand thinning, particularly prior to germination.

Strategy II: Fertilize cool season species with nitrogen in the fall to encourage increased density.

Strategy III: Apply a selective postemergence herbicide according to recommendations for your area when yellow rocket is in an actively growing vegetative state, with spring being best if it has invaded cool season grass species.

WILD CARROT

Daucus carota L.

Description

The foliage of this biennial is very similar to the domestic carrot, being multi-branched and somewhat lace-like. The root of wild carrot, although edible, is quite tough and fibrous. Like other biennials, wild carrot produces an aggressive leafy rosette during the first year. Wild carrot is somewhat more adaptable to mowing pressure than yellow rocket. Flowers and viable seed cannot be produced under mowed conditions of less than 45 mm.

Cultural Strategies

Wild carrot rarely produces a dense stand of plants in established turf and can be physically removed quite easily.

Strategy I: During physical removal, make sure to remove a significant portion of the taproot, at least two-thirds.

Strategy II: Apply a selective postemergence herbicide according to recommendations for your area.

BLACK MEDIC

Medicago lupulina L.

Description

Black medic is considered an annual, biennial, or perennial depending on the source. It is a trifoliate species like clover and Oxalis, but differs in that the middle leaflet is extended on a pedicel or stalk. In most turf situations, black medic persists as a perennial legume through stoloniferous growth, but produces flowers and viable seed under most mowed conditions. The stolons rarely root at nodes. Individual leaflets usually have a short spur at the tip.

Cultural Strategies

Black medic is not persistent or aggressive when turf is adequately fertilized with nitrogen. Maintaining proper nitrogen fertility, with a balance between nitrogen and phosphorus, is beneficial. Maintain a ratio of 2:1 for N and P_2O_5.

Strategy I: Increase nitrogen fertility level and correct for low phosphorus if soil test results indicate a need.

Strategy II: Apply a selective postemergence herbicide according to recommendations for your area.

MANAGING PERENNIAL BROADLEAF WEEDS

Perennials are weeds that live more than two years and are represented by the largest group of species causing problems in turf. Many perennials are very easy to manage, while others are virtually impossible to manage even with herbicides.

Many perennials spread by seed and vegetative plant parts. Often viable seed are produced below the cutting height routinely used for a given turf species and/or the seeds are disseminated by the wind.

The presence of abundant plants of a given species often is a good indicator of certain soil conditions or improper management. For example, legumes indicate low nitrogen fertility, sheep sorrel and cinquefoils indicate low pH conditions, yarrow is a good indicator of poor or droughty soil conditions, and creeping buttercup indicates shade and poor drainage. Therefore, modifying soil conditions or changing lime and fertility programs can have a significant positive impact on management programs for perennial weeds.

WILD GARLIC

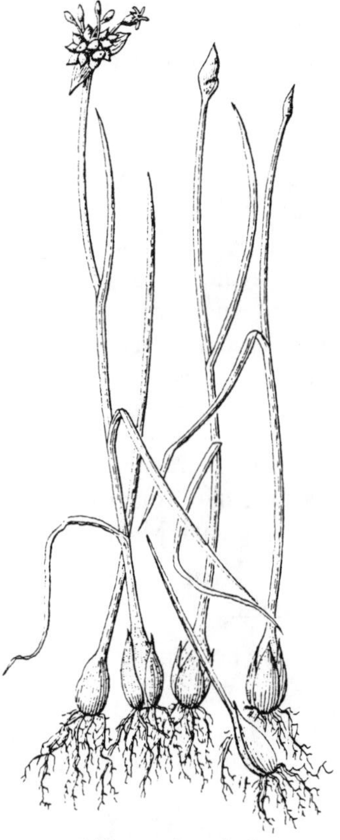

Allium vineale L.

Description

Although wild garlic and wild onion can both be found in turf, wild garlic is significantly more common. Wild garlic leaves are smooth, slender, hollow, and arise from bulbs or bulblets. The leaves have a strong odor that is the result of allyl sulfide contained in the leaf. Wild garlic grows rapidly in early spring and is a significant problem when found in dormant warm-season turfgrasses. Wild garlic reproduces from seed, bulbs, and both aerial and below ground bulblets.

Cultural Strategies

The vigorous, early spring growth requires frequent mowing, perhaps before the turf is in need of mowing or has even broken dormancy in the case of warm-season turfgrasses. Control spring diseases that reduce turfgrass competitiveness.

Strategy I: Close and frequent mowing, particularly in early spring.

Strategy II: Apply ester-based selective broadleaf herbicides according to recommendations for your area.

Strategy III: Apply a nonselective translocatable herbicide if turf is dormant with a wetting agent according to label directions.

DANDELION

Taraxacum officinale L.

Description

Dandelion has deeply lobed leaves arranged in a rosette. Lobes may be opposite or alternate and the tips may point away from or toward the apex. The large fleshy taproot contains vegetative buds mostly near the soil surface that can develop new plants. The flower is bright yellow turning to white and is wind disseminated.

Cultural Strategies

Care must be exercised when physically removing dandelions so a majority of the taproot is removed, at least two-thirds. If not, vegetative buds on the upper portion of the taproot will produce rosettes and 2 to 3 plants will grow where one was removed.

Strategy I: Maintain dense turf through adequate nitrogen fertility, disease and insect control to reduce voids in the turf in the spring into which dandelion seed can be disseminated.

Strategy II: Physically remove, being careful to include all reproductive vegetative plant parts.

Strategy III: Apply a selective postemergence broadleaf herbicide when dandelions are actively growing. Fall applications often are most effective. Follow recommendations for your area.

WHITE CLOVER

Trifolium repens L.

Description

This perennial stoloniferous legume has compound leaves, usually palmately divided into 3 leaflets. Flowers are white, globular, and appear from mid-May through September. White clover can flower and produce viable seed even at very low mowing heights of 6 mm. Low nitrogen nutrition (a third of that typically recommended for a species) and other causes for a thin turf make it particularly susceptible to invasion by white clover.

Cultural Strategies

Physical removal of white clover is not a good management strategy as stolons spread further than expected and portions can be left behind. Good liming and nitrogen fertilization practices that maintain turf density are desired.

Strategy I: Nitrogen fertility level is of greatest importance. When the nitrogen level for a particular turf species is not adequate, white clover or other low-growing legumes commonly invade.

Strategy II: Physical removal is recommended only under conditions of very close mowing of 6–8 mm where pluggers can be used effectively.

Strategy III: Apply a selective postemergence broadleaf herbicide before clover begins to flower in combination with increased nitrogen fertility. Do not apply herbicides when white clover is in flower or under moisture stress, as poor control will result. Follow recommendations for your area.

PLANTAIN

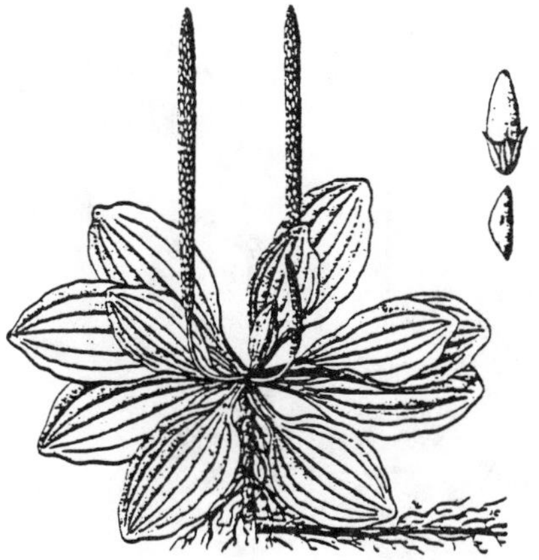

Plantago major L., *Plantago rugelii* L.

Description

This perennial is a commonly occurring weed that has a broad tolerance to soil types. It grows in a rosette with large, broad leaves that have nearly parallel prominent veins. Leaves often are oriented parallel to the ground and the margins often are wavy. Seedstalks grow erect, with seeds forming along more than half the length. Common plantain is green with surface hairs on the leaves, while Rugel's has a purplish cast to the leaves and petioles.

Cultural Strategies

Common plantain persists in thin turf by competing for space with large, broad, prostrate leaves that shade out new tillers of turfgrass that emerge. Maintain turf density with good management or physically remove.

Strategy I: Proper pH and adequate fertility levels should be used. Common plantain can compete with turf particularly well where pH is high (> 8.0). Areas of lime spills or where irrigation water has a high pH may have more serious common plantain encroachment problems. Acidify soil, 6.5 to 7.0, if soil testing so indicates.

Strategy II: See Dandelion.

Strategy III: Apply a selective postemergence broadleaf of herbicide in mid-spring or fall. Plantain resumes seasonal growth later in spring than dandelion, and often is poorly controlled with herbicides when applied prior to dandelion bloom.

BUCKHORN

Plantago lanceolata L.

Description

This perennial has long, slender, strap-like leaves growing in a rosette and is frequently found in thinned turf. Perhaps only dandelion is more abundant. The leaves are dark green with nearly parallel venation. The seed stalk is very fibrous and difficult to cut with mowers, particularly reels, and the seeds are formed in a terminal capsule. The modified taproot has many adventitious buds and new plants can form near the sodline.

Cultural Strategies: See Common Plantain.

Strategy I: See Dandelion.

Strategy II: See Dandelion.

Strategy III: See Common Plantain.

MOUSE-EAR CHICKWEED

Cerastium vulgatum L.

Description

This low-growing perennial spreads by stolons and seed. Mouse-ear chickweed can adapt to mowing heights as low as 6 mm and even produce viable seed under such conditions. The leaves are opposite, oval, and quite hairy, resembling a mouse's ear. The stolons also are hairy with leaves attached directly. Flowers are small, 5-petaled, and white.

Cultural Strategies

Although mouse-ear chickweed grows quite well in full sun, it also can compete in shaded sites. Use turfgrass species adapted to shade and improve soil drainage. For physical removal, see White Clover.

Strategy I: Decrease shade and improve drainage.

Strategy II: See White Clover.

Strategy III: See White Clover.

MALLOW

Malva neglecta Wallr.

Description

This perennial most commonly persists in areas where the cutting height is above two inches (50 mm) and frequently is less than weekly. The leaves are lobed, grow upright, and are borne on sprawling branches that emerge from a deep taproot. In regularly mowed turf, the leaves remain small and resemble ground ivy. However, mallow does not have the degree of upper leaf surface hairiness that ground ivy possesses. Also, mallow stems do not root at the nodes.

Cultural Strategies

Mallow will not invade turf mowed at least weekly where the turf stand is dense. More commonly found where the height of cut is greater than 3 inches (75 mm).

Strategy I: Maintain close mowing of turf (< 35 mm) through proper height adjustment and frequency schedule.

Strategy II: See Dandelion.

Strategy III: See Dandelion.

GROUND IVY

Glechoma hederacea L.

Description

This perennial is a strongly competitive, low-growing weed that spreads by vigorous stolons. Adapted to sun or shade, this species can form thick mats that crowd out existing turf. The leaves are small, roundish, slightly lobed, and have short bristle-like surface hairs somewhat resembling mallow. They are opposite on low-growing squarish stems that root at the nodes. Flowers are funnel-shaped, blue to purple in color, and are attractive, particularly when ground ivy is used as a ground cover planting.

Cultural Strategies

Ground ivy is difficult to manage without herbicide use. Turf often is invaded by ground ivy that originates in mulched landscape plantings. Therefore, physical removal must include that found in the turf as well as in other plantings.

Strategy I: See Dandelions.

Strategy II: See Dandelions.

Strategy III: Apply a selective postemergence broadleaf herbicide in mid-spring or fall. Repeat herbicide applications every 3 weeks for 3 applications, if necessary, according to recommendations for your area.

SHEEP SORREL

Rumex acetosella L.

Description

This perennial spreads by a strongly branching rhizome system which can give rise to a very dense mat of leaves that crowd out existing turf. Red sorrel has arrow-shaped leaves 1 to 3 inches (25 to 75 mm) long with spreading basal lobes. They are alternate on the stem and have a bitter taste, although they are sometimes eaten in salads. Flowers of red sorrel are borne on separate plants, yellow on male and reddish on female. Flowers seldom occur in mowed turf (< 35 mm).

Cultural Strategies

Maintaining proper pH and adequate nitrogen fertility. Inspecting acid-loving nursery stock prior to planting for red sorrel contamination is a good preventive practice. Do not attempt physical removal, as rhizome fragments often are left behind.

Strategy I: Lime to a pH of 6.7 to 7.2 for good turf vigor.

Strategy II: Physical removal is not particularly effective.

Strategy III: See Ground Ivy.

CANADA THISTLE

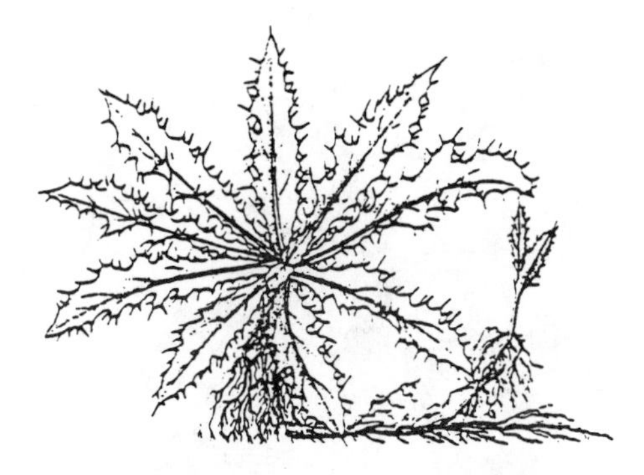

Cirsium arvense L. (Scop.)

Description

This thistle is the most troublesome in turf. Spreading by vigorous rhizomes, Canada thistle competes with turf even when mowed. It commonly invades turf from mulched landscape plantings. Considered a noxious weed in most states, Canada thistle in turf not only reduces aesthetic quality, but can cause physical discomfort for those who come in contact with it. Ridged spines form at the ends of lobes along the leaves. As plants emerge from rhizome buds, they form rosettes that are capable of maintaining a low-growth habit.

Cultural Strategies

Managing Canada thistle in turf begins with using certified seed or sod during turf establishment and making sure that plants are not introduced when planting nursery stock. If preventive measures are followed, encroachment by Canada thistle is limited mostly to situations where fields border turf areas.

Strategy I: Use certified seed and sod during establishment and manage Canada thistle in landscape plantings.

Strategy II: Use a selective postemergence broadleaf herbicide. Repeat applications at 3-week intervals may be required, according to recommendations for your area.

CHICORY

Cichorium intybus L.

Description

Chicory grows in a rosette and closely resembles dandelion. However, it has blue flowers borne along a tough, fibrous stalk, quite unlike the succulent tubular flower stalk of the dandelion. Vegetative buds occur on the chicory flower stalk. Under mowed conditions these buds can produce aerial rosettes, as chicory does not commonly flower when mowed below 40 mm on a weekly basis. The deep, fibrous taproot can be dried, ground, and used as a supplement in coffee (chicory coffee). The upper surface of chicory leaves are rougher to the touch than dandelion.

Cultural Strategies: See Dandelion

Strategy I: See Dandelion.

Strategy II: See Dandelion.

Strategy III: See Dandelion.

CURLY DOCK

Rumex crispus L.

Description

This perennial has long, strap-like leaves originating from buds at the top of a strong fleshy taproot. The leaves are arranged in a rosette with each leaf having a "wavy" margin, hence the common name curly dock. Under mowed conditions, curly dock rarely produces viable seed, but does persist due to the taproot. Curly dock is most competitive during periods of moisture stress when the turf species wilt, as curly dock rarely exhibits permanent wilt.

Cultural Strategies: See Wild Carrot.

Strategy I: See Wild Carrot.

Strategy II: See Wild Carrot.

BULL THISTLE

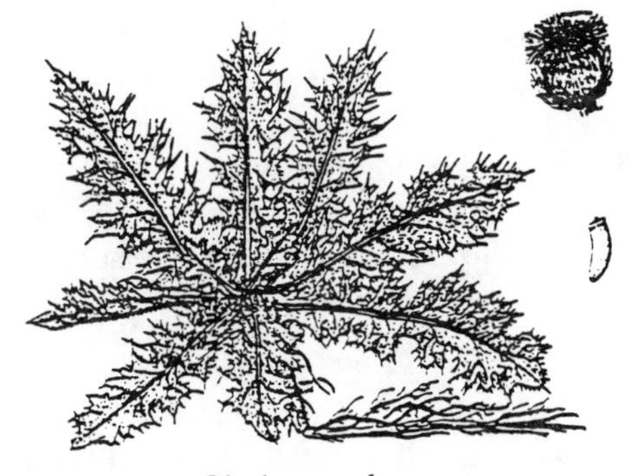

Cirsium vulgare

Description

Musk and bull thistles are botanically biennials, but tend to persist as perennials in turf situations. They grow in rosettes, are quite spiny, and commonly develop during establishment of cool-season grasses. Not particularly competitive in established turf, these species are easily identified and can be physically removed.

Cultural Strategies

By maintaining the shoot density, using proper seeding rates, and selecting adapted species, bull thistles generally are not serious weed problems.

Strategy I: See Wild Carrot.

Strategy II: See Wild Carrot.

HEALALL

Prunella vulgaris L.

Description

Healall is a widely adapted broadleaf perennial weed that is somewhat difficult to manage. Spreading by fibrous stolons, healall competes strongly with weakened turf and in shaded conditions where it tends to be dark green. The leaves and stems are hairy, the upper leaf surface has a "puckered" appearance, and the stem is squarish. Flowers are purplish, borne on compact spikes with overlapping green bracts.

Cultural Strategies

Select turf species that are well adapted to shaded conditions and maintain maximum shoot density through proper cultural practices.

Strategy I: See Dandelion.

Strategy II: See Dandelion.

Strategy III: See Ground Ivy.

OX-EYE DAISY

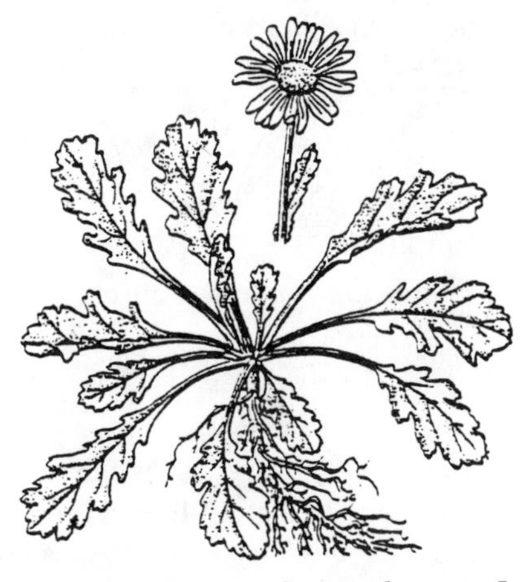

Chrysanthemum leucanthemum L.

Description

This perennial spreads by fleshy rhizomes and can compete with low maintenance turf (such as along roadsides). Ox-eye daisy has typical chrysanthemum leaves, deeply lobed, but is not a daisy even though the flower is daisy-like in shape and color. Leaves are dark green, even when nitrogen fertility is low.

Cultural Strategies

Ox-eye daisy does compete in turf that is even moderately managed. It is much more prevalent in low maintenance areas like roadsides and other rights-of-way.

Strategy I: See Dandelion.

Strategy II: See Dandelion.

Strategy III: See Dandelion.

HAWKWEED

Hieracium pratense Tausch.

Description

Hawkweed forms rosettes of very hairy leaves borne on an aggressive rhizomes system. The leaves are long and strap-like, with some hairs being quite long. When turf is thin, hawkweed will grow in clusters; while in denser turf, individual rosettes tend to appear. Hawkweed persists through vegetative reproduction in most turf situations, as mowing eliminates the flowering capability.

Cultural Strategies: See Dandelion.

Strategy I: See Dandelion.

Strategy II: See Dandelion.

Strategy III: See Dandelion. Include surfactant with postemergence herbicide treatment to improve retention on leaf surface.

THYME-LEAF SPEEDWELL

Veronica serpyllifolia L.

Description

This perennial spreads by slow growing, but aggressive stolons that compete strongly with turfgrasses. The leaves are dark green, oblong, with a waxy, shiny surface and occur opposite along the stem. Usually thyme-leaf speedwell roots into the soil at each node as the stolon grows. The leaves orient themselves parallel to the soil surface and cannot be removed by close mowing. Thyme-leaf speedwell can be a serious broadleaf problem, even on putting greens and other closely mowed areas.

Cultural Strategies

There is no selective chemical management for thyme-leaf speedwell, and physical removal is tedious. When digging it out and pulling it up, be sure to remove all stolon segments.

CREEPING SPEEDWELL

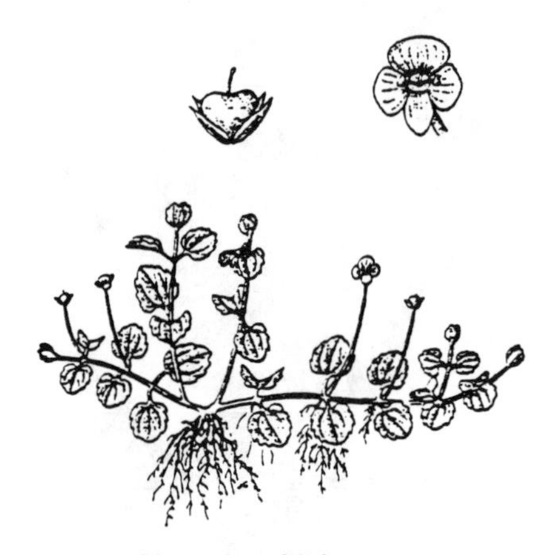

Veronica filiformis L.

Description

This speedwell is the most common one found in cool season grasses. It spreads aggressively by stolons that frequently grow over the top of a turf without rooting at the nodes. The leaves are pale green, lobed slightly, roundish with upper surface hairs. Rapid stolon development results in dense patches of creeping speedwell even when turf growth is good. Although not restricted to shaded sites, creeping speedwell is more competitive in the shade. In late April and May, bluish to purple flowers form in abundance.

Cultural Strategies

Creeping speedwell requires a diligent combination of cultural and chemical programs to effect good management.

Strategy I: Reduce shade where possible and establish turfgrass species best adapted to the prevailing conditions. Improve air and soil drainage to allow turf to become more competitive.

Strategy II: Apply selective postemergence herbicides according to extension recommendations for your area in mid-spring (prior to flowering) and continue every 3 to 4 weeks. Repeat applications are necessary, as creeping speedwell recovers from herbicide "knock-down" through vegetative bud break along the stolons.

Strategy III: Apply wettable or flowable formulation of DCPA when speedwell is actively growing. This preemergence herbicide has specific postemergence activity on this particular speedwell species. The granular formulation will not be effective.

VIOLET

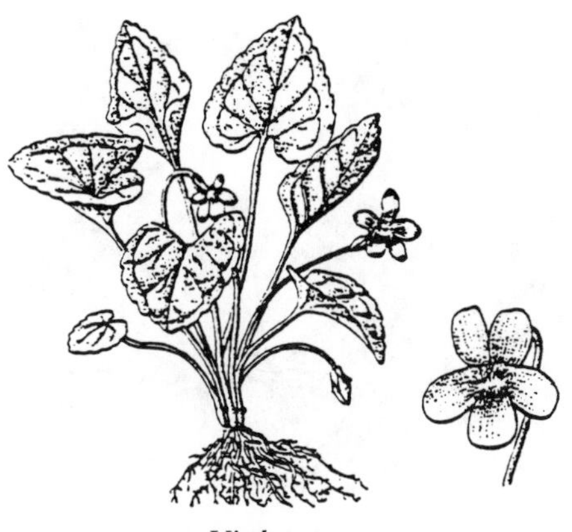

Viola spp.

Description

Numerous species of wild violets occur in turfs. Characterized by a vigorous root system, heart-shaped leaves, and purple to blue flowers, this perennial is extremely difficult to manage. Spreading effectively through turf by rooting stolons, very dense patches of stemless plants form. Although more competitive in shade, wild violets compete in sunny turf sites as well. Violets often invade turf areas from other locations in the landscape where they have been used as a ground cover.

Cultural Strategies

Wild violets should not be used as a ground cover unless they can be physically restrained from invading turfed portions of the landscape.

Strategy I: Physically remove all vegetative portions of the wild violet plants. This requires some digging. Remove wild violets from nonturf areas or contain them where they are desired by using physical barriers.

Strategy II: Repeated applications of selective postemergence broadleaf herbicide combinations will reduce wild violet competition, but management is extremely difficult to achieve. Follow extension recommendations for your area.

STITCHWORT

Stellaria graminea L.

Description

This perennial spreads primarily by rapidly growing stolons that often form dense mats. The leaves are opposite and similar to an elongated teardrop with a somewhat shiny appearance. The squarish stem roots at the node, but more often grows over the top of the turf, particularly when mowing is infrequent. White flowers may form, but usually do not occur under mowed conditions. Stitchwort can closely resemble common chickweed under certain growing conditions.

Cultural Strategies

Management can be achieved by mowing more frequenly, reducing height of cut, and proper herbicide use.

Strategy I: Increase mowing frequency to at least weekly and, where possible, lower the height of cut to at least 35 mm.

Strategy II : Apply selective postemergence broadleaf herbicides when stitchwort is actively growing. Addition of nonionic surfactant is often beneficial. Follow herbicide recommendations for your area.

BINDWEED

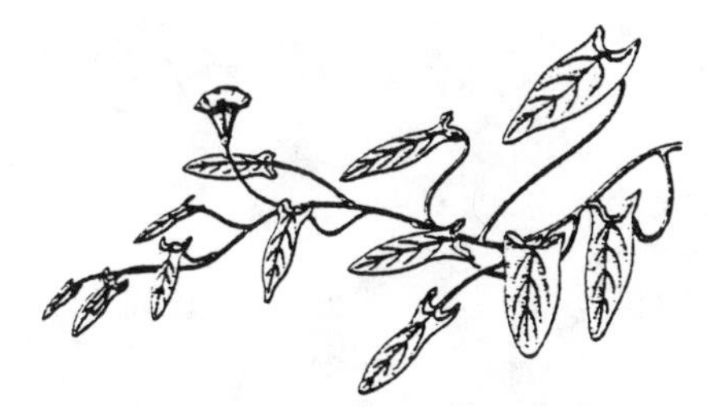

Convolvulus arvenis L.

Description

This commonly occurring weed can invade turf from other portions of the landscape or neighboring fields due to a vigorous, vining growth habit. Field bindweed has spade-like leaves that form alternately along vines that can grow over obstacles like fences, shrubs, or other ornamentals. It can grow very low and adapt to frequent mowing at heights of cut between 1 and 2 inches (25 to 50 mm). Flowers can form along the vine and they are white to pinkish and about one-half inch (13 mm) across.

Cultural Strategies

Field bindweed is very difficult to manage, and like wild violets, must be physically removed at the source of the encroachment which usually is from the ornamental plantings in the landscape.

Strategy I: Physically remove from hedge rows, fence lines, and ornamental plantings. Be sure to remove as much of root system as possible.

Strategy II: Apply selective postemergence broadleaf herbicides when weed is actively growing. Repeated applications are usually required. Apply according to recommendations for your area.

BELLFLOWER

Campanula rapunculoides L.

Description

Used as an effective ground cover, this species often escapes into turf areas where it is commonly mistaken for wild violets. The growth habit and leaf shape (heart-shaped) are similar to violets, but bellflower is more likely to form dense patches. The flower color also is similar to violets, but the flower resembles a bell rather than being open, like violets.

Cultural Strategies: See Wild Violet.

Strategy I: See Wild Violet.

Strategy II: See Wild Violet.

CINQUEFOIL

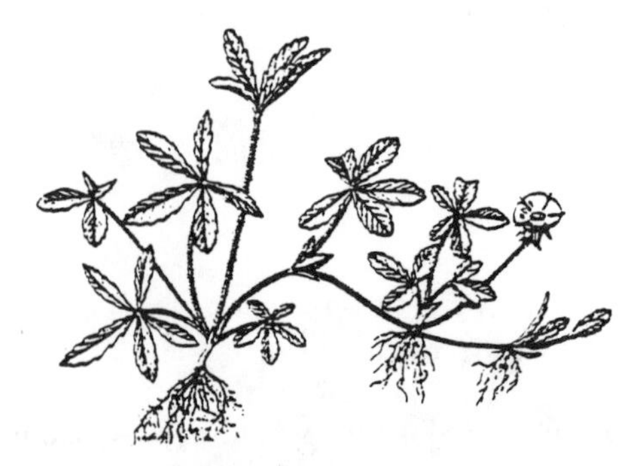

Potentilla norvegica L.

Description

Several cinquefoil species occur in turf, but the most common one is the five-leaflet type. Cinquefoil leaflets are palmately divided and arise from fibrous, hairy creeping stems. The leaflet shape is similar to that of a strawberry, and the three-leaflet cinquefoil, that grows more upright, is very similar to the cultivated strawberry. The five-leaflet type can compete effectively in frequently mowed situations, but does not compete strongly when nitrogen fertility is adequate for good turf growth. Cinquefoil often is found when the soil type is gravelly and the pH is below 6.0. The undersurface of the leaflets appears silvery due to pubescence.

Cultural Strategies

Maintaining turf shoot density through adequate nitrogen fertilization and liming significantly decreases cinquefoil competition.

Strategy I: Increase nitrogen fertilization and lime to pH range of 6.7 to 7.2.

Strategy II: See Dandelion.

Strategy III: See Dandelion.

YARROW

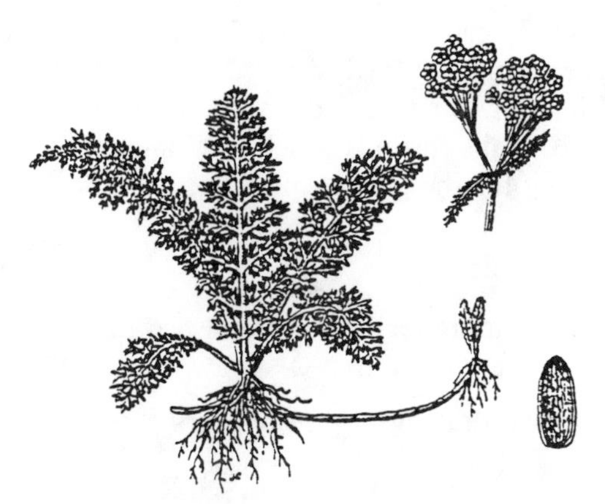

Achillea millefolium L.

Description

Yarrow, under nonmowed conditions, can grow 2 to 3 feet (0.6 to 0.9 m) tall. However, when mowed frequently it grows more prostrate, spreading by fleshy rhizomes, with individual plants forming modified rosettes. The leaves are compound, finely divided (fern-like), and appear woolly due to grayish-fine hairs. When the leaves are crushed they have an aromatic scent. Although rarely flowering in mowed turf, in less frequently mowed areas a white umbel flower is produced that is somewhat similar to wild carrot.

Cultural Strategies

Yarrow is less competitive when frequently mowed below 30 mm and nitrogen fertility is adequate for good turf growth.

Strategy I: Physically remove by digging, being sure to remove all rhizome segments. Improve soil physical conditions by turf cultivation and modification with an organic matter source.

Strategy II: Apply a selective postemergence broadleaf herbicide in late spring or fall. Repeated applications usually are necessary. Addition of a wetting agent to the herbicide treatment improves leaf surface coverage. Follow extension recommendations for your area.

MOSS

Scientific Name

Selagimella spp. (See Plate 2–72)

Description

Moss encompasses a relatively primitive form of plant life that has many species. Forming dense mats of finely divided vegetation that hugs the ground, moss grows well in cool, shaded, and poorly drained areas. Moss does not form true roots, but forms rhizoids which are filamentous structures that do not provide anchoring for the plants equal to a true root.

Cultural Strategies

Moss management is more readily accomplished through cultural practices than through herbicide applications.

Strategy I: Improve drainage by recontouring surface and/or installing subsurface drainage system.

Strategy II: Remove understory growth and trim trees and ornamentals to reduce shade and to allow better air movement and diffuse sunlight to penetrate.

Strategy III: Increase overall fertility level, particularly in fall and early spring when shade conditions are lessened.

Strategy IV: Apply pulverized limestone at 75 to 100 pounds per thousand square feet (38 to 50 kg/100 m^2) directly to the mossy areas. Rake out moss after it has become dehydrated.

BETONY

Stachys floridana Shuttlew.

Description

This perennial in southern turf often grows more like a winter annual than a perennial. It grows when temperatures are cool and moisture is prevalent. The leaves are opposite on squarish stems and have serrated margins. Vegetative reproduction occurs from rapidly spreading underground tubers. As hot weather begins, betony is not competitive with warm-season turfgrasses and disappears only to return from tubers when cool weather returns.

Cultural Strategies

Insignificant infestations of betony can be physically removed, using caution to remove the tubers.

Strategy I: Physical removal.

Strategy II: Application of triazine herbicides to existing and established betony in centipedegrass, carpetgrass, zoysiagrass, and St. Augustinegrass. Follow extension recommendations for your area.

BEGGARWEED

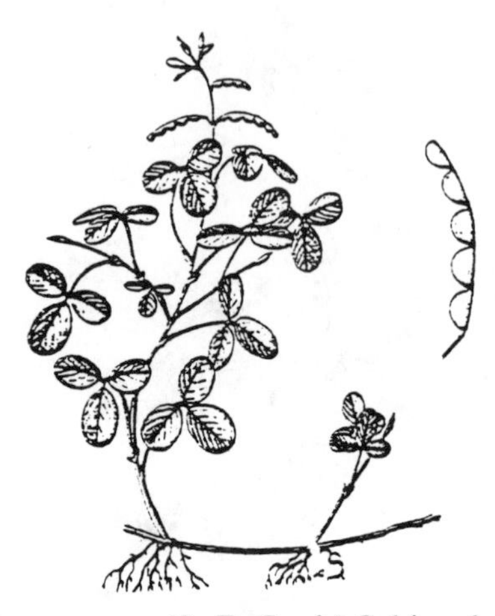

Desmodium canum (J. F. Guel.) Schinz & Thellung

Description

Closely resembling clover, beggarweed spreads by rhizomes that give rise to three-leaflet plants. The middle leaflet is extended on a short stalk much like the leaves of black medic. Beggarweed produces pinkish flowers that form jointed pods which can break up and be disseminated by people and animals.

Cultural Strategies

Beggarweed can persist under close mowing (< 20 mm) and physical removal can be difficult.

Strategy I: Maintain a dense turf stand through adequate nitrogen fertility and irrigation sufficient to eliminate moisture stress.

Strategy II: Spot treatment with a nonselective translocated herbicide according to extension recommendations for your area.

BUR CLOVER

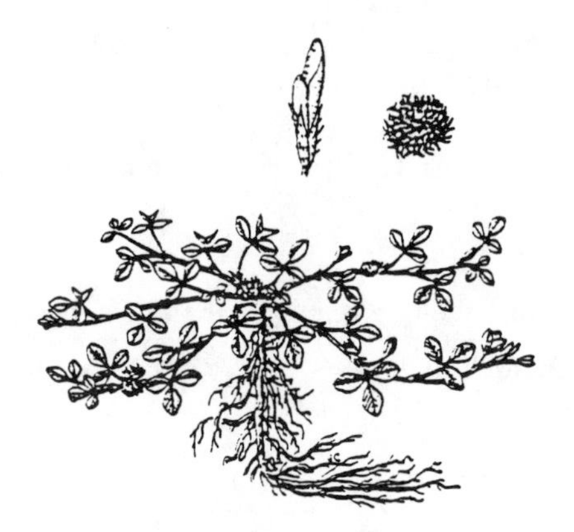

Medicago minima (L.) Bartal.

Description

This species spreads by stolons which allows for competition even under close mowing conditions (< 25 mm). Bur clover produces yellow flowers that mature into spiny seed pods which are easily transported by man and animals upon contact. The leaflets may have reddish-brown spots on the upper surface; otherwise they closely resemble black medic.

Cultural Strategies: See White Clover.

Strategy I: See White Clover.

Strategy II: See White Clover.

Strategy III: See White Clover.

ENGLISH DAISY

Bellis perennis L.

Description

Originally used as an ornamental bedding plant, this weed has become a serious turf weed in some parts of the country. Growing in modified rosettes, English daisy forms clusters of vegetation in thin turf. The leaves vary from smooth to hairy, are narrow at the base and are slightly lobed. Typical daisy flowers are produced on 3 to 4 inch (7.5 to 10 mm) flower stalks.

Cultural Strategies

Physical removal must include removal of the extensive taproot.

Strategy I: See Dandelion.

Strategy II: See Dandelion.

Strategy III: See Dandelion.

OXALIS (CREEPING)

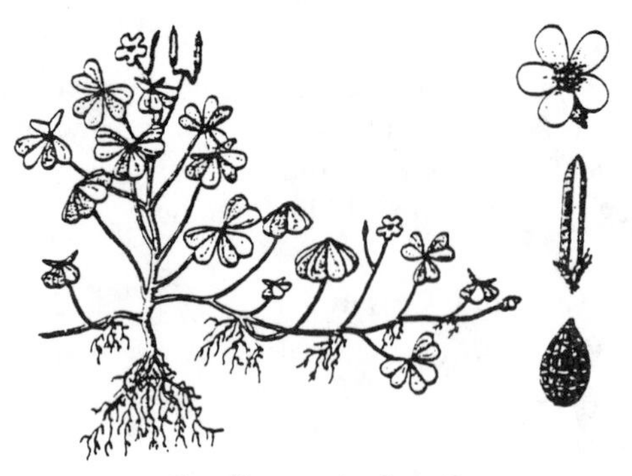

Oxalis corniculata L.

Description

This perennial oxalis is quite similar to the summer annual yellow woodsorrel, spreading by stolons. Creeping oxalis leaves are trifoliate, heart-shaped, and have a reddish-purple tint. The stolons are slightly hairy and root at the nodes. This species maintains a prostrate growth habit and is competitive in closely mowed situations (20–30 mm). As with other oxalis, creeping oxalis forms yellow flowers that mature into seedpods which explode when dried.

Cultural Strategies: See Yellow Woodsorrel.

Strategy I: See Yellow Woodsorrel.

Strategy II: See Yellow Woodsorrel.

Strategy III: Apply a selective postemergence broadleaf herbicide in late spring and fall. Repeated applications are usually required. Make applications after turf has been mowed to improve coverage. Follow extension recommendations for your area.

WEED MANAGEMENT IPM

As with other pests, Integrated Pest Management (IPM) for weeds is a viable management philosophy that has been practiced for years. Managing weed problems is, perhaps, more complex than managing other pests that afflict turf in that the pest is also a plant rather than some other organism. Therefore, successful weed management requires skillful manipulation of plant competition through proper cultural practices in addition to exercising outstanding control of insect and disease pests. Lack of control of these pests often leads to turf losses which then result in weed encroachment.

In this chapter, it has been shown that weeds are good indicators of certain soil conditions and that some species often occur as a result of mismanagement of the turf. Such knowledge provides the basis for developing a scouting plan.

The first part of the plan should include an historical account of the property (including records of previous weed problems, cultural attempts to manage, herbicide use and success rate, changes in species predominance). After the historical record is assembled and understood, the next step is to scout the property. This requires an appropriate transect of the land accompanied with a detailed record of those weeds identified and their distribution and abundance. Scouting provides the information needed to develop a customized and efficient weed management strategy. Particular attention should be given to those weed species that have escaped previous control procedures. Any unknown weeds should be sampled and identified by any and all appropriate means possible. It is not necessary to identify and record all weed species encountered during scouting. From a scouting perspective, in areas where control has been limited, a simple notation reflective of the fact that an array of broadleaf weeds are present is sufficient. In any event, the scouting report form that is developed should provide the opportunity to generalize, but also allow for specific notation of troublesome weeds and those that require particular management attention. For example, the occurence of several summer annual weeds can be reported as "summer annual weeds." However, a notation that goosegrass is the predominant species of the group signals the important fact that the site is probably compacted. Therefore, core cultivation will undoubtedly improve turf competition and enhance the effectiveness of any herbicide that might be chosen as a chemical control strategy.

Scouting for weeds can be done in any number of ways. Walking or riding the turfgrass site in a zig-zag fashion, stopping periodically to sample and identify existing weeds, will begin to build a plant inventory of the site. Once this inventory of weeds and their distribution is assembled, underlying reasons for the particular array of weeds might be discerned. Such information should provide the basis for a management strategy that will maximize the competitive nature of the desired turfgrass species and enhance the effectiveness and efficiency of any herbicide that might be chosen as a chemical control agent.

REFERENCES

Neal, Joseph C. 1993. Turfgrass Weed Management—An IPM Approach. Cornell Weed Management Series No. 8.

Muenscher, Walter C. 1980. Weeds. 2nd Edition. Cornstock Publishing Associates, Cornell University Press. Ithaca, NY.

Shurtleff, M. C., T. W. Fermanian, and R. Randell. 1987. Controlling Turfgrass Pests. Prentice-Hall, Inc. Englewood Cliffs, NJ.

Lorenzi, H. J. and H. S. Jeffery. 1987. Weeds of the United States and Their Control. Van Nostrand Reinhold Company, Inc. New York, NY.

Zimdahl, R. L. 1993. Fundamentals of Weed Science. Academic Press, Inc. San Diego, CA.

International Turfgrass Society Research Journal Vol. 7. 1993. R. N. Carrow, N. E. Christians, R. C. Shearman (ed.). Intertec Publishing Corp. Overland Park, KS. pp. 238–310.

Table 1.1. Trade Names and Use Characteristics of Common Turfgrass Herbicides[a,b]

Common Name	Some Trade Names	Pre-emergence[d]	Post-emergence[d]	Non-selective[d]	Selective[d]
Asulam	Asulox		X		X
Atrazine[c]	X	X		X	
Benefin	Balan, LESCO 2.5 Benefin Granular	X			X
Benefin + oryzalin	XL	X			X
Benefin + trifluralin	Team	X			X
Bensulide	Bensumec, Betasan, Lescosan, Pre-San, PROTURF Weedgrass Preventer, Regalsan	X			X
Bentazon	Basagran T/O. LESCOGRAN		X		X
Bentazon + atrazine[c]	Prompt		X		X
Cacodylic acid	Phytar 560, Weed Ender		X	X	
Chlorsulfuron	LESCO TFC		X		X
CMA (CAMA)	C410 Crabgrass Killer, CGK-89, Grasstron II, M-M Crabgrass Killer, Super Dal-E-Rad Calar		X		X
2,4-D	2,4-D Granules, 2,4-D Low-Volatile Ester, LESCO A-4D, See 2,4-D LV 4, Standard 2,4-D Amine, Weedar 64, Riverdale AM-40, Weedone LV4, Weed Pro 4# Amine		X		X
2,4-D + dicamba	Bonide Lawn Weed Killer, 81 Selective Weedkiller, Majestic Green Banvel + 2,4-D, Pro Triple D Lawn Weed Killer, Turfgo Four Power Plus		X		X
2,4-D + dichlorprop	2D + 2DP Amine, LESCO Granular Herbicide, PROTURF Fluid Broadleaf Weed Control, Turf D + DP, Weedone DPC, Weedone Amine		X		X
2,4-D + dichlorprop + dicamba	Super Trimec		X		X
2,4-D + mecoprop	Cleary's MCPP-2,4-D, Expedite Broadleaf Herbicide, Lebanon Dandelion Killer and Lawn Weed Control, 2 MCPP + 2D Amine, 2 Plus 2		X		X
2,4-D + MCPP + dicamba	Brushfire, Brush-out, Brush-Whacker, LESCO Bentgrass Selective Herbicide, LESCO Three-Way Selective Herbicide, Liquid Weedaway, Norkem 400T, Riverdale Triplet Selective Herbicide, SNS-2000, Three-Way Lawn Weed Killer, Trex-San, Trex-San Bent, Trimec Bentgrass Formula, Trimec Classic, Trimec 992, Trimec Southern, Turfgo MecAmine-D		X		X

Table 1.1. Trade Names and Use Characteristics of Common Turfgrass Herbicides[a,b] (Continued)

Common Name	Some Trade Names	Pre-emergence[d]	Post-emergence[d]	Non-selective[d]	Selective[d]
2,4-D + MCPP + dichlorprop	Riverdale Tri-Ester, Riverdale Triamine, Triamine Granular Weed Killer, Triamine Jet-Spray Spot Weed Killer		X		X
2,4-D + MCPP + MSMA + dicamba	Trimec Plus		X		X
2,4-D + triclopyr	Chaser		X		X
DCPA	Cornbelt Crabgrass Control, Dacthal, Pretrol, Pre-Weed, Pro Crabgrass Preventer/Dacthal, Pro Super Dacthal 686, Seed-trol, Stand-off, Vegetable, Turf and Ornamental Weeder	X	X		X
Dicamba	Banvel, Pro-Turf K-O-G Weed Control		X		X
Diclofop[c]	Illoxan		X		X
Diquat	Aquatate, Castaway, LESCO ProTrim, Mantek, Reward, Vegetrol, Watrol		X	X	
Dithiopyr	Dimension	X	X		X
DSMA	Cleary's Methar 30, Drexel DSMA Slurry		X		X
Ethofumesate	Prograss	X	X		X
Fenoxaprop	Acclaim		X		X
Fluazifop-P-butyl	Fusilade Turf and Ornamental		X		X
Glyphosate	Avail, Bareback, Expedite Grass and Weed Herbicide, Hoedown, Kem-til, Quick Claim, Roundup, Striker, Trailblazer		X	X	
Imazaquin	Image	X	X		X
Isoxaben	Gallery	X			X
MCPA	MCPA-4 Amine		X		X
MCPA + MCPP + dicamba	Eliminate, Riverdale Tri-Power, Trimec Encore		X		X
MCPA + MCPP + dichlorprop	Riverdale Triamine II, Riverdale Tri-Ester II		X		X
Mecoprop (MCPP)	Certi-CM, Chemweed-265, Cleary's MCPP, Lescopex, MCPP Low-Volatile Ester, Mecomec, Milpro 360, Riverdale MCPP-4 Amine, Selecticide, Super Lescopex, Turfgo MCPP-4K		X		X
Metolachlor	Pennant	X			X
Metribuzin	Sencor 75 Turf Herbicide		X		X

Metsulfuron methyl	DMC Weed Control		X		X
MSMA	Daconate 6, Daconate Super, Drexar 530, Majestic Green Sandbur and Nutsedge Killer, MSMA, MSMA Turf, 912 Herbicide, 120 Herbicide, Super Crabgrass Killer, Weed Hoe		X		X
MSMA + cacodylic acid	Monside, Broadside		X	X	
Napropamide	Devrinol	X			X
Oryzalin	Surflan, Weed Stopper	X			X
Oxadiazon	Chipco Ronstar	X			X
Oxadiazon + bensulide	PROTURF Goosegrass/Crabgrass Control	X			X
Pendimethalin	LESCO PRE-M, PROTURF Southern Weedgrass Control, PROTURF Turf Weedgrass Control, PROTURF Weedgrass Control	X			X
Potassium salts of fatty acids	DeMoss, Moss and Algae Eraser, SharpShooter		X	X	
Prodiamine	Barricade	X			X
Pronamide[c]	Kerb	X	X		X
Sethoxydim	Vantage		X		X
Siduron	Crabgrass Preventer and Weed Killer, Lebanon Crabgrass Control 4.6% Tupersan, Pro Crabgrass Preventer-Tupersan, Pro Super Tupersan, Tupersan	X			X
Simazine	Simazine, Princep, Regal Wynstar	X	X		X
Triclopyr	Turflon Ester		X		X
Triclopyr + clopyralid	Confront		X		X

[a] Source: *Grounds Maintenance,* January 1993.
[b] No endorsement of named products by the author or publisher is intended, nor is criticism implied for products that are not mentioned.
[c] Restricted-use pesticide.
[d] Based on usage at label rates.

Table 1.2. Hard-to-Control Broadleaf Weeds and Their Controls[a,b,c]

	Bedstraw	Catnip	Curly dock	Ground ivy	Hawkweed	Healall	Knawel	Knotweed	Mallow	Pennywort	Pineappleweed	Sorrel, red or sheep	Speedwell	Spurge	Vervain	Violet	Wild carrot	Wild garlic	Wild onion	Yarrow	Yellow woodsorrel (oxalis)
Pre-emergence controls:																					
Atrazine								X					X	X							X
Benefin + oryzalin								X						X[f]			X[f]				X[f]
Benefin + trifluralin														X[f]							X[f]
DCPA													X	X[e]							
Dithiopyr											X		X	X							X
Imazaquin							X					X						X	X		
Isoxaben								X	X[f]	X	X		X	X[e]							X
Oryzalin								X						X[f]							X[f]
Oxadiazon																					X
Pendimethalin								X						X							X
Simazine							X				X		X								
Post-emergence controls:																					
Bentazon + atrazine													X								
Chlorsulfuron	X[e]		X[e]					X[e]	X		X		X[e]			X[e]	X	X[e]	X	X	
2,4-D		X	X	X	X	X		X	X	X					X		X	X	X		
2,4-D + dicamba	X		X	X	X	X	X	X	X	X		X	X	X	X		X	X	X	X	X
2,4-D + dichlorprop			X	X	X	X		X	X	X		X[e]	X	X	X	X[e]	X	X	X	X[e]	X[e]
2,4-D + dichlorprop + dicamba	X		X	X		X		X	X			X	X	X			X	X	X	X	X
2,4-D + MCPP			X	X	X	X		X	X	X		X	X	X	X			X	X	X	X

2,4-D + MCPP + dicamba	X		X	X		X		X	X		X	X	X		X	X	X	X	X
2,4-D + MCPP + dichlorprop	X		X	X		X		X	X	X	X	X	X		X	X	X	X	X
2,4-D + MCPP + MSMA + dicamba	X		X	X		X		X	X		X	X	X		X	X	X	X	X
2,4-D + triclopyr		X	X	X[e]			X		X			X	X[e]	X[e]	X[e]			X	X[e]
DCPA												X[e]							
Dicamba	X		X[e]	X[e]	X			X	X		X		X		X	X[e]	X[e]	X[e]	X
DSMA																			X
Glyphosate	X		X									X							
Imazaquin							X				X			X[g]		X	X		
MCPA			X					X							X				
MCPA + MCPP + dicamba	X		X	X	X	X		X	X	X	X	X	X		X	X	X	X	X
MCPA + MCPP + dichlorprop			X	X	X	X		X	X	X	X	X	X	X	X	X	X	X	X
MCPP			X	X				X		X		X	X					X[e]	
Metribuzin[d]	X							X				X	X						
Metsulfuron methyl			X												X	X	X	X	X
MSMA																			X
Triclopyr			X	X										X	X			X	X
Triclopyr + clopyralid			X	X	X				X		X			X[e]					X[e]

[a] Source: *Grounds Maintenance,* January 1993.
[b] The user should check the label for species and cultivar tolerance and selectivity.
[c] Some labels specifically list weeds controlled by genus and species. Other labels use common names only. This can be confusing when two species have similar common names. Consult your distributor if you are unsure about a specific weed problem.
[d] Apply to dormant turf.
[e] May require a second application for control or application at a higher labeled rate.
[f] For suppression or partial control only.
[g] Labeled, but control is marginal.

Table 1.3 Tolerances of Established Turfgrasses to Herbicides[a]

Herbicide: Confirm species tolerance before using any product, whether listed here or not.	Turfgrass Species											
	Bahia-grass	Creeping bentgrass	Bermuda-grass	Annual bluegrass	Kentucky bluegrass	Buffalo-grass	Centipede-grass	Fine fescue	Tall fescue	Perennial ryegrass	St. Augustine-grass	Zoysia-grass
Asulam			T[g]								T	
Atrazine[b]			T[e]			T[h]	T				T	T
Benefin[b]	T[d]		T		T		T	T	T	T	T	T
Benefin + oryzalin[b]	T		T				T		T		T	T
Benefin + trifluralin[b]	T	T[f]	T		T		T	T	T	T	T	T
Bensulide	T	T	T		T		T	T	T	T	T	T
Bentazon[b]	T	T	T		T		T	T	T	T	T	T
Bentazon + atrazine[b,c]						T				T	T	
Chlorsulfuron	T	T[f]	T		T			T				
CMA			T		T[e]				T			T
Dicamba	T	T[e]	T	T	T	T[e]	T	T	T	T		T
Diclofop			T[c]									
Diquat			T[h]				T[h]				T[h]	T[h]
Dithiopyr[b]	T	T	T		T	T	T	T[g]	T	T	T	T
DSMA			T		T							T
Ethofumesate		T[b,c]	T[h]		T[b,f]					T		
Fenoxaprop		T[b,f,g]		T	T			T	T	T		T
Isoxaben[b]	T	T	T		T	T	T	T	T	T	T	T
Imazaquin[b]			T				T				T	T
MCPP[b]		T[e]	T		T[e]			T	T			
Metolachlor	T		T				T				T	T
Metribuzin			T[b,f]									
Metsulfuron methyl			T		T			T			T	
MSMA			T		T							T

Napropamide	T		T				T	T	T		T	
Oryzalin[b]	T		T			T	T		T[i]		T	T
Oxadiazon			T		T	T			T	T	T[c]	T[c]
Pendimethalin[c]	T	T[b]	T	T[b]	T		T	T	T	T	T	T
Prodiamine[b]		T	T		T		T	T	T	T	T	T
Pronamide			T[b]									
Sethoxydim[c]			T			T	T					
Simazine[b]			T[e]				T				T	T
Siduron[b]		T[e]			T			T	T	T		T
Triclopyr					T			T	T	T		
Triclopyr + clopyralid					T		T	T	T	T		

[a] Source: *Grounds Maintenance,* January 1993.
[b] Not labeled for one or more of the following: greens, native soil greens, tees or collars.
[c] Some products or formulations not labeled in all states.
[d] Adequate tolerance when used according to label.
[e] Injury may occur to certain species or cultivars. Read label before using.
[f] Not labeled for closely mowed turf.
[g] Certain species or varieties only.
[h] When dormant.
[i] Growing in warm-season turf areas.

Note: Blanks indicate that the product can cause severe injury, that is not labeled for use on the particular turf species, or that there is insufficient information to document the level of tolerance. The user should be sure to confirm specific species and cultivar tolerances by checking the label on the product under consideration.

Table 1.4. Grassy Weeds and Sedges and Their Controls[a,b,c]

	Annual bluegrass	Bahia-grass	Barnyard-grass	Crabgrass, large or smooth	Dallis-grass	Fall panicum	Field sandbur	Foxtail, green or yellow	Goose-grass	Nutsedge, yellow or purple	Tall fescue
Pre-emergence controls:											
Atrazine	X		X	X				X			
Benefin	X		X	X				X	X		
Benefin + oryzalin	X[d]		X	X			X	X	X		
Benefin + trifluralin	X		X	X				X	X		
Bensulide	X		X	X		X		X	X[g]		
DCPA	X[e]		X[e]	X		X[e]	X[e]	X	X[e]		
Dithiopyr	X		X	X				X	X		
Ethofumesate	X		X	X				X		X[f]	
Isoxaben	X[e]		X[e]	X[e]				X[e]	X[e]		
Imazaquin	X[f]										
Metolachlor	X			X					X[e]	X[h]	
Napropamide	X			X					X		
Oryzalin	X[d]		X	X			X	X	X		
Oxadiazon				X			X		X		
Oxadiazon + bensulide				X					X		
Pendimethalin	X[e]		X	X		X		X	X		
Prodiamine	X			X				X	X[e]		
Pronamide	X										
Siduron			X	X				X			
Simazine	X		X	X		X		X	X		
Post-emergence controls:											
Asulam				X			X		X		
Atrazine	X										
Bentazon										X[h]	
Bentazon + atrazine										X[h]	
Chlorsulfuron											X
CMA			X	X	X		X	X	X	X	
2,4-D + MCPP + MSMA + dicamba				X	X				X	X	

Diclofop									X		
Diquat	X[e]										
DSMA		X		X	X		X			X	
Dithiopyr				X							
Ethofumesate	X										
Fenoxaprop			X	X		X		X	X		
Fluazifop-P-butyl			X	X		X	X	X	X		
Glyphosate	X	X	X	X	X	X	X	X	X	X	X
Imazaquin	X[f]			X[f]			X			X	
MCPA										X	
MCPA + MCPP + dichlorprop										X	
MSMA		X	X	X	X		X	X		X	
MSMA + cacodylic acid			X	X	X		X	X	X	X	
Metribuzin	X								X		
Metsulfuron methyl		X						X	X		
Pronamide	X										
Sethoxydim		X[e]	X	X		X	X	X	X		
Simazine	X										

[a] Source: *Grounds Maintenance,* January 1993.
[b] The user should check the label for species and cultivar tolerance and selectivity.
[c] Some labels specifically list weeds controlled by genus and species. Other labels use common names only. This can be confusing when two species have similar common names. Consult your distributor if you are unsure about a specific weed problem.
[d] When a winter annual.
[e] May require a second application or application at a higher rate.
[f] For suppression or partial control only.
[g] Labeled but control is marginal.
[h] Labeled for yellow nutsedge only.

Note: All chemicals mentioned in this article should be applied in accordance with the manufacturer's directions on the label as registered under the Federal Insecticide, Fungicide and Rodenticide Act. Mention of trademark or proprietary product does not constitute a guarantee or warranty of the product by the author or the publisher. Nor does it imply approval to the exclusion of other products that may also be suitable. The use of certain pesticides effective against turfgrass weeds, diseases, insects, small animals and related pests may be restricted by some local, state or federal agencies. Be sure to check the current status of the pesticide you are considering before use. Before using any pesticide, READ THE ENTIRE LABEL.

CHAPTER 2

Turfgrass Diseases and Their Management

Peter H. Dernoeden

INTRODUCTION

Most turfgrass diseases are caused by pathogenic fungi that can invade leaves, stems, and roots of plants. The second leading cause of turfgrass disease is plant parasitic nematodes that attack the roots. There are very few bacterial or virus diseases of turf. Fungi and plant parasitic nematodes use a combination of physical pressure and enzymes to enter plants, invade tissues, and/or disrupt metabolic processes. As a result of the injurious effects of a pathogen (disease-causing agent), a plant may exhibit various responses known as symptoms. Examples of symptoms include leaf spots; root and stem rots; bronzing, yellowing or other changes in leaf color; and death of leaves, tillers, or entire plants. Symptoms may be unique to a particular pathogen and can be used to easily identify certain diseases. Unfortunately, symptoms can overlap or share commonality with several different diseases. Fungal pathogens sometimes produce visible structures known as signs. Examples of signs include mushrooms; white powdery mildew; white, fluffy mycelial growth; pink gelatinous mycelial growth; red or black pustules on leaves; and sclerotia. Mycelium is the vegetative body of a fungus, and it is composed of a network of fine tubes that often appear cottony. Sclerotia are compact, hardened masses of mycelium that are produced by some fungi to serve as survival structures. Sclerotia are variously shaped, may have distinctive colors, and often can be seen without the aid of a hand lens. The presence of sclerotia assists greatly in the identification of gray snow mold, red thread, southern blight, Rhizoctonia diseases (especially *R. zeae*), and other pathogens. It is through the use of a combination of symptoms and signs that disease problems are diagnosed. Many signs, such as spores, can only be seen with a microscope. Only those signs and symptoms that can be seen with the naked eye or the aid of a hand lens will be emphasized here.

Time of year, turfgrass species, and environmental conditions also provide very important clues for disease diagnosis. For example, brown patch and Pythium blight are seldom a problem in cool-season grasses when night temperatures fall below 65°F (18°C), and Pythium blight rarely causes severe injury to mature stands of Kentucky bluegrass. Conversely, dollar spot is more damaging during the cooler night temperatures of late spring and fall than during the hottest weeks of summer. It is important to realize that not all grasses are susceptible to all diseases. Summer patch is strictly a high temperature, summer disease of Kentucky bluegrass, fine leaf fescues and annual bluegrass. Species such as perennial ryegrass, creeping bentgrass, and tall fescue are resistant to summer patch. In summary, diseases are diagnosed using a combination of signs, symptoms, weather conditions, and a knowledge of those diseases that are most likely to appear on a given turf species in any particular season of the year. This chapter will focus on disease identification by providing a guide of key factors that includes: field symptoms, distinctive and easily observed signs, predisposing environmental conditions, and grass species likely to be affected. Approaches to reduce injury through cultural, mechanical, biological, and chemical methods are outlined for each disease.

MONITORING DISEASE AND ESTABLISHING THRESHOLDS

Successful disease management is contingent on early disease detection and a proper diagnosis. Knowledge of the environmental requirements for disease development, symptoms of disease, pathogen signs, and the diseases likely to affect any particular turfgrass species are the primary factors scouts use to detect diseases. Visual monitoring is essential to early detection and selection of a tactic that will effectively address each disease.

Turfgrass managers are responsible for monitoring, establishing acceptable thresholds of disease damage, and selecting the most appropriate management technique(s) (e.g., cultural, biological, chemical, or no action). Monitoring can only be effectively achieved on relatively small, well-defined sites such as a golf course; athletic field; park, campus, church or public lawns; or selected lawns in a neighborhood. Visually monitoring vast areas or large numbers of lawns is not usually economically feasible or logistically possible. In the latter situation, chronically affected sites or "hot spots" are preferentially monitored. Hot spots are most often associated with low and frequent mowing; full sun or heat sink sites; low-lying areas; wet shade; high traffic areas; and heavily thatched sites. Every lawn care organization branch manager, golf course superintendent, landscape specialist, and athletic field manager should know those sites (i.e., a particular neighborhood or lawn, a low-lying green, etc.) that are first or most likely to develop disease problems. Some golf course superintendents install environmental monitoring stations and employ computerized disease forecast models to assist in predicting diseases. Although these devices have some limitations, they provide very useful information that assists in monitoring several diseases. There is, however, no substitute for vigilant visual scouting.

Monitoring

The frequency of monitoring depends on the level of management, and the economic or aesthetic value of the site. For example, greens should be scouted every other day during the growing season, and daily during periods of high temperature stress, high humidity, or extended wet and overcast weather. Lawns or athletic fields with chronic disease problems may require weekly to bimonthly scouting during the growing season.

Thresholds

The disease threshold refers to that level of disease injury that requires implementation of a management tactic. This threshold is based on the turfgrass species (i.e., its likelihood to be severely damaged); the prevailing environmental conditions (i.e., the likelihood the weather will remain conducive to disease activity); the economic or aesthetic value of the site; and the cost of chemical treatment versus renovation of damaged turf sites. In Table 2.1 the major turfgrass species are ranked according to their susceptibility to disease. It should be noted that any attempt to rank species is artificial, since there often is great variation in disease susceptibility among cultivars and regions. Intensively managed sites, particularly greens, are rendered more susceptible to disease by virtue of high traffic combined with extremely low and frequent mowing. Manicured Kentucky bluegrass lawns or athletic fields are more likely to sustain disease damage than tall fescue or zoysiagrass. Bermudagrass or bentgrass greens have a 0% Pythium blight threshold and a dollar spot threshold of less than 0.5% blighting. A Kentucky bluegrass lawn maintained by a lawn care company may have a dollar spot threshold of less than 10% blighting. The brown patch threshold for the same lawn may be much greater (say 25%). This is because dollar spot generally is much more destructive than brown patch in a Kentucky bluegrass turf. Conversely, a tall fescue lawn affected with net-blotch or red thread would not be a candidate for chemical treatment since this species is likely to rapidly recover from these diseases. Fungicides are generally not warranted after a turf has been extensively damaged. In that situation, or wherever a disease is a chronic problem, it is wiser to

Table 2.1. Disease Proneness As Related to Turfgrass Species[a]

NORTHERN, WESTERN,[b] AND TRANSITION ZONE REGIONS	
I. Very High	**III. Moderate**
Annual bluegrass	Chewings fescue
Colonial bentgrass	Creeping red fescue
Creeping bentgrass	**IV. Relatively Low[c]**
Velvet bentgrass	Bermudagrass
II. High	Blue sheep fescue
Kentucky bluegrass	Buffalograss
Perennial ryegrass	Hard fescue
	Tall fescue
	Zoysiagrass
SOUTHEASTERN STATES	
I. High to Moderately High	**II. Relatively Low[d]**
Bermudagrass	Bahiagrass
St. Augustinegrass	Carpetgrass
Zoysiagrass	Centipedegrass

[a] There can be great variation in susceptibility among cultivars, regions, or levels of turfgrass management.
[b] Turfgrass diseases are generally less severe in arid or semi-arid regions of the western U.S.
[c] Each species may suffer one or more debilitating diseases and each is rendered more disease-prone as nitrogen level or other management inputs are increased.
[d] At least one disease can be extremely destructive to each species.

renovate the turf with disease resistant, regionally adapted species and cultivars.

Thresholds also are based on the disease "history" of the site. Turf managers should maintain detailed records of all pest problems. Records should be organized by management units such as lawns in a particular neighborhood, individual athletic fields, or by green, tee, fairway, etc. The turfgrass species and cultivar(s), if known, should be recorded as well as the following: (a) date of the appearance of each disease; (b) the duration of disease activity; (c) level of damage tolerated before a management tactic was imposed; and (d) comments regarding the success or failure of the tactic employed. Disease histories for each turfgrass unit provide invaluable information for future disease management programs.

RELATIONSHIP OF ENVIRONMENTAL CONDITIONS AND CULTURAL PRACTICES TO DISEASE

The relationship among environmental conditions, turfgrass plants, cultural practices and pathogens as they interact with one another are the key factors involved in disease development. Of these factors, the environment is the most important determinant of a disease outbreak. For serious disease problems to occur, the environment must simultaneously exert diverging conditions on both plant and pathogen. Those environmental conditions must be conducive for growth of the pathogen, while at the same time be unfavorable for plant growth. For example, snow mold pathogens can only be destructive when low temperatures slow or stop growth of the turf. As long as there is sufficient moisture and favorable temperatures (32 to 50°F or 0 to 10°C), snow mold fungi can actively grow and parasitize grass plants. Similarly, brown patch is most damaging to cool-season grasses in summer when high temperatures reduce turf vigor. Conversely, brown patch is damaging to warm-season grasses in fall because lower temperatures retard their growth and vigor at this time.

The intensity and nature of turfgrass management also greatly influence the severity, and sometimes the types of diseases that occur. Without question, diseases are much more severe on golf courses, and fungicide use on these sites is common. This is due not only to intense traffic, but more importantly, to low and frequent mowing, frequent irrigation, and other intensive cultural practices routinely performed on golf courses. Because of greater flexibility in maintenance tactics, many lawn diseases can be managed with little use of fungicides. The key to any effective disease management program begins with planting regionally adapted, disease resistant turfgrass species and cultivars. Planting improved cultivars alone does not eliminate disease problems. However, proper mowing, fertilization, irrigation, and cultivation practices (i.e., aerification, verticutting, overseeding, etc.) can greatly minimize the severity of many diseases.

Species and Cultivars

A stand of turf normally is composed of plants representing a single or a few species. Planting a single cultivar of Kentucky bluegrass, whose seed is mostly genetically identical, or planting vegetatively propagated warm-season grasses results in monostands. Where a monostand exists, little or no genetic variation in disease resistance occurs. As a consequence, if one plant is susceptible to a particular disease, all plants are susceptible. Except where warm-season grasses are grown or bentgrass or bermudagrass green turf, monoculture is not a major factor in disease susceptibility. This is because it is common practice to blend and mix turfgrass species and cultivars to provide a broad genetic base and thereby reduce the probability of a disease epidemic. Hence, blending cultivars of cool-season grasses, particularly Kentucky bluegrass, is an extremely important first step in managing turfgrass to reduce potential disease problems. The best source of information regarding regionally adapted turf species and cultivars is the Land Grant university of your state and its Cooperative Extension Offices.

CULTURAL PRACTICES

The susceptibility of a particular turf to disease is primarily governed by environmental and cultural factors. Weather conditions obviously cannot be manipulated, but cultural practices can be adapted to reduce disease problems. The management factors that most influence turf diseases include: mowing, irrigation, fertilization, thatch depth, soil compaction, and soil pH.

Mowing

Mowing favors infection by creating wounds through which a pathogen can easily enter plants. Mowing also spreads spores and fungal mycelium of some pathogens. Hence, when foliar diseases are active, especially brown patch, dollar spot and Pythium blight, it is wise to avoid mowing until the leaves are completely dry. Clipping removal from greens, and poling or dragging greens and fairways to speed leaf drying, may help to reduce some diseases such as dollar spot, brown patch, and Pythium blight. For lawns, however, clipping removal probably would have little effect on disease severity. Indeed, it could be argued that clippings provide nutrients that promote turf vigor and thereby indirectly reduce some diseases and help speed recovery of turf from injury.

Height of cut is a major factor in disease susceptibility. Close mowing exacerbates most, if not all, turf diseases. This is particularly true for Helminthosporium diseases, rusts, summer patch and other root diseases, anthracnose, brown patch, dollar spot, and injury caused by plant parasitic nematodes. Cultivars of bluegrass that are apparently resistant to summer patch may be predisposed to this disease by low mowing. It is important to note that the term "resistance" means only that a plant is less susceptible to a disease. Resistance is a relative concept that does not imply that a cultivar is immune to a particular disease. There are few examples of turfgrass immunity to common fungal diseases.

The continuous removal of the youngest, most photosynthetically productive tissues, below recommended heights, causes depletion of carbohydrate reserves in the grass plant. These reserves play a key role in active disease resistance processes and are needed by plants to recover from injury. Increasing mowing height increases leaf area, which results in an increase in carbohydrate synthesis. High mowing also helps to moderate soil temperature and to stimulate deeper root penetration. One of the most important cultural recommendations for most disease problems is to increase mowing height when turf injury is first observed. For example, studies have shown that summer patch was less damaging to Kentucky bluegrass maintained at 3.0 inches (7.6 cm) as compared to 1.5 inches (3.8 cm). When severely damaged bluegrass turf that had been maintained at 1.5 inches (3.8 cm) was allowed to grow to 3.0 inches (7.6 cm) it recuperated more rapidly from summer patch than turf maintained continuously at a 1.5 inch (3.8 cm) height.

Irrigation

Irrigation provides moisture critical to spore germination and fungal development. The timing, duration, and frequency of irrigation may greatly affect disease intensity. Light, frequent irrigations discourage root development and predispose turf to injury when extended periods of drought occur. Light and frequent irrigations that keep soils moist during periods of high temperature stress greatly increase summer patch of bluegrasses and creeping red fescue. This is because the fungus that causes summer patch (*Magnaporthe poae*) grows much more rapidly along roots when soils are moist and very warm (>85°F or 29°C in the root zone). Studies also have shown that summer patch was much less severe when 3.0 inch (7.6 cm) high turf (4% injury) was subjected to deep (i.e., water penetrates soil throughout the root zone) and infrequent (i.e., water only applied during early signs of wilt) irrigation when compared to 3.0 inch (7.6 cm) high turf (16% injury) or 1.5 inch (3.8 cm) high turf (36% injury) subjected to light and frequent irrigation. The light and frequent irrigation treatments in that study kept soil moist for extended periods, which favored root infection by *M. poae* during the warm summer months. Frequent cycles of wetting and drying of thatch enhance spore production by Helminthosporium leaf spot pathogens (i.e., *Drechslera* and *Bipolaris* spp.). Deep irrigation helps to encourage bacteria that keep leaf spot pathogens from producing enormous spore populations in thatch. Sometimes, however, light and frequent irrigation is recommended in situations where a disease or an insect pest already has caused extensive damage to roots. For example, studies have shown that the severity of necrotic ring spot, another root disease of Kentucky bluegrass, is reduced by light daily watering.

Subjecting turf to extreme drought stress appears to enhance the level of injury in stands affected by Helminthosporium diseases, take-all, stripe smut, powdery mildew, dollar spot, necrotic ring spot, and fairy rings. Conversely, excessive irrigation restricts root development and encourages numerous diseases. Turfgrasses grown under wet conditions develop succulent tissues and thinner cell walls, which are presumably more easily penetrated by fungal pathogens. Algae, mosses, black layer and the germination of weed seed are encouraged by wet or waterlogged soils, particularly where stand density is poor. Early morning or afternoon irrigation is often recommended during summer to ensure that plant tissues are dry by nightfall. This practice reduces the number of nighttime hours when leaves are wet and thereby helps to minimize the intensity of foliar blighting diseases such as Pythium blight and brown patch.

Fertility

Proper soil fertility improves plant vigor and the ability of plants to resist or recover from disease. Excessive use of nitrogen encourages snow molds, Helminthosporium leaf spot, brown patch, powdery mildew, Nigrospora blight, Pythium blight, and other diseases. Excessive use of nitrogen promotes tissue succulence and thinner cell walls which, as previously mentioned, are more easily penetrated by fungal pathogens. Excessively applied nitrogen also may divert carbohydrates and important biochemicals synthesized by the plant to resist pathogen attack into growth processes devoted to leaf production. Diversion of these plant chemicals into developing tissues, rather than to plant defense, may result in shortages that adversely affect the capacity of the plant to resist fungal invasion. Conversely, turfgrasses grown in nutrient poor soils lack vigor and are more prone to severe damage from dollar spot, red thread, and rust diseases. Application of nitrogen to diseased turf under these conditions stimulates foliar growth at a rate that exceeds the capacity of these fungi to colonize new tissues.

In general, slow-release nitrogen fertilizers are recommended where chronic disease problems occur. The controlled release of these nitrogen sources results in a moderate rather than rapid or excessive growth rate. Selected composted animal waste or sludge nitrogen sources show promise for managing diseases such as dollar spot, red thread, brown patch, and necrotic ring spot. While slow-release nitrogen sources are preferred for disease management, there are some exceptions to this generalization. Rapid release, water soluble nitrogen sources may be recommended during autumn months to help speed recovery of turf injured by summer diseases or environmental stresses. Furthermore, water soluble nitrogen from ammonium sulfate or ammonium chloride reduces the severity of take-all of bentgrass and spring dead spot of bermudagrass, and ammonium sulfate can reduce pink snow mold and summer patch damage.

There is little documented evidence that phosphorus, potassium, or micronutrients alone significantly impact on turf diseases. Deficiencies in these nutrients, however, can weaken plants and predispose them to more disease injury than normally would have occurred. For example, take-all disease of bentgrass frequently is more severe in greens where phosphorus is deficient. It is likely that Pythium root rot and other root diseases may be more severe in phosphorus deficient soils. More importantly, N + K and N + P + K fertilizers (especially when applied in a 3:1:2 ratio) have been shown to reduce the severity of take-all, stripe smut, spring dead spot, red thread, and Fusarium patch, when compared with nitrogen alone.

Thatch

Many turfgrass pathogens survive as resting structures or as saprophytes (organisms living on dead organic matter) in thatch. Thatch also provides fungi with moisture. As previously noted, some pathogens, such as *Helminthosporium* spp., produce enormous numbers of spores in thatch, particularly when the thatch is subjected to frequent cycles of wetting and drying. Stripe smut, Helminthosporium melting-out, and summer patch are diseases that appear to be favored by excessive thatch accumulation.

Traffic and Soil Compaction

Traffic, like mowing, produces wounds that are easily invaded by some fungal pathogens. Compaction caused by intense traffic impedes air and water movement into soil and eventually restricts root function. This results in a decline in plant vigor and disease resistance. Summer patch, Helminthosporium leaf spot, and anthracnose tend to be more damaging on highly trafficked or compacted soil sites. Core cultivation can alleviate compaction and reduce the severity of these diseases. Conversely, plant parasitic nematode populations tend to be higher in light-textured, sandy loam soils than in clays or compacted soils.

Soil pH

Soil pH also may affect disease development in turfgrasses. In general, most turfgrass pathogens are able to grow at any pH encountered by turf. High soil pH, however, favors take-all and pink snow mold. Factors that encourage take-all under alkaline soil conditions remain imperfectly understood. It is likely, however, that this pathogen achieves a competitive advantage over antagonistic microorganisms in alkaline soils. Acidifying nitrogen fertilizers reduce the severity of take-all, pink snow mold, spring dead spot, and summer patch. Soil pH, however, does not directly impact many diseases, so it is prudent in most situations to maintain or adjust soil pH to a range of 6.0 to 7.0.

Adherence to sound cultural practices is basic to reducing disease severity in turf. The turfgrass environment, however, is not static, and managers must continually modify cultural practices to encourage vigorous growing conditions. To maintain optimal conditions, the turfgrass manager must routinely monitor the nutrient and pH status of soil, adjust mowing heights, judiciously irrigate, overseed with disease resistant cultivars, manipulate different sources of nitrogen fertilizer, and control thatch and soil compaction. The basic principles of cultural disease management of turfgrasses are outlined in Table 2.2. Shade is a key factor that limits root growth and contributes to a decline in turf vigor. Lawns and greens shrouded in trees and untrimmed ornamentals are chronically weak turfs. Hence, heavy trimming and sometimes the removal of numerous trees would be sound mechanical measures for improving turf vigor and thereby alleviating some diseases. Mechanical approaches to alleviating greens diseases are listed in Table 2.3.

BIOLOGICAL CONTROL OF TURFGRASS DISEASES

The development of biological control agents for turfgrass disease management lags behind advances in insect pest and weed control. It generally is easier to identify diseases of insects and weeds, isolate the disease agents, grow them in a laboratory, and direct sprays of propagules (infecting units, such as spores

Table 2.2. Basic Principles for Managing Turfgrass to Minimize Damage Caused by Common Fungal Pathogens

1. Plant disease-resistant species and cultivars. Recommendations for regionally adapted cultivars can be obtained from most Cooperative Extension Service offices.
2. Use a balanced N-P-K fertilizer at the appropriate times. In the absence of a soil test, N-P-K should be applied at least once annually in a ratio of 3:1:2. Most of the total annual nitrogen (about 75%) should be applied to cool-season grasses during autumn months. Conversely, most nitrogen should be applied to warm-season grasses in late spring and during summer months.
3. At least 50% of all nitrogen applied per year should be from a slow-release nitrogen source. Some examples of slow-release nitrogen sources include: isobutylidene diurea (IBDU), methylene urea, sulfur coated urea, and composted sludges or animal waste products.
4. Maintain a high mowing height within a species' adapted range. Raise the mowing height during periods of environmental stress or disease outbreaks. To reduce disease damage in lawns, it is best to maintain Kentucky bluegrass or perennial ryegrass at 2.5 to 3.0 inches (6.3 to 7.6 cm), and all fescue species at 3.0 to 3.5 inches (7.6 to 8.9 cm) in height. Bermudagrass lawns should be maintained at a height of 1.0 to 1.5 inches (2.5 to 3.8 cm); centipedegrass and zoysiagrass 1.5 to 2.0 inches (3.8 to 5.1 cm); St. Augustinegrass 2.5 to 3.0 inches (6.3 to 7.6 cm); and bahiagrass 3.0 to 3.5 inches (7.6 to 8.9 cm).
5. Avoid mowing turf when leaves are wet and foliar mycelium is evident.
6. Irrigate deeply to wet soil to a depth of 4 to 6 inches (10–15 cm) when the turf first exhibits signs of wilt. Avoid frequent and light applications of water, except when root systems are severely injured or shallow.
7. Test soil every 2 or 3 years for phosphorus and potassium levels as well as soil pH. Adjust soil pH to a range of 6.0 to 7.0.
8. Avoid application of broadleaf herbicides or plant growth regulators when diseases are active.
9. Alleviate soil compaction and reduce thatch to about 0.5 inch (1.3 cm) in depth through core cultivation, verticutting, or a combination of these methods.
10. Overseed or renovate chronically damaged sites with disease resistant species and regionally adapted cultivars.

Table 2.3. Mechanical Approaches to Alleviating Diseases on Greens

1. Irrigate at dawn to remove leaf surface exudates and physically knock down mycelium of some pathogens.
2. Remove dew and leaf surface exudates to speed leaf drying by dragging, poling, or whipping.
3. Avoid mowing wet foliage when foliar mycelium is evident.
4. Keep mowers adjusted and blades sharp; use walk-behind greensmowers and increase the height of cut whenever possible.
5. Core cultivate or verti-drain compacted sites.
6. Employ water injection or core cultivation in combination with wetting agents to alleviate localized dry spots or fairy ring damage.
7. Never roll greens when soils are wet.
8. Remove trees and brush to improve air movement and sunlight penetration; electric fans may improve air movement in some situations.
9. Prune tree roots around greens to reduce competition for water and nutrients.

or mycelium) to sites infested with these pests. Using bacteria or fungi to control a fungal turfgrass pathogen is much more complicated. The microscopic size of the target pest, and the complex environmental forces affecting both the biological control agent and the pathogen, are significant barriers to progress. Pathologists have demonstrated that several fungi and bacteria provide suppression of some turfgrass diseases in controlled laboratory studies. In field studies, however, these agents generally have failed to provide a significant level of disease suppression. An exception is the fungus *Typhula phacorrhiza*, which reduced gray snow mold blighting by 40% to 70% in field tests. Most field failures are due to the inability of biological control agents to survive, compete, or reproduce in populations large enough to provide disease suppression. Biological disease control research being conducted in turf with the bacterium *Enterobacter cloacae*, as well as other bacteria and fungi, however, is very promising.

Maintaining turf vigor through sound cultural practices, planting disease resistant cultivars, and enhancing microbial activity in soil are the most effective means of biologically reducing a turfgrass disease. Application of composted sludges, manures, and agricultural wastes sold as organic fertilizers provides organic matter and nutrients, which encourage the growth and proliferation of microorganisms and some invertebrates (e.g., earthworms) in soil. Indeed, prior to the advent of synthesized fertilizers and fungicides, when greenskeepers relied on composts for nutrients, some diseases such as dollar spot were probably suppressed. Composted topdressings were prepared by mixing equal parts of well-rotted manure, rotted grass clippings and soil. The mixture was allowed to compost for a 2- to 3-year period prior to being screened, mixed with sand, and spread on greens.

Microorganisms that use composts as an energy source proliferate and provide several benefits. They degrade organic matter, which provides nitrogen for plant uptake, and they may form mycorrhizal relationships, which enhance phosphorus availability. More importantly, from a disease viewpoint, they act as antagonists of plant pathogens. These microbes may produce antibiotic substances that interfere with the growth of pathogens or they may simply outcompete pathogens for nutrients, water, or space. Antagonistic microorganisms can block infection sites on roots and may trigger a plant's own disease resistance mechanisms.

Application of certain composted sludges, manures, and agricultural wastes as fertilizers has been shown to reduce the severity of necrotic ring spot, dollar spot, and brown patch. In some regions, however, brown patch, summer patch, and large patch of zoysiagrass suppression was not achieved with these same organic fertilizers. The greatest potential for disease suppression benefits appears to be in sandy or other low nutrient soils, or sites where turf is introduced to previously wooded sites or other nontraditional agricultural lands.

COLLECTING AND SENDING DISEASED SAMPLES

As you know, there are a multitude of pathogens that attack turfgrasses. Fungi and plant parasitic nematodes cause most disease problems in turf. Many fungal diseases are easily distinguishable because of unique signs or symptoms that accompany their activity. For example, the red thread fungus produces a pink gelatinous mycelium and characteristic red, antler-like sclerotia, making this disease easy to diagnose. Unfortunately, many diseases do not always produce textbook symptoms. Certain signs, such as foliar mycelium, can be deceiving. During early infection stages, fungi that cause dollar spot, brown patch, and Pythium blight may produce large amounts of foliar mycelium. Most turfgrass managers equate foliar mycelium with Pythium blight. In many situations, this is an error in judgment. These three diseases can, however, be easily separated by microscopic observation of the mycelium or through the use of a disease identification kit. Other pathogens, particularly those that attack roots, can be extremely difficult to identify. Take-all, summer patch, necrotic ring spot, spring dead spot, and bermudagrass decline are root and stem diseases. With few exceptions, these pathogens cannot be identified by mycelial characteristics. To be absolutely certain, a pathologist must isolate the pathogen on an agar medium. Isolation techniques, however, are not consistently successful. Once isolated, these fungi must be manipulated to produce their spore producing structures. These techniques often require a 4 to 8 week waiting period, and frequently they are unsuccessful. Many fungal diseases, or the activity of parasitic nematodes, may produce symptoms that mimic environmental stress or insect pest damage. The similar symptomatologies of numerous diseases, as well as environmental stresses and insect pests, makes diagnosis a challenge for even the experts. When perplexed or exasperated, turf managers eventually send samples to a laboratory to obtain a precise answer for unknown maladies.

Most Land Grant universities provide plant disease diagnostic services. Many of these laboratories, however, are not staffed by pathologists trained to diagnose turfgrass diseases. Few, if any, labs are able to identify virus diseases of turfgrasses. Some university and all private labs charge a fee. It is prudent to choose a laboratory that you have confidence in, but it is equally important that samples are properly collected and shipped. Since time is always an essential factor in treating turf diseases, samples must be transported as rapidly as possible. Either direct transport or overnight express shipment is preferable. Most turf pathologists at universities have little or no daily responsibilities in plant disease labs or clinics run by the Cooperative Extension Service. Hence, if you want a particular specialist to handle your problem, it is wise to contact that individual by telephone prior to shipping samples. Otherwise, the turf disease specialist may be out-of-town or have other commitments that preclude a rapid handling of your sample. If shipping by express mail, never ship on Friday or Saturday, as samples arriving over a weekend or holiday will not be processed until the following Monday or workday.

The pathologist should receive fresh samples in good condition. Samples must be collected while the disease is actively injuring or causing a decline of the turf. Samples collected even a few days after disease activity has subsided may yield negative or misleading information. Avoid sending dead plants or samples from turf areas recently treated with fungicides. Samples also should be clearly marked on the *outside* of packaged soil or turf plugs, and should be accompanied with a sample submission form from the lab or a letter that provides the information outlined below. The form or letter should be placed in a separate plastic bag.

1. Turf Species and Site Affected
   a. Green, tee, fairway, lawn, athletic field, etc.
   b. Full sun, partial shade, heavy shade
   c. Soil often wet or dry
   d. Soil type and pH; is soil compacted?
   e. Name of cultivar and age of turf, if known
   f. Mowing height and frequency

2. Symptoms and Environmental Conditions
   a. Describe symptoms (e.g., circular patches, diffuse thinning, color changes) and distribution (e.g., widespread, localized, scattered)
   b. Describe weather conditions just prior to and during the period when injury became evident
   c. Date symptoms first appeared and date of sample collection

3. Cultural and Chemical
   a. List all pesticides applied to the site within 21 days prior to appearance of symptoms
   b. Note the total amount of N-P-K applied in the last 12 months
   c. Irrigation frequency, amounts and time of day

4. Polaroid Photographs
   a. Close-up photos and photos taken from a standing position are particularly useful

The preferred methods of collecting soil samples for plant parasitic nematode assays and turf samples for fungal disease diagnoses are outlined below.

Nematode Assay

All too often, turfgrass managers send a shovelful of soil from a single site to a lab and request a nematode assay. This type of sample is unacceptable since the results will likely provide meaningless information. A good sample from a single green, lawn, field, etc., will require several minutes of work. It is best to collect soil a day after rain or irrigation when it is moist, but not wet or muddy. About one pint (500 cc) of soil is necessary from each site, rather than a composite sample of two or more greens, tees, lawns, etc. Where parasitic nematodes are a chronic problem, soil should be collected in late spring or early summer (about mid-June) and again in late summer (about mid- to late

September). Nematode counts from monitored sites will provide baseline information on population densities of parasitic nematodes for each site. This baseline information is particularly useful for interpreting problems with turfs that are poorly rooted and tend to wilt rapidly each summer.

Samples should be collected in the upper 3 inches (7.6 cm) of soil since this area represents the zone where most turfgrass roots are found. If roots are confined only to a 1 to 2 inch (2.5 to 5.1 cm) soil depth, this would be the appropriate zone from which to sample. The best tool for the job is the common "soil probe," which extracts cores that range from 0.75 to 1.0 inch (1.9 to 2.5 cm) in diameter. Soil cores should be randomly taken from the interface between damaged or declining turf and healthy turf. The thatch layer, and plant stems and leaves should be discarded. Only soil and roots are needed. Samples from dead areas, or where few plants persist, will provide false population density information. This is because plant parasitic nematodes are obligate parasites that feed only on living plant tissue. Hence, parasitic nematodes naturally migrate outward from severely thinned areas, and populations generally are highest in the region where unhealthy and healthy plants meet. About 15 to 20 randomly selected soil cores, or one pint (500 cc) of soil, will provide a satisfactory sample size. These soil cores should be uniformly mixed and placed in a plastic bag or specially-lined bags provided by the testing laboratory. Avoid using thin-paper lunch bags, as moist soil is likely to cause them to fall apart during transport. It is very important that site information be marked on the outside of the bag with a waterproof marker. Do not place small scraps of paper that identify the collection site inside the bag. These paper scraps deteriorate rapidly, and will probably be unreadable within a few hours of contact with moist soil.

Once collected, the soil should be kept out of direct sunlight. Soil placed on the dashboard or in the trunk of a car will heat up rapidly, killing the nematodes. If samples must be held for a few days, they should be refrigerated or placed in a cool and dark room. Extremes in temperature, high or low, will kill nematodes. This must be avoided because only living nematodes can be effectively extracted from soil. In the laboratory, various methods for extraction are used. Many labs use a glass funnel technique. Soil is placed in a funnel lined with porous paper. The funnel base is clamped, and the soil is flooded with water. Within 24 hours, most of the living nematodes will swim from the soil through the paper and be trapped at the base of the funnel. Water in the base of the funnel stem is drawn off into a shallow dish. The dish is placed under a dissecting or stereo microscope, and the nematodes are identified to genus and counted. Unfortunately, there are no reliable data correlating nematode number per sample and expected degree of turf injury in the field. For example, a count of 800 root knot nematodes/250 cc of soil may cause no visual damage to a green that is well drained and deeply rooted; whereas, 100 root knot nematodes/250 cc soil from a green where rooting is shallow may result in severe thinning of the stand during hot, dry, or windy periods. The nematologist, however, generally will be able to provide a good management recommendation based not only on the total number of plant parasitic nematodes, but also on the type of nematodes that are present.

Other Diseased Turf Samples

Sod pieces the size of a book or cup cutter size plugs are ideal for situations where a fungal or bacterial disease, environmental stress, or chemically damaged tissue are suspected. As noted previously, most labs are not equipped or staffed to identify viruses that inhabit turfgrasses. Known virus diseases of turf are diagnosed by host symptoms and not by serological or DNA-based techniques. Bacterial diseases are relatively rare in turf, and are primarily diagnosed by sectioning tissue and searching for streaming oozes of bacterial cells. Bacteria are identified based on colony growth characteristics on specialized bacteriological media; staining (i.e., Gram stain) character; and ability to grow in the absence of oxygen (anaerobic growth).

Two or more plugs should be carefully selected from areas where both early and advanced symptoms are evident. Each plug should contain dead tissue (if present), and some healthy appearing plants. The majority of the sample, however, should contain various stages of unhealthy, declining or blighted tissue. Plugs should be taken to a soil depth of about 2 inches (5.1 cm). Send only soil that is clinging to roots; don't send deep plugs where there is a conspicuous cleavage zone beyond which roots do not penetrate. The only exception would be if black-layers are evident from greens. Plugs should be collected when soil is moist, but well drained and not in a muddy or soggy condition. Samples should then be wrapped in an inner, damp paper towel, followed by an outer wrapping of aluminum foil. Some labs prefer foil wrapping of roots and soil only, thereby leaving foliage exposed and unwrapped. Do not box plugs that are loose or unbound in paper bags or newspaper. Aluminum foil is preferred because it helps ensure that soil and roots remain intact during transport. Plugs wrapped in paper or crowded into large plastic bags normally arrive as a loose mass of soil mixed on top and within the turfgrass canopy. Broken plugs with soil mixed into diseased foliar tissue greatly impedes the pathologist, and may be cause for an improper diagnosis. Before packing the foil-wrapped plugs, ensure that field identification information is securely fixed to the *outside* of the foil. Obviously, if diseased turf samples are being hand-delivered they would not have to be as carefully wrapped. However, samples shipped express or overnight mail should be foil-wrapped and packed with newspaper or appropriate packing material to prevent tumbling and breakup of plugs. Package the samples for transport as if they were glass or some other breakable object.

As noted for nematode assays, samples should not be shipped unless they will arrive on a work day. Never send samples by expensive overnight mail on a Friday or Saturday or near a holiday. If you want a specific turf specialist to handle your sample, be sure to contact the individual by phone or at least leave a message alerting the specialist that a sample is enroute and when to expect its arrival. Samples that spend a weekend in a hot post office or delivery truck, or lay unopened on a desk for a few days generally are badly damaged and cannot be used for disease diagnostic purposes.

In the laboratory, most samples are incubated overnight in a humidity chamber, particularly if a foliar pathogen is suspected. For most diseases a diagnosis generally is made within 48 hours of receiving a fresh, properly collected and shipped sample. If the pathologist must first attempt to isolate a possible pathogen on a sterile media, the process may require a waiting period of one or more weeks. Other techniques may involve months of waiting. All techniques that begin with an attempt to isolate a pathogen are tedious and time-consuming to perform. During the interim, however, a turf pathologist is likely to suggest cultural and/or chemical approaches to alleviate the disease stress and further turf damage.

WINTER DISEASES

Snow protects dormant turfgrass plants from desiccation and direct low temperature kill, but also provides a microenvironment conducive to the development of some low-temperature, pathogenic fungi. There is no shortage of fungi capable of damaging turf during cold periods between late fall and early spring. The most common low temperature fungal diseases are pink snow mold and gray snow mold. Other diseases known to be active under snow cover or during winter months include red thread, Helminthosporium leaf spot, yellow patch, and Pythium snow blight. Crown hydration, freezing, desiccation, and ice damage are also important problems, especially in northern tier states, Rocky Mountain states and Canada. Cold weather environmental stress damage sometimes can be confused with snow molds.

Snow mold fungi are remarkable because they are active at temperatures slightly above freezing. Snow molds are damaging when turf is dormant or when its growth has been retarded by low temperatures. Under these conditions, turfgrasses cannot actively resist fungal invasion. Although these fungi are known as snow molds, they can attack turf with or without snow cover. In general, these diseases develop when temperatures are cool (32–55°F or 0 to 13°C) and there is an abundance of surface moisture.

Pink Snow Mold or Fusarium Patch

Pathogen: *Microdochium nivale* (Fr.) Samuels and Hallett
Primary Hosts: Most turfgrasses, especially annual bluegrass and bentgrasses
Predisposing Conditions: Prolonged periods of snow cover or cold, wet, and overcast weather

Pink snow mold affects a wide range of turfgrass species under snow including perennial ryegrass, bluegrasses, bentgrasses and the fescues (Plates 2–1, 2–2, and 2–3). This disease is generally most destructive to annual bluegrass and bentgrasses. Conditions favoring pink snow mold include low to moderate temperatures; abundant moisture; prolonged deep snow; snow fallen on unfrozen ground; lush turf stimulated by late season application of excessively high amounts of nitrogen fertilizer; and alkaline soil conditions. Prolonged periods of cold, rainy weather are particularly conducive to disease development on greens. When this disease occurs in the absence of snow it is referred to as Fusarium patch.

Symptoms of pink snow mold appear as small water-soaked patches 2 to 3 inches (5.1 to 7.6 cm) in diameter. Most fully developed patches are 3 to 8 inches (7.6 to 20.3 cm) in diameter, but some patches may range from one or two feet in diameter and coalesce. Large patches are most likely to appear in lawns, greens, or fairways when the Fusarium patch phase precedes a heavy snow. The pink coloration of diseased turf at the edge of the patches is produced by the pinkish color of the mycelium. Plants eventually collapse and die, and mycelium mats the leaves. Matted leaves have a tan color, but on close inspection they may display a pinkish cast.

On greens, Fusarium patch often appears as patches 1 to 3 inches (2.5 to 7.6 cm) in diameter. These patches generally have a reddish-brown or yellow color and may increase in size to 6 or more inches (>15 cm) in diameter. Yellow patches bordered by a reddish-brown or pinkish periphery can easily be confused with yellow patch (also known as cool temperature brown patch). Gray-colored smoke rings also may be associated with Fusarium patch. Mycelium on the leaf blades produces fruiting bodies called sporodochia, upon which large amounts of spores are borne. These white or salmon-pink sporodochia are very tiny and appear as flecks on necrotic (dead) tissue. These flecks may be seen with a hand lens when the disease is active, but cannot be seen on dried tissue. The spores are easily spread by water, machinery, and foot traffic. Therefore, blighting can appear in streaks or even straight lines when spores are carried on wheels. When damage occurs under snow, the extent of injury usually is more severe than without snow cover. After snow recedes, patches are bleached white and may or may not have a pink fringe. Normally, most plants in affected patches under snow are killed.

Management

Snow mold injury can be reduced by using a balanced N-P-K fertilizer in fall and avoiding excessive late season applications of water soluble nitrogen. Ammonium sulfate may be suggested as a nitrogen source where soils are alkaline and pink snow mold is common. Avoid the use of limestone where soil pH is above 7.0 since soil alkalinity may encourage pink snow mold. In the Pacific Northwest, sulfur and phosphorus nutrition has been shown to reduce Fusarium patch. Sulfur, however, must be used with caution on greens in most other regions. Continue to mow late into the fall to ensure that snow will not mat a tall canopy. On golf courses, snow fences and windbreaks should be used to prevent snow from drifting onto chronically damaged greens. Divert skiers and snow mobiles around greens to avoid snow compaction.

Mercury-based fungicides were widely used for snow mold prevention, but they are no longer commercially available. Pentachloronitrobenzene (PCNB, Terraclor), iprodione (Chipco 26019), propiconazole (Banner), and triadimefon (Bayleton) also provide good control of pink and gray snow molds. Fungicidal control is best achieved with a preventive application prior to the first major snow storm of the year. Subsequent applications to greens or other prone locations should be made during mid-winter thaws and spring snow melt in areas where the disease is chronic. During extremely wet winters, the disease can cause extensive injury to lawns. It is best to spot apply fungicides on lawns where the disease has developed in localized pockets. Widespread blighting of lawns on some occasions may require blanket fungicide treatment. In most regions of the U.S., snow mold prevention with fungicides is only warranted for golf course turf, bowling greens and grass tennis courts.

Gray Snow Mold or Typhula Blight

Pathogens:	*Typhula incarnata* Lasch ex Fr., *T. ishikariensis* Imai
Primary Hosts:	All turfgrasses
Predisposing Conditions:	Prolonged periods of snow cover, especially when snow falls on unfrozen ground

Gray snow mold or Typhula blight is a serious disease with a wide host range. Symptoms initially appear as light brown or gray patches 2 to 4 inches (5.1 to 10.2 cm) in diameter (Plates 2-1 and 2-4). Patches may enlarge to 2 feet (60 cm) in diameter and coalesce. Gray snow mold may occur with or without snow cover; however, damage usually is minimal in the absence of snow. Like pink snow mold, Typhula blight is more damaging under prolonged deep snow, particularly when heavy snow accumulates on unfrozen ground.

Gray snow mold fungi begin the disease cycle as saprophytes colonizing dead organic matter. Under snow, however, the fungus penetrates living leaves, sheaths and may ultimately invade the crown. Normally, *Typhula* spp. do not completely kill crowns so plants generally recover during the spring. Conversely, *M. nivale* (the pink snow mold fungus) more frequently invades crown tissues and kills turf. *Typhula* spp. survive unfavorable environmental conditions as sclerotia. Sclerotia of *Typhula* spp. are compact masses of fungal mycelium covered with a dark colored, protective rind. Sclerotia are wrinkled, rounded, chestnut brown or black in color and often are less than 1/8 inch (3.2 mm) in diameter. Sclerotia also may appear as tiny, rounded specks or as elongated, blackened or reddish-brown crusts embedded in or clinging to necrotic tissue. When cool, moist weather conditions return in late fall, these sclerotia germinate to produce fungal mycelium or a specialized fruiting body upon which spores are borne. Both *Typhula* species that attack turf produce similar symptoms. Sclerotial color is one of the primary characteristics pathologists use to differentiate between *Typhula* spp. After sclerotia are produced in large numbers in late winter or spring, which may give the affected gray patches a speckled appearance, the disease subsides. Hence, spring application of a fungicide after sclerotia appear is of little, if any, disease management value.

Management

See the pink snow mold section for cultural approaches for managing both pink and gray snow mold. The fungus *Typhula phacorrhiza* applied on grain inoculum was shown to suppress gray snow mold on creeping bentgrass greens by 44% to 70%. Although not commercially available, it is believed that this biological control agent can be formulated into pellets and applied by standard fertilizer spreaders.

Chloroneb (Terraneb SP), flutolanil (ProStar), iprodione (Chipco 26019), propiconazole (Banner), and triadimefon (Bayleton) provide effective control of the disease. Preventive fungicide applications are more beneficial in situations when gray snow mold is chronic. Curative applications in spring after sclerotia are produced provide little benefit.

Yellow Patch or Cool Temperature Brown Patch

Pathogen: *Rhizoctonia cerealis* Van der Hoeven
Primary Hosts: Annual bluegrass and creeping bentgrass greens
Predisposing Conditions: Cold, wet periods with abundant surface moisture

Yellow patch, sometimes called cool temperature brown patch, develops during prolonged cool and overcast or moist periods from late fall to early spring. Yellow patch is a disease of bentgrass, annual bluegrass, and sometimes it occurs in Kentucky bluegrass and perennial ryegrass turf (Plates 2–5 and 2–6). It is most frequently observed on greens where it produces rusty-brown and yellow rings, or circular yellow patches, a few inches (6 to 20 cm) to one or more feet (> 30 cm) in diameter. This disease can be confused with pink snow mold and diagnosis should be confirmed if in doubt. Damage generally is superficial, but significant thinning of turf may occur during prolonged wet and overcast weather during late winter and early spring.

Management

The disease may be less severe if surface drainage is improved or if standing water is squeegeed off greens during rainy weather. A broad-spectrum fungicide, such as chlorothalonil (Daconil 2787) or iprodione (Chipco 26019) will prevent severe thinning, but no fungicides or cultural practices are known to prevent the formation of rings and patches.

Low Temperature Pythium or Snow Blight

Pathogens: *Pythium graminicola* Subramanian, *P. iwayamai* Ito, others
Primary Hosts: Annual bluegrass and creeping bentgrass greens
Predisposing Conditions: An abundance of surface water or saturated soils during cool or cold periods.

The disease is favored by abundant moisture, particularly when snow falls on unfrozen soils. It can potentially affect many grasses, but is currently recognized as a disease primarily restricted to greens. Creeping bentgrass and annual bluegrass greens can be damaged in winter by several *Pythium* species. This disease often follows drainage patterns and is most injurious in low areas of poorly drained greens. Damage appears as small tan, gray, or yellow to orange spots or patches. Injury often appears first at the outer peripheral areas of the green surface, near the interface between mineral soil and the sandy greens mix. There may or may not be any foliar mycelium evident. While the disease occurs in greens grown on mineral soils, it is most common in high sand content greens. These *Pythium* spp. cause both a foliar blight and stem or root rot. In late winter or early spring, dead patches may mimic old dollar spot damage; i.e., patches are 1 to 3 inches (2.5 to 7.6 cm) in diameter and leaves appear grayish-tan.

Management

There are no known cultural measures for managing this disease. Improving surface drainage may help to reduce disease severity. Check underneath winter blankets frequently for snow blight development. Where the disease is chronic, apply fosetyl-aluminum (Aliette), or other *Pythium*-targeted fungicides with systemic (i.e., penetrates tissue) activity, in combination with mancozeb (Dithane M-45, Fore). Apply fungicides prior to the first heavy snow and again at snow melt, or during a mid-winter thaw, on greens with a past history of the disease.

DISEASES INITIATED IN FALL OR SPRING THAT MAY PERSIST INTO SUMMER

The diseases described in this section normally develop in response to cool and moist weather. Many of these diseases, however, can persist or develop anew during the summer. For example, red thread can be active at almost any time of year when weather conditions are wet and overcast. Red thread is one of the first diseases to appear in spring and can be active in late winter during snow melt. Similarly, anthracnose basal rot can be severe in early spring, but it also frequently develops in late summer. The seasonal appearance of most diseases can never be precisely predicted since environmental conditions are impossible to predict. Furthermore, there usually are exceptions to the rule, and the precise conditions that trigger most diseases are not well understood.

Red Thread and Pink Patch

Pathogen:	Red Thread = *Laetisaria fuciformis* (McAlpine) Burdsall Pink Patch = *Limonomyces roseipellis* Stalpers & Loerakker
Primary Hosts:	Bentgrasses, bluegrasses, bermudagrasses, fescues and ryegrasses
Predisposing Conditions:	Extended overcast and rainy weather

Red thread and pink patch are common diseases of turfgrasses, and their development is favored by cool (65–75°F or 18–24°C), wet and extended periods of overcast weather in spring and fall. Pink patch was once believed to be a form of red thread, but it is actually a distinct disease with the same general hosts and symptoms of red thread. Pink patch, however, usually does not develop in the absence of red thread. *Limonomyces roseipellis* frequently is complexed with *L. fuciformis* on the same plant and sometimes on the same leaf. Both diseases may occur during warm and drizzling weather in summer, very cold weather in the presence of abundant surface moisture, or at snow melt in winter. These diseases may become widespread among turfgrass species during mild winters.

Red thread and pink patch often occur together, and are most damaging to perennial ryegrass, common-type Kentucky bluegrasses, creeping red fescue, and other fine leaf fescues (Plates 2–7, 2–8, 2–9, and 2–10). Improved cultivars of Kentucky bluegrass, as well as tall fescue, bentgrasses, and bermudagrass also may be affected, but these grasses do not usually sustain a significant level of injury if sufficiently fertilized with nitrogen. Red thread, however, is becoming more commonplace on lawns serviced by lawn care companies and even on golf course fairways. Red thread therefore should not be always thought of as a disease of poorly nourished turf.

Symptoms of red thread and pink patch are concentrated in circular patches 2 inches (5.1 cm) to 2 feet (60 cm) in diameter. Patches frequently coalesce to involve large, irregular shaped areas. From a distance, affected turf has a straw-brown, tan, or pinkish color. The signs of red thread are distinctive and unmistakable. In the presence of morning dew or rain, a coral pink or reddish layer of gelatinous fungal growth easily can be seen on leaves and sheaths. The infected green leaves of these plants soon become water-soaked in appearance. When leaves dry, the fungal mycelium becomes pale pink in color and is easily seen on the straw-brown or tan tissues of dead leaves and sheaths. Bright red, hard and brittle strands of fungal mycelium called "red threads" or sclerotia can invariably be observed extending from leaf surfaces, particularly cut leaf tips. These red threads fall into the thatch and serve as resting structures for the fungus by surviving long periods that are unfavorable for growth of the pathogen. The visible sign of pink patch appears as a gelatinous mass of pinkish mycelium associated with a water-soaked appearance of leaves. The pink patch fungus, however, does not produce "red threads." While these two fungi can be easily distinguished with a microscope, the presence or absence of sclerotia is the primary means of their identification in the field.

Management

Red thread and pink patch are generally, but not always, most injurious to poorly nourished turfs. Frequently, either disease is best controlled by an application of 0.5 to 1.0 lb nitrogen/1000 ft^2 (25 to 50 kg N/ha). Application of nitrogen during periods too cool for turf growth will not aid in reducing disease severity. This is because nitrogen alleviates red thread or pink patch injury by stimulating plant growth and vigor. The nitrogen-stimulated plants are able to replace damaged tissues more rapidly than these fungi can inflict injury. Nitrogen plus potassium has been shown to alleviate red thread more effectively than nitrogen alone.

These diseases, however, may develop in well-fertilized turf. The blighting that occurs often opens the stand to invasion by weeds. Hence, fungicide use may be necessary in some situations. Fungicides that control red thread include chlorothalonil (Daconil 2787), iprodione (Chipco 26019), flutolanil (ProStar), triadimefon (Bayleton), and vinclozolin (Curalan, Touche, Vorlan). Pink patch may be more effectively controlled with iprodione, triadimefon, or vinclozolin.

Helminthosporium Leaf Spot and Melting-Out

Pathogen:	*Drechslera* spp. and *Bipolaris* spp.
Primary Hosts:	Bentgrasses, buffalograss, bermudagrasses, bluegrasses, fescues, perennial ryegrass
Predisposing Conditions:	Varies among turfgrass hosts and fungal species

Many of the fungi that cause leaf spotting and melting-out diseases of turfgrasses were once assigned to the genus *Helminthosporium*. These fungi are now more appropriately referred to as species of *Drechslera* or *Bipolaris*, but the common names of the diseases they cause remain Helminthosporium leaf spot, melting-out or net-blotch (Plates 2–11, 2–12, 2–13, and 2–14). A list of the most common of these diseases and pathogens is provided in Table 2.4.

Among the most important spring and autumn diseases of Kentucky bluegrass is leaf spot, caused by *Drechslera poae*. This disease is not as devastating as it once was, due to the development and widespread use of resistant bluegrass cultivars. 'South Dakota,' 'Kenblue,' 'Park,' and other "common" types of Kentucky bluegrass are very susceptible to leaf spot. The common types, which generally survive extreme environmental stresses, are still used today as components of bluegrass blends in some regions of the U.S. because they lend genetic diversity to the stand.

Drechslera poae is a cool weather pathogen of bluegrasses that is most active

Table 2.4. Common Hosts and Names of Diseases Caused by Fungi Previously Known as Helminthosporium

Host	Common Name of Disease	Pathogen
Kentucky bluegrass and other bluegrasses	Leaf spot, melting-out	*Drechslera poae* (Baudys) Shoemaker
	Leaf spot, melting-out	*Bipolaris sorokiniana* (Sacc.) Shoemaker
Tall fescue	Net-blotch	*D. dictyoides* (Drechs.) Shoemaker
Creeping red fescue	Leaf spot, melting-out	*B. sorokiniana*
	Net-blotch	*D. dictyoides*
Perennial ryegrass	Leaf spot, melting-out	*B. sorokiniana*
	Net-blotch	*D. dictyoides*
	Brown blight	*D. siccans* (Drechs.) Shoemaker
Bentgrasses	Red leaf spot	*D. erythrospila* (Drechs.) Shoemaker
	Leaf spot	*B. sorokiniana*
Bermudagrass	Leaf spot, melting-out	*B. cynodontis* (Marig.) Shoemaker

during the spring (especially March, April, and May), autumn (especially September and October), and throughout mild winter periods. *Drechslera poae* and other diseases in this group may develop in two phases: the leaf spot phase and/or the melting-out phase. In the leaf spot phase, distinct purplish-brown, oval-shaped leaf spot lesions with a central tan spot are produced on the leaves of affected bluegrass plants. In a heavily infected stand, the turf appears yellow or red-brown in color when observed from a standing position. During favorable disease conditions, numerous lesions coalesce to encompass the entire width of the blade, causing a generalized dark-brown blight and dieback from the tip. Leaf spot lesions initially are associated with older leaves, which die prematurely as a result of the invasion. If environmental conditions favorable for disease continue, particularly overcast, cool and drizzling weather, successive layers of leaf sheaths are penetrated, and the crown is invaded. Once the crown is invaded the disease enters the melting-out phase. During this phase, entire tillers are lost, and the turf loses density. Hence, it is the melting-out phase that is most damaging to the stand.

Net-blotch disease of tall fescue, perennial ryegrass, and creeping red fescue is caused by another Helminthosporium, *Drechslera dictyoides. Drechslera dictyoides* also is a cool, wet weather pathogen that attacks turf primarily during cool and moist periods of spring and fall. Symptoms initially appear as minute, purple-brown specks on fescue leaves. As the disease progresses, a dark brown, net-like pattern of necrotic lesions develops on leaves of tall fescue and creeping red fescue. These net-blotches may coalesce, and leaves will turn brown or yellow, and die back from the tip. On leaves of perennial ryegrass, numerous oblong, dark-brown lesions are produced by both *D. dictyoides* and *D. siccans* (i.e., brown blight). Under ideal environmental conditions, these fungi may invade stems and cause a melting-out of the stand. Both diseases can be active during relatively warm, rainy periods of winter.

Bipolaris spp. and *D. erythrospila* attack grasses during warm weather from late spring through summer. These pathogens, and the diseases they cause, are discussed in the summer section.

Management

Cultural practices that minimize injury from leaf spot, melting-out, net-blotch, or brown blight are the same regardless of the pathogen. Renovation or overseeding resistant cultivars is the simplest and best approach to managing these diseases in Kentucky bluegrass. Unfortunately, homeowners continue to buy inexpensive Kentucky bluegrass seed mixes, which invariably contain high percentages of susceptible, common-type cultivars. These diseases are clearly much more destructive under low mowing. The leaf lesions reduce the photosynthetic capacity of plants and low mowing therefore exacerbates the condition. Hence, when these diseases become evident, it is important to increase mowing height immediately. Avoid spring application of high rates of water-soluble nitrogen fertilizers. These fungi produce huge populations of spores when thatch is subjected to frequent wetting and drying cycles. Irrigation water should therefore be applied deeply and infrequently. Thatch should be controlled by verticutting and/or core cultivation when layers exceed 0.5 inches (1.3 cm) in depth. Broadleaf, phenoxy herbicides and plant growth regulators aggravate *Helminthosporium* diseases and their use should be avoided when these diseases are active. Fungicides that effectively control leaf spot diseases include chlorothalonil (Daconil 2787), iprodione (Chipco 26019), mancozeb (Dithane M-45, Fore), and vinclozolin (Curalan, Touche, Vorlan). Fall application of PCNB (Terraclor) to Kentucky bluegrass has been shown to control leaf spot during mild winters and early spring periods.

Take-All Patch

Pathogen: *Gaeumannomyces graminis* (Sacc.) Arx & Olivier var. *avenae* (Turner) Dennis
Primary Hosts: Bentgrasses
Predisposing Conditions: Cool, wet weather in fall and spring, followed by hot and/or dry summer conditions

Take-all patch (formerly known as Ophiobolus patch) is almost exclusively a disease of bentgrass turf (Plates 2-15, 2-16, 2-17, and 2-18). It has been observed in annual bluegrass on rare occasions. Take-all was first reported in Holland in 1937 on a bentgrass putting green, but its occurrence in the U.S. was first documented in western Washington in 1960. It was not until the 1970s that the disease was reported in the eastern U.S. Take-all is now known to occur anywhere where bentgrass is grown. The pathogen attacks roots and stems, and there are no distinctive leaf spot or sheath lesions.

Take-all is most common on newly constructed golf courses, particularly those carved out of woodlands, peat bogs, or other areas that have not supported crops or grasses for decades. This disease can be especially damaging to greens renovated with methyl bromide. The disease tends to spread more rapidly and with greater severity in sandy soils. Take-all may appear as early as the spring immediately following a fall seeding. It generally becomes most severe in the second year following seeding. The pathogen actively attacks roots during cool and wet periods. Symptoms of the disease are most conspicuous from late April, throughout summer, and may recur in autumn. Bentgrass affected by take-all in the spring may recover by summer; however, if irrigation is withheld, those areas affected in the spring are the first to die from drought stress. Initially, the circular patches of take-all are only a few inches in diameter and reddish-brown in color. Patches may increase to 2 feet (60 cm) or more in diameter, particularly on chronically affected bentgrass sites. Most patches range from 6 to 18 inches (15 to 46 cm) in diameter, but they may develop in tight clusters that give the appearance of a single large, 2 to 3 foot (60 to 90 cm) diameter patch. When the disease is active, the perimeter of the patch usually assumes a bronzed appearance, and the turf eventually turns a bleached or tan color. Patches also frequently appear reddish-brown in color and bronzing may be absent. The small, circular patches increase in size over a number of years and dead bentgrass in the center of the patch may be colonized by weeds if herbicide use is restricted. Sometimes, small horseshoe-shaped crescents are associated with take-all. On rare occasions, turf turns brick-red and thins out in a nonuniform pattern. Because the fungus attacks the root system, turf in affected areas is easily detached and is reminiscent of the type of damage caused by white grubs. Patches may coalesce resulting in large, irregular areas of dead turf.

Over time the disease will naturally decline, presumably due to a buildup of microorganisms that antagonize or in some other way prevent *G. graminis* var *avenae* from damaging roots. The decline phenomenon may occur within 3 years from the time that first disease symptoms are observed. During the decline phase, patches appear chlorotic (yellow) and turf may thin-out, but significant damage does not occur. Take-all may persist indefinitely where soils are alkaline or irrigation water has a high pH.

Management

There is an interesting relationship between soil pH and take-all patch. This disease can occur over a wide range of soil pHs, but it is most severe where soils are in the neutral to alkaline range (i.e., pH >7.0). Acidification of soil with ammonium forms of nitrogen fertilizer is the primary cultural approach to controlling take-all. The early studies that were used to establish this method of control involved excessively high levels of nitrogen (i.e., 8 to 12 lb N/1000 ft^2 per year or 400 to 600 kg N/ha/yr). It is now known that as little as

3.0 lb N/1000 ft^2 per year (150 kg N/ha/yr) from either ammonium chloride or ammonium sulfate significantly reduces, but does not necessarily completely eliminate take-all. One of the aforementioned N-sources should be used as the exclusive N-source for at least 2 years and perhaps longer where the disease is more persistent. Sulfur is used to suppress this disease in the Pacific Northwest; however, ammonium-based N fertilizer remains the best and safest choice for take-all management in most regions of the U.S.

Phosphorus (P) and potassium (K) also have been linked to reducing take-all severity. Phosphorus should be applied even when soil tests indicate moderate (>50 lb P/acre or >2500 kg P/ha) or high P levels. For best results, an ammonium-based nitrogen fertilizer should be applied with P and K in a 3:1:2 ratio. Ordinary superphosphate (0–18–0) is the preferred P source since it also contains some sulfur in levels that are safe to use on bentgrass greens. A total of 3 to 5 lb N/1000 ft^2 (150 to 250 kg N/ha) from one of the aforementioned N-sources should be applied annually for at least 2 years. The use of ammonium-based fertilizers will provide very good winter color to turf, but will also encourage growth, and therefore increased mowing into early winter. Furthermore, the use of lime or topdressing soil with a pH above 6.0 should be avoided, and thatch should be controlled through core cultivation and/or verticutting.

Take-all decline can occur within 3 to 7 years after seeding bentgrass. This may account for why take-all is seldom a problem in bentgrass stands older than 10 years. Take-all can recur indefinitely where soil or well water pH is very high. Hence, irrigation water should be analyzed for pH where take-all is a problem. In extreme cases, acid injection of irrigation water may be required to speed take-all decline.

Acidification of the rhizosphere (i.e., the root surface and microenvironment) is believed to be the primary factor responsible for alleviating take-all with ammonium forms of nitrogen. It has been suggested that acidification of soil water either directly reduces growth of the take-all fungus or favors growth of other microorganisms, which effectively compete with or in some other way antagonize *G. graminis* var. *avenae.* More recently, manganese oxidation by microbes in soil has been linked to increased take-all. Manganese fertilization, however, does not reduce take-all because microbes readily convert this micronutrient to an unavailable form. Acidification apparently reduced the ability of microbes to oxidize manganese, and the resulting increase in manganese availability for root uptake assists plants in their defense against the take-all fungus.

Fenarimol (Rubigan), propiconazole (Banner) or triadimefon (Bayleton) applied in late fall or early winter for preventive control of snow mold should provide some protection against take-all. For additional protection, these fungicides should be applied 2 to 3 times on a 21-day interval at the onset of symptoms in spring. The most effective take-all fungicide was PMA, but it is no longer available.

Necrotic Ring Spot

Pathogen: *Leptosphaeria korrae* Walker & Smith
Primary Host: Kentucky bluegrass
Predisposing Conditions: Cool, wet weather in spring and fall, followed by high temperature or drought stress in summer

Necrotic ring spot (NRS) is a disease of Kentucky bluegrass, annual bluegrass, and creeping red fescue (Plates 2-19 and 2-20). Although first described in 1986, NRS probably has been around for some time, but has been confused with summer patch. *Leptosphaeria korrae* primarily attacks roots during cool, wet weather in spring and fall. In some regions, disease symptoms may be apparent all year to include mild winter periods. In contrast, summer patch is a high-temperature summer disease that normally appears in July or August. Frequently, symptoms of NRS do not become evident until summer, when environmental stresses kill plants whose root systems were damaged by *L. korrae* in spring. The confusion between NRS and summer patch results because both diseases share common symptoms, such as rings of dead grass with living turf in the center, and both can appear in summer. Initially, leaves of NRS-affected plants display a purple color or wilted appearance, and plants may be stunted in a circular pattern. During early stages of the disease symptoms are inconspicuous and often are not detected. Diagnosis is most often based on the development of dead rings or thinned-out turf in circular patches. In chronically affected areas, NRS patches tend to be large, often greater than 1 foot (30 cm) in diameter, and a distinct frog-eye symptom is common. Conversely, patches associated with summer patch tend to be less than 1 foot (30 cm) in diameter, but a few can exceed 18 inches (46 cm) in width.

Sodded lawns in newly cleared woodland sites appear most likely to be affected, but NRS also occurs where lawns are seeded. To date, NRS has been observed primarily in Kentucky bluegrass turf grown in the Northeast, upper Midwest, Rocky Mountain states, and Pacific Northwest regions. Its occurrence is believed to be less common than that of summer patch in transition zone regions of the U.S.

Management

Because NRS is a root disease, frequent irrigation is required during summer months when heat and drought severely stress injured plants. Daily, light applications of water are recommended to cool turf and assist the dysfunctional root system where the disease is chronic and severe. Use of slow-release and organic fertilizers such as sewage sludges or animal waste products help to reduce disease severity. The Kentucky bluegrass cultivars 'Wabash,' 'Vantage,' 'I-13,' and 'Adelphi' were reported to have some NRS resistance. An early spring application of fenarimol (Rubigan), or thiophanate (CL 3336, Fungo 50) reduces NRS severity, but fungicide use may not completely control this disease. The severity of this disease may begin to decline naturally about 5 years after it first appears.

Spring Dead Spot

Pathogen:	*Leptosphaeria korrae* Walker & Smith, *Leptosphaeria narmari* Walker & Smith, *Ophiosphaerella herpotricha* (Fr. : Fr.) Walker, and others
Primary Host:	Bermudagrass
Predisposing Conditions:	Cool to cold and wet weather in fall and spring

Spring dead spot (SDS) is perhaps the most damaging disease of bermudagrass turf (Plates 2–21, 2–22, and 2–23). In the U.S., this disease is caused by several root pathogens including *L. korrae* and *O. herpotricha*. *Gaeumannomyces graminis* (Sacc.) Arx & Olivier var. *graminis* is associated with both spring dead spot and bermudagrass decline in southern states. As the name implies, SDS injury becomes apparent in the spring. The actual infection may begin as early as autumn, but root injury by the pathogen becomes rapid just prior to spring green-up. As bermudagrass breaks dormancy, circular patches of tan or brown, sunken turf 2 inches (5.1 cm) to 3 feet (90 cm) or greater in diameter become conspicuous. Rhizomes and stolons from nearby, healthy plants eventually spread into and cover the dead patches. This filling-in process is slow, a period which may last four to eight weeks or longer following spring green-up. The slowness of the filling-in process is believed to be due to toxic substances generated in the soil below the dead patches. Weeds commonly invade the dead patches. These weeds should be controlled to reduce competition with the bermudagrass, which helps to speed the recovery process.

Spring dead spot is extremely destructive to both low and high management bermudagrass. Recovery, however, is very slow in turf maintained with low levels of nitrogen. The disease is most commonly associated with bermudagrass turf older than 3 years, but it may appear the spring following sprigging with stolons from sites previously affected with SDS. Spring dead spot injury is most likely to occur where thick thatch layers exist and where heavy applications of nitrogen fertilizers were applied during late summer or fall.

Management

Cultivars of bermudagrass with greater low temperature hardiness such as 'Midiron,' 'Midfield,' and 'Vamont' are more resistant and tend to recover more rapidly from SDS. 'Tufcote' and most bermudagrass hybrids are very susceptible to this disease. It is important to eliminate weeds from diseased sites as their presence slows recovery of the bermudagrass. Ammonium sulfate or ammonium chloride (applied at 1.0 lb N/1000 ft^2 or 50 kg N/ha) and potassium chloride (applied at 1.0 lb K/1000 ft^2 or 50 kg K/ha) applied on monthly intervals from mid-May to mid-August speeds recovery of turf injured by SDS, and helps to alleviate disease severity over time. It is important, however, to cease nitrogen application 6 weeks prior to the anticipated dormancy of bermudagrass. Irrigate frequently during dry periods in summer to encourage re-growth by stolons and rhizomes. Control thatch and alleviate compaction by core cultivation. Fenarimol (Rubigan) applied once in mid- to late September or about 30 days prior to anticipated dormancy alleviates SDS, but does not provide complete control.

Large Patch of Zoysiagrass

Pathogen: *Rhizoctonia solani* Kuhn
Primary Host: Zoysiagrass
Predisposing Conditions: Cool, rainy periods in spring and fall

Large patch of zoysiagrass (sometimes referred to as zoysia patch) primarily occurs in 'Meyer' zoysiagrass grown on golf course fairways and sod farms in transition zone regions of the U.S. (Plate 2–24). The disease appears during extended rainy overcast periods, particularly in early spring and late fall. Evidently, zoysiagrass is rendered susceptible to *R. solani* as its growth slows in response to cool temperatures prior to fall dormancy and at spring green-up. The disease is characterized by huge circular patches that range from 2 to 10 feet (60 cm to 3 m) or more in diameter. Patches are brown or yellow-orange in color. At the periphery of patches, the turf may exhibit a brilliant-orange firing, particularly when the disease appears in the fall. Unlike brown patch in cool-season grasses, *R. solani* attacks basal portions of zoysiagrass leaf sheaths in the thatch region, producing small reddish-brown or black lesions. Eyespot lesions may appear on basal stems and stolons. Leaves are blighted, and stems may be infected and tillers killed. Turf within affected areas thins out, and 85% to 90% of the shoots may be killed. Turf within these large, almost dead-appearing patches eventually recovers but the process is very slow. As temperatures increase in spring, the disease subsides and stolon growth results in a slow improvement in turf density. These symptoms are similar to those observed in St. Augustinegrass and centipedegrass affected by brown patch in the spring or fall.

Management

Increasing mowing height to 1.5 to 2.0 inches (3.8 to 5.1 cm) is perhaps the most effective means of reducing disease progress and enhancing turf recovery. Core cultivation and verticutting affected sites following spring green-up stimulates stolon growth and helps to reduce thatch. Thatch reduction is important because most damage to tillers occurs primarily within the thatch layer. Redistribution of soil from cores assists in thatch degradation. Do not apply any nitrogen fertilizer until disease activity has ceased. Research in Kansas indicates that most nitrogen fertilizers, including urea and composted turkey litter, help to stimulate recovery, but they do not appear to suppress disease development. Spring application of water-soluble N-sources can enhance zoysia patch. The total amount of nitrogen applied to healthy zoysiagrass should not exceed 2.0 lb N/1000 ft^2 per year (100 kg N/ha/yr). Large patch also is enhanced by overwatering in the spring and fall, and tends to be most severe in poorly drained sites. Avoiding mowing when leaves are wet may help to reduce disease severity.

Fungicides assist in blight reduction, but generally do not prevent the disease. A fall application in mid-to-late September is recommended where the disease is chronic. Flutolanil (ProStar), iprodione (Chipco 26019), propiconazole (Banner), and triadimefon (Bayleton) have been shown to provide satisfactory disease suppression and their use is associated with more rapid turf recovery from the disease.

Pythium Root Rot

Pathogens:	*Pythium aristosporum* Vanterpool, *P. graminicola* Subramanian, *P. vanterpooli* Kouyeas & Kouyeas, others
Primary Host:	Creeping bentgrass and annual bluegrass greens
Predisposing Conditions:	Cool and wet weather, particularly in spring

There are several *Pythium* species that invade roots and crowns of most turfgrasses. Pythium root rots, however, are primarily a problem in creeping bentgrass and annual bluegrass grown on greens (Plate 2–25). These root rots are puzzling because they can occur in all soils and at nearly any time of year. Pythium induced root dysfunction was initially described as a disease of greens on older golf courses that were renovated by introducing high sand content soil mixes. Pythium root rot is most likely to appear during or following long cool periods when soils are excessively wet or saturated. Damage most frequently occurs during late winter and spring. Indeed, several of the low-temperature tolerant *Pythium* species associated with snow blight have been implicated as agents of these root rots. Infected plants turn yellow or reddish-brown, tissues may have a water-soaked appearance, and turf dies out in irregular patterns. These symptoms of Pythium root rot mimic Helminthosporium melting-out and anthracnose. There is no foliar mycelium, and diagnosis should be confirmed by a trained diagnostician as rapidly as possible during early stages of the disease. The presence of some Pythium oospores in stem or root tissue, however, does not always indicate that severe turf damage will occur.

Management

Pythium root rot appears to be more common in northern regions of the U.S. The distribution, as well as most other epidemiological aspects of this disease, are unknown. Mowing height must be increased to 0.25 inches (0.6 cm) or higher on greens. Mow affected greens no more frequently than every other day with a walk-behind greensmower. Syringe as frequently as needed during dry periods, but avoid heavy irrigation. Overseeding is often necessary. Control with chemicals is erratic, and once extensive damage appears, the turf seldom responds to fungicide treatment. Because of its true systemic properties, fosetyl aluminum (Aliette) generally is recommended in tank mix combination with mancozeb (Dithane M-45, Fore). Should there be little or no response in 5 to 7 days, a follow-up spray of another Pythium-targeted fungicide is suggested.

Dollar Spot

Pathogen: *Sclerotinia homoeocarpa* F.T. Bennett
Primary Hosts: Most turfgrass species
Predisposing Conditions: Warm days, cool nights, and heavy dew formations

Dollar spot is widespread and extremely destructive to turfgrasses (Plates 2–26, 2–27, 2–28, and 2–29). The taxonomy of *S. homoeocarpa* is unclear, and this fungus may be referred to as an unknown species of either *Moellerodiscus* or *Lanzia*. Dollar spot is known to attack most turfgrass species including annual bluegrass, bentgrasses, fescues, Kentucky bluegrass, perennial ryegrass, bermudagrasses, zoysiagrasses, centipedegrass and St. Augustinegrass. The symptomatic pattern of dollar spot varies with turfgrass species and cultural practices. Under close mowing conditions, as with intensively maintained bentgrass, bermudagrass or zoysiagrass, the disease appears as small, circular, straw-colored spots of blighted turfgrass about the size of a silver dollar (4 cm diam.). With coarser textured grasses that are suited to higher mowing practices, such as Kentucky bluegrass or perennial ryegrass, the blighted areas are considerably larger, straw-colored patches 3 to 6 inches (7.6 to 15.2 cm) in diameter. Affected patches frequently coalesce and involve large areas of turf. Grass blades often die back from the tip, and have straw-colored or bleached-white lesions that are shaped like an hourglass. The hourglass banding on leaves often is made more obvious by a definite narrow brown, purple, or black band that borders the bleached sections of the lesion from the remaining green portions. Hourglass bands may not appear on warm-season grasses. On bermudagrass and other warm-season grasses, the lesions may be oblong or oval-shaped, but there is a brown staining of tissue where the tan lesion and green tissue meet. When the fungus is active and leaf surfaces are wet, a fine, white, cobwebby mycelium covers the diseased patches during early morning hours. The disease is favored by warm and humid weather, and when night temperatures are cool enough to permit early and heavy dew formation. In cool-season grasses, disease severity usually peaks in late spring to early summer and again in late summer to early fall. In the upper Midwest, however, the disease tends to be most damaging during autumn. In southern states, dollar spot can be a chronic problem in bermudagrass during the summer. In some regions, dollar spot can remain active during mild periods throughout fall and into early winter.

Management

Dollar spot tends to be most damaging to poorly nourished turfs, particularly if soils are dry and humidity is high or a heavy dew is present. On greens, removal of dew and leaf surface exudates by poling, dragging, or whipping can be beneficial. Mowing greens early in the morning speeds surface drying, and has been linked to reduced dollar spot. Maintain a balanced N-P-K fertility program. In poorly nourished turf, an application of nitrogen (50% water-soluble plus 50% slow-release) will stimulate growth and mask the disease. Any subsequent nitrogen applications should be in a slow-release form. Increasing mowing height is among the most effective cultural approaches to minimizing dollar spot injury. Core cultivation, to alleviate soil compaction and control thatch, assists in reducing dollar spot. Avoid drought stress, and irrigate deeply during early morning hours. Irrigating between 5 A.M. and 8 A.M., when dew is present on leaves, does not extend the fungal infection period. Hence, where water use is restricted to the hours between sundown and sunup, predawn irrigation will not promote disease and will not violate local watering laws. Most fungicides that are labeled for dollar spot will provide effective disease control. Tank mixing a fungicide with 0.1 to 0.2 lb nitrogen per 1000 ft^2 (5 to 10 kg N/ha) from urea is sometimes associated with improved dollar spot control.

Stripe Smut and Flag Smut

Pathogen:	Stripe smut = *Ustilago striiformis* (Westend.) Niessl
	Flag smut = *Urocystis agropyri* (Preuss) Schrot
Primary Host:	Kentucky bluegrass
Predisposing Conditions:	Overcast, cool and wet weather in spring and fall

Stripe smut (*U. striiformis*) and flag smut (*U. agropyri*) are diseases that occur primarily in mature Kentucky bluegrass stands (Plate 2–30). Stripe smut is occasionally a problem in bentgrass and perennial ryegrass turf. Symptoms are most conspicuous during the cool, moist seasons of spring and fall. Infected plants are often stunted and pale green or yellow in color. Narrow, silvery or gray-black streaks appear in the leaves. These streaks are fruiting structures (sori) in which large masses of black spores (teliospores) are produced. When sori mature, the cuticle and epidermis rupture, and the leaves shred and curl, which releases the teliospores. Both pathogens are obligate parasites and once infected, plants will remain colonized by these fungi until they die or they are treated with an effective fungicide.

Although both pathogens are systemic and persistent in surviving plants, during summer months infected plants may appear amazingly healthy if properly maintained. In spring or autumn, however, badly infected stands may appear chlorotic (yellow) and in need of nitrogen fertilizer. From early winter until spring green-up, leaves that had been shredded by matured fruiting bodies in the fall develop a gray-brown, desiccated appearance. Affected stands may appear winter dormant many weeks prior to bitterly cold weather.

Management

Although leaf shredding symptoms may not be evident, stripe and flag smuts can be very damaging to infected plants during periods of heat and drought stress. If properly irrigated and fertilized, however, badly smutted stands often survive and exhibit only a decrease in turf quality and some thinning during stressful summer months. These smut diseases most commonly occur in mature (2 to 4 years and older) stands that have been managed with high levels of nitrogen fertilizer. 'Merion,' 'Windsor,' and 'Fylking' Kentucky bluegrasses are among the most susceptible cultivars. Recently introduced cultivars are less susceptible to these diseases, which has led to a reduction in the occurrence of stripe and flag smuts. Using a balanced N-P-K fall fertilizer program, increasing the mowing height in summer, and deep irrigation at the first sign of drought stress are effective cultural practices that greatly minimize smut injury. Severely infected stands can be effectively revitalized with a single, spring or fall application of a sterol-inhibiting fungicide such as propiconazole (Banner) or triadimefon (Bayleton).

Powdery Mildew

Pathogen: *Erysiphe graminis* DC.
Primary Host: Kentucky bluegrass
Predisposing Conditions: Shade

Powdery mildew is a disease generally confined to shaded sites. The presence of grayish-white mycelium on the upper leaf surfaces is a conspicuous, diagnostic sign of the disease. The lower, older leaves of the plant generally are more heavily infected than upper, younger leaves. In severe infestations, leaves appear to have been dusted with ground limestone or flour. The abundant surface mycelium absorbs nutrients from the epidermal cells and the leaves turn yellow. Eventually leaves and tillers may die and the turf exhibits poor density. The fungus seldom kills plants; however, it weakens plants and can predispose them to injury from environmental stresses or other diseases. Some Kentucky bluegrass cultivars infected with powdery mildew can lose their resistance to leaf spot.

Powdery mildew can be found at almost any time of year, but peak activity normally occurs when days are warm and nights are cool. The white coating of mycelium and spores is, however, most prevalent on leaves during cool, humid and cloudy periods of late summer and early fall. Spores are produced in abundance on leaf surfaces and they are disseminated by air currents and equipment to adjacent, healthy leaves. The spores germinate rapidly, even in the absence of dew or water.

Management

Because shade is the primary predisposing factor for powdery mildew, reducing shade and improving air circulation are sound, but often impractical approaches to reduce damage. Planting or overseeding with shade-tolerant cultivars, increasing mowing height, avoiding drought stress, and using a balanced N-P-K fertilizer program will promote turfgrass growth and help to minimize injury from powdery mildew. Fungicides may be applied in situations where the disease is yellowing plants and thinning the stand. Some effective fungicides include propiconazole (Banner) and triadimefon (Bayleton).

Anthracnose

Pathogen: *Colletotrichum graminicola* (Ces.) Wils.
Primary Hosts: Annual bluegrass, creeping bentgrass and creeping red fescue
Predisposing Conditions: Prolonged periods of overcast and rainy weather or hot and dry weather

Anthracnose is a serious disease of annual bluegrass and creeping bentgrass turf on golf courses (Plates 2-31, 2-32, 2-33, and 2-34). Anthracnose also has been reported to attack creeping red fescue, perennial ryegrass, and Canada and Kentucky bluegrasses. The fungus may cause either a foliar blight or a basal rot. Foliar blighting generally occurs during periods of high temperature and drought stress. Plants affected by anthracnose also may be invaded by other disease agents, including those causing leaf spot, summer patch or Leptosphaerulina blight. Affected turf initially develops a reddish-brown color and thins-out in irregularly shaped patterns several feet or more in diameter.

On greens, basal rot of annual bluegrass and creeping bentgrass occurs during both cool periods in spring, and during warm and moist periods of summer. The disease often is associated with soil compaction, heavy traffic, and low nitrogen fertility. Infected plants may initially appear as orange or yellow spots about the size of a dime (2 cm diam.). The dime spot symptom is especially common in annual bluegrass during cool and wet periods from fall to spring. Individual plants may have both green healthy appearing tillers, and yellow-orange infected tillers. The central, or youngest leaf, is last to show the yellow-orange color change. Removal of all sheath tissue to expose the stem base reveals a water-soaked black rot of crown tissues where roots and new buds are produced. Black aggregates of fungal mycelium often are present on infected stolons of creeping bentgrass. Spore bearing structures (acervuli) with short, black hairs called setae and the black mycelial aggregates can be seen on dying tissues with a hand lens. Once acervuli develop on sheath or leaf tissue, the basal rot phase is advanced and plants generally die. When the disease develops on greens during warm and moist weather, turf develops an orange or reddish-brown color and thins-out in large, irregularly-shaped areas. Symptoms under these conditions mimic melting-out, red leaf spot disease, and Pythium root rot. Acervuli are commonly observed on stem tissues of bentgrass and annual bluegrass, but only occasionally on green leaves or sheaths of infected plants. Hence, managers are advised to look on stem bases for the mats, which appear as black "fly specks" during the early stages of the disease. Once abundant acervuli appear on necrotic leaves, it generally is too late to arrest the disease with fungicides.

Management

Basal rot is very difficult to control once turf shows signs of thinning. To alleviate basal rot, use walk-behind greensmowers and increase the height of cut immediately, and divert traffic by moving cups frequently. In the fall, after symptoms have dissipated, core cultivate and overseed. A modest application of nitrogen (0.15 to 0.25 lb N/1000 ft^2 or 7.5 to 12.5 kg N/ha) combined with a fungicide, such as chlorothalonil (Daconil 2787) tank mixed with either fenarimol (Rubigan), propiconazole (Banner), thiophanate (CL 3336 or Fungo 50) or triadimefon (Bayleton) should help reduce, but not eradicate the disease. Where basal rot is a chronic problem on greens, tees, or fairways, fungicides should be used preventively in combination with an improved fertility program. Mowing height should be immediately increased and water from irrigation should be applied only as needed to prevent wilt. Moderate nitrogen levels (3.0 lb N/1000 ft^2/yr or 150 kg N/ha/yr) are associated with less foliar blighting by anthracnose, especially when fungicides such as propiconazole (Banner) or triadimefon (Bayleton) are used preventively. For fairways and other large areas it is best to control annual bluegrass and renovate with less susceptible grasses such as Kentucky bluegrass, perennial ryegrass, zoysiagrass, etc.

Ascochyta and Leptosphaerulina Leaf Blights

Pathogens:	*Ascochyta* spp. *Leptosphaerulina trifolii* (Rostr.) Petr.
Primary Hosts:	Bentgrasses, bluegrasses, fescues and perennial ryegrass
Predisposing Conditions:	Overcast and wet weather from spring to fall

Both of these diseases are common and can cause extensive foliar blighting (Plates 2-35 and 2-36). Turf usually recovers fairly rapidly; however, *Leptosphaerulina* may cause severe damage in some inexplicable situations. Both diseases occur from spring to fall, and they are triggered by prolonged periods of humid or wet and overcast weather. These diseases can be found in association with most turfgrasses, but are particularly troublesome in Kentucky and annual bluegrasses, perennial ryegrass, and creeping red fescue. Tall fescue frequently suffers a tip dieback due to *Ascochyta* blight in the spring. Leptosphaerulina blight also occurs in creeping bentgrass and zoysiagrass grown on golf courses.

Ascochyta and *Leptosphaerulina* produce a myriad of symptoms, which largely appear as nonuniform blighting. Both cause a tip dieback, and their brownish to black and rounded fruiting bodies are invariably found embedded in blighted leaves. *Leptosphaerulina* blighted leaves often develop a reddish-brown color prior to turning completely brown. Both pathogens restrict their activity to leaf blighting, and do not appear capable of attacking stem or root tissue. Blighting ceases once the weather shifts from rainy to sunny and dry conditions.

Management

Fungicides are seldom recommended for these diseases, since significant blighting often occurs during wet periods before the disease can be either properly diagnosed or treated. Improving cultural practices (i.e., increasing mowing height and fall application of balanced N-P-K fertilizer) and perhaps fall overseeding are recommended. On golf courses, fungicides such as chlorothalonil (Daconil 2787), iprodione (Chipco 26019), or vinclozolin (Curalan, Touche, Vorlan) may be suggested to arrest blighting.

Yellow Tuft or Downy Mildew

Pathogen: *Sclerophthora macrospora* (Sacc.) Thirum., Shaw & Naras.
Primary Hosts: All turfgrasses
Predisposing Conditions: Extended periods of cool and wet weather

Yellow tuft disease is caused by a downy mildew fungus, and infects all turfgrass species (Plate 2–37). A severe infection mars the appearance and the playability of greens. Yellow tuft attacks all grasses and has rendered Kentucky bluegrass sod temporarily unsalable. Infected plants generally are not killed by the parasite and the disease is primarily a problem on greens. On bentgrass greens and tees, the disease appears as yellow spots 0.25 to 0.5 inch (0.6 to 1.3 cm) in diameter. In Kentucky bluegrass, and other wider bladed grasses, the yellow spots are 1 to 3 inches (2.5 to 7.6 cm) in diameter. Each spot consists of 1 or 2 plants that have 20 to 30 or more tillers, giving plants a tufted appearance. Roots of infected plants are short and the tufts are easily detached from the turf. During cool and moist periods in late spring and autumn, plants develop a yellow color, at which time the infected plants are yellow tufted. The yellowing is the result of heavy spore production by the fungus. These spores swim (zoospores), and this accounts for why the disease is more severe in low lying areas that puddle. Zoospores are produced in lemon-shaped structures called sporangia. Sporangia develop on leaf surfaces from sub-stomatal cavities below the leaf epidermis. During early morning hours, when leaves are wet, the pearly white sporangia can be seen on the upper leaf surfaces of infected plants with a hand lens. During most summer months infected plants appear amazingly healthy.

In St. Augustinegrass, the disease is called downy mildew instead of yellow tuft, and the symptoms are different. The disease appears as white or yellow-green, linear streaks that run parallel to the leaf veins. Leaves turn yellow and there may be some browning of leaf tips. Excessive tillering does not occur. The disease is disfiguring and St. Augustinegrass growth may be stunted.

Plants infected with *S. macrospora* can persist in excess of 2 years. Over time, new shoots may escape systemic invasion by the fungus and eventually downy mildew-free tillers replace the original plants. Escape of tillers helps to explain the ephemeral nature of yellow tuft symptoms in older stands. Seedlings are most susceptible to infection by *S. macrospora*, which accounts for why the disease is most commonly observed in the spring following fall seeding. The disease can recur in older greens during years marked by excessively wet weather.

Management

Improving drainage may help to alleviate yellow tuft, since the disease is most severe in low lying areas where water collects and the swimming spores are able to move easily to uninfected plants. Yellow tuft can be controlled chemically with either fosetyl aluminum (Aliette) or metalaxyl (Subdue).

DISEASES INITIATED DURING SUMMER THAT MAY PERSIST INTO AUTUMN

Many of the diseases in this section are initiated in response to high temperature and humidity. Some are most severe during wet and overcast periods, while others are more damaging during periods of drought. A root pathogen may damage turf during cool and moist conditions, but symptoms may not develop until periods of heat or drought stress. Hence, in some regions take-all, necrotic ring spot, and injury induced by plant parasitic nematodes may be more apparent in summer despite the fact that most damage was inflicted to the root systems earlier in the year. Furthermore, several of these diseases can occur in some turfgrasses at different times of year in different regions. For example, brown patch attacks cool-season grasses in the summer, but it often is more damaging to warm-season grasses in southern states in the spring and fall.

Bermudagrass Decline

Pathogen: *Gaeumannomyces graminis* (Sacc.) Arx & Olivier var. *graminis*
Primary Host: Bermudagrass
Predisposing Conditions: Hot and wet periods in summer

Bermudagrass decline has been recognized in Florida for many years, where it typically occurs on greens during hot and wet summer periods. Lower leaves are the first to turn yellow and the chlorosis (yellowing) progresses to younger leaves emanating from the crown. Roots on stolons typically are shortened and blackened. Turf normally dies out in a nonuniform manner and distinct, circular patches generally are not evident.

In more northern regions, *G. graminis* var. *graminis* and *Bipolaris cynodontis* can simultaneously attack bermudagrass during wet periods following spring green-up. Affected plants exhibit a brilliant chlorosis (yellowing) and turf thins out in irregularly-shaped areas. Stolons tend to dieback from the younger tips, and yellowing progresses backward to the crown. Roots on stolons are reduced to blackened nubs that can no longer anchor stolons in soil. Mild symptoms consisting of yellow spots of 1 to 2 inches (2.5 to 5.1 cm) in diameter may appear in late summer.

Management

Effective control measures for this disease have not as yet been determined. Mowing height, however, should be increased until turf has fully recovered. It has been suggested that cultural and chemical approaches to spring dead spot management may help to alleviate bermudagrass decline.

Brown Patch or Rhizoctonia Blight

Pathogen(s):	*Rhizoctonia solani* Kuhn, *R. zeae* Voorhees, *R. oryzae* Ryker & Gooch
Primary Hosts:	Most species
Predisposing Conditions:	High night temperatures, high humidity, and long periods of leaf surface wetness

Brown patch, also known as Rhizoctonia blight, is caused by *Rhizoctonia solani* and is a common, summertime disease of cool-season turfgrasses (Plates 2-38 to 2-43; Plate 2-62). In southeastern states, *R. zeae* and *R. oryzae* also may produce symptoms that are typical of brown patch. *R. solani* attacks nearly all grasses used as turf. In northern regions, it is most damaging to tall fescue, perennial ryegrass, creeping bentgrass, and annual bluegrass. *R. solani* attacks sheaths of zoysiagrass during autumn or spring and causes a disease known as large patch. In southern states, *R. solani* is most likely to damage bermudagrass, St. Augustinegrass, and centipedegrass in the spring or fall.

The symptoms of brown patch vary according to host species. On closely mown turf, affected patches are roughly circular and range from 3 inches (7.6 cm) to 3 feet (90 cm) or greater in diameter. The outer edge of the patch may develop a 1 to 2 inch (2.5 to 5.1 cm) wide smoke ring. The smoke ring is blue-gray or black in color and is caused by mycelium in the active process of infecting leaves. On high-cut turfs, smoke rings are usually not present, and patches may have an irregular rather than circular shape. Close inspection of leaf blades reveals that the fungus primarily causes a blight or dieback from the tip. This gives diseased turf its brown color. *R. solani* produces distinctive and often greatly elongated lesions on tall fescue leaves. The lesions are a light, chocolate brown color, and are bordered by narrow, dark-brown bands. On perennial ryegrass, smaller leaf lesions are produced and tip dieback commonly occurs. On bentgrass or bermudagrass, distinct lesions may not be evident because the leaf blades are too fine textured to observe lesions. During early morning hours, when the disease is active, a cobweb-like mycelium can develop in sparse to huge amounts on leaves laden with dew. Late in the season, distinctive circular patches may not appear. In perennial ryegrass, for example, the turf may simply exhibit a nonuniform thinning-out and there may be little or no foliar mycelium evident in the morning.

In bermudagrass, centipedegrass, and St. Augustinegrass, *R. solani* attacks the base of leaf sheaths where they join stolons. Leaves turn yellow in St. Augustinegrass and to a reddish color in centipedegrass before dying. In these warm-season grasses, the disease tends to be more severe in the fall and can cause extensive injury during cool, wet periods prior to entering winter dormancy. In transition zone regions (i.e, more northern regions of bermudagrass adaptation), brown patch is seldom a problem in the more cold-tolerant bermudagrass cultivars. As previously noted, zoysiagrass is most likely to be damaged by *R. solani* in the spring and fall.

Sclerotial color can be used to separate *R. zeae* from other *Rhizoctonias*. Sclerotia of *R. zeae* are distinctively orange, round, and 1/32 to 1/16 inch (0.8 to 1.6 mm) in diameter. *R. zeae* sclerotia are produced in large numbers in thatch at the base of blighted plants. *R. solani* sclerotia or bulbils are very difficult to find and usually are brown to black and appressed to dead tissue. *R. zeae* often is more destructive than *R. solani*, and is more difficult to control.

Environmental conditions that favor disease development are day temperatures above 85°F (29°C) and high relative humidity. A night temperature above 68°F (20°C) and periods of leaf surface wetness exceeding 10 hours are the most critical environmental requirements for disease development on cool-season grasses. This disease becomes extremely severe during prolonged, overcast wet periods in summer as long as average daily temperatures remain above 68°F (20°C). *R. solani*, however, can be quite active at lower temperatures. This explains its ability to damage several warm-season grasses in the fall and spring. *Rhizoctonia zeae* and *R. oryzae* are more commonly found in southern regions, and they may require higher temperatures to become destructive. *Rhizoctonia zeae*, however, may be active during cooler periods than is commonly recognized. For example, *R. zeae* may produce yellow rings in annual bluegrass in the Midwest when air temperatures are in a moderate range. In transition zone regions, *R. zeae* can cause yellow circular patches or small yellow blotches in perennial ryegrass in early summer.

Management

Avoiding nitrogen when the disease is active, increasing mowing height, and irrigating early in the day (i.e., between dawn and 8 A.M.) are some cultural practices that may help alleviate brown patch. Spring or summer applications of fertilizers, in particular water-soluble N sources, may enhance disease injury from brown patch in cool-season grasses. Conversely, late season application of nitrogen increases disease severity in warm-season grasses in southern states. Organic nitrogen sources have been reported to reduce brown patch severity in northern regions. Where brown patch is chronically severe, however, organic and other nitrogen fertilizers may have little or no impact on brown patch. Fall applications of organic nitrogen fertilizers have been associated with less brown patch when compared to spring applications of water soluble nitrogen. Hence, fall application of nitrogen in cool-season grasses is encouraged, while use of nitrogen in the spring or summer is discouraged. Spoon feeding greens with small amounts of nitrogen (e.g., 0.1 lb N/1000 ft^2 or 5 kg N/ha) intermittently during summer probably has little or no influence on brown patch. When conditions warrant high rates of nitrogen, phosphorus and potassium should be used with nitrogen in a 3:1:2 ratio. Dragging, poling, or mowing greens early in the morning speeds leaf drying and may help to reduce disease activity and improve the residual effectiveness of fungicides. Chlorothalonil (Daconil 2787), iprodione (Chipco 26019), flutolanil (ProStar), mancozeb (Dithane M-45, Fore), thiophanates (CL 3336, Fungo 50) and vinclozolin (Curalan, Touche, Vorlan) effectively control brown patch caused by *R. solani*. Fungicides often are much less effective against *R. zeae*.

Pythium Blight

Pathogens: *P. aphanidermatum* (Edson) Fitzp., *P. ultimum* Trow, *P. myriotylum* Drechs., others
Primary Hosts: Annual bluegrass, bentgrasses, and perennial ryegrass in northern regions, and tall fescue and bermudagrass in southeastern states
Predisposing Conditions: Hot and humid weather

Pythium blight develops rapidly during nighttime and is among the most destructive turfgrass diseases (Plates 2–44 to 2–47). During periods of high relative humidity, night temperatures above 70°F (21°C) and abundant surface moisture, the disease progresses with remarkable speed. Huge areas of turf can be destroyed within 24 hours, particularly if there are thunder showers at night. This disease is often first observed in areas that are shaded, low lying and adjacent to water where air circulation is poor.

A general misconception is that Pythium blight is a common, widespread disease. Although *Pythium* spp. can cause damping-off of any seedling species, it rarely, if ever, attacks mature lawns comprised of Kentucky bluegrass, tall fescue, fine fescue or zoysiagrass. Pythium blight is most likely to attack creeping bentgrass, annual bluegrass, or perennial ryegrass grown under the intensive management (i.e., frequent night irrigation, low mowing and high nitrogen fertility) conditions commonly found on golf courses. It occasionally damages tall fescue and bermudagrass in the southeastern U.S. Pythium blight can cause severe damage to hybrid bermudagrass greens in the Gulf Coast region in the fall.

On closely mown bentgrass greens, Pythium species kill turf in circular patches, rings, or streaks that follow the water drainage pattern. During morning hours when the disease is active, bentgrass turf displays an orange-bronze color and there may be a gray smoke ring or grayish-white mycelium on the periphery of affected patches. In low lying areas where water collects, the patches are brown and all plants usually are killed. In perennial ryegrass, infected foliage develops an oily or dark-gray color, and leaf blades have a water-soaked appearance. Blades later collapse, mat together, and turn brown.

A cottony web of mycelium covers the grass leaves and is visible during early morning hours when leaves are wet. *Pythium* spp. are capable of producing an abundance of mycelium in just a few hours. Mycelium bridges leaf blades and is responsible for the cottony appearance seen on affected turf. The brown patch fungus (*R. solani*), however, also may produce copious amount of foliar mycelium. *Pythium* spp. primarily spreads over a turf by rapid mycelial growth or by movement of mycelial fragments and motile spores in rain or irrigation water. These fungi also are effectively spread by equipment when driven across wet foliage, which is covered with mycelium.

Management

Water management can greatly influence disease severity. It is therefore helpful to irrigate early in the day to avoid moist foliage at nightfall. Improving water and air drainage will help reduce disease development, but these cultural measures often are expensive and difficult to achieve. Avoiding the use of lime in alkaline soils and avoiding the application of nitrogen fertilizers during summer stress periods help to reduce disease incidence and severity. A fall fertilization program using a balanced N-P-K fertilizer improves turf vigor and density. Cultural practices, however, will likely have only minimal beneficial effects on Pythium blight during high disease pressure periods. Severely damaged stands should be converted to less susceptible species by overseeding or by complete renovation. Increasing mowing height and watering early enough in the day to ensure dry leaf surfaces at nightfall may help to reduce the rate of pathogen spread.

While fungicides generally are not used in lawn care for Pythium blight control, they are considered a necessity on golf courses in several regions of the U.S. Before the advent of improved fungicides in the early 1980s, the disease was combatted with short residual chemicals such as chloroneb (Terraneb SP) and ethazole (Koban). Metalaxyl (Subdue) was registered for use on turf in 1981 and if it is not used excessively it can provide over 20 days of control. Metalaxyl can be used either preventively or curatively. The widespread reliance and continuous usage of metalaxyl on golf courses, however, has led to reduced effectiveness and in some cases the selection of *Pythium* spp. biotypes resistant to this fungicide. Reduced residual effectiveness also may be attributable to a buildup of microorganisms that degrade the active ingredient of the chemical. Propamocarb (Banol) and fosetyl-Al (Aliette) are other fungicides that provide long, residual Pythium blight control. The latter are most effective when applied as preventive treatments. Chloroneb and ethazole provide a rapid knock-down and often are recommended for curative treatments.

To avoid the buildup of fungicide-resistant biotypes, and the reduction of residual effectiveness of compounds due to microbial buildup, Pythium-targeted fungicides should always be rotated or applied in tank-mix combinations whenever economically feasible. Resistance problems can be delayed or avoided by tank mixing reduced rates of metalaxyl + propamocarb + fosetyl-Al. Consult the label for directions when using lower rates of any of these fungicides in tank-mix combinations. Tank mixing metalaxyl, propamocarb or fosetyl-Al with mancozeb (Dithane M-45, Fore) also reduces the probability that resistant *Pythium* biotypes will dominate. Alternating sprays of systemics with contacts (e.g., chloroneb, ethazole, or mancozeb), although the latter may only provide 3 to 7 days of control, also helps to reduce these potential problems from occurring. Most Pythium-targeted fungicides can yellow putting green turf when applied during hot weather.

Summer Patch

Pathogen:	*Magnaporthe poae* Landschoot & Jackson
Primary Hosts:	Annual bluegrass, Kentucky bluegrass, creeping red fescue
Predisposing Conditions:	High temperature stress, moist soils and low mowing

Summer patch is a destructive disease of Kentucky bluegrass, creeping red fescue, and annual bluegrass turfs (Plates 2–48 to 2–54). Symptoms of summer patch initially appear as wilted, dark-green or pale areas of turf. These areas rapidly turn into straw-brown, dead patches which resemble those of dollar spot. These patches soon increase in size and may become crescent-shaped or remain circular. Healthy turf may persist in the center of patches, producing rings or "frog-eye" symptoms. In some regions, the frog-eye symptom is only occasionally observed, while the circular patch with only a few or no living plants in the center is more common. Affected regions may coalesce, and large areas of turf can be destroyed within a 7 to 10 day period. There are no distinctive leaf lesions associated with this disease, but leaves generally die back from the tip, and plants at the periphery of affected patches display a bronze or copper color when the disease is active. The copper colored plants at the edge of patches are only evident for a few days, and they are most conspicuous under low mowing.

In annual bluegrass maintained under green conditions, patches range from a few inches (7.6 to 15.2 cm) to about 1 foot (30 cm) in diameter. Death of plants may be nonuniform rather than in discrete patches, particularly when annual bluegrass is mixed with bentgrass. Affected plants develop a reddish-brown, bronze, or yellow color before dying. Creeping bentgrass plants adjacent to dying or dead annual bluegrass plants are unaffected and fill into dead areas vacated by the bluegrass. While *M. poae* may be found on bentgrass roots, it currently is not recognized as being highly pathogenic to 'Penncross' or other commonly used cultivars of creeping bentgrass. Bentgrass greens grown in extremely hot and humid areas such as south Florida and similar climates may be damaged by this pathogen. In creeping red fescue, dead patches generally are 3 to 6 inches (7.6 to 15.2 cm) in diameter. Turf initially develops a bronze or reddish-brown color, but dead leaves invariably turn a straw-brown color. Depressions in the turf called crater pits are common and creeping red fescue frequently does not completely fill into the dead pit areas.

Summer patch most commonly occurs in susceptible species that are 2 years of age or older. The disease may appear in the summer following a fall seeding, but generally is not as severe then as in subsequent summers. To date, the disease principally has been a problem in Kentucky bluegrass, annual bluegrass, and creeping red fescue, but it also may attack hard, blue sheep, and

Chewings fescues. Environmental conditions play a significant role in the predisposition of turf to the disease. Summer patch generally appears in late June or early July when daytime air temperatures above 88°F (31°C) prevail. It is most severe on sunny, exposed slopes or other heat-stressed areas of lawns, such as those adjacent to paved walks and driveways. The disease most frequently occurs during periods of drought stress that were preceded by wet weather in late spring or early summer. Turf allowed to enter drought-induced dormancy after its root system was extensively injured is severely damaged. Mysteriously, the disease may flair up following rainy periods in late summer and September. Low and frequent mowing are most conducive to severe summer patch development. Other predisposing factors include: spring applications of high levels of nitrogen fertilizer, accumulation of thatch, frequent light irrigations or rain storms, and soil compaction. The most important environmental factors required for disease development are for the soil to be moist and for root zone temperatures to exceed 78°F (26°C).

Management

Low mowing (especially <2.0 inches or <5.1 cm) is the major cultural practice that exacerbates summer patch. For home lawns, increase mowing height to 3.0 inches (7.6 cm) in late spring, and apply water deeply and only at the onset of visual wilt. Similarly, the height of cut on greens should be increased to the maximum acceptable level, preferably above 0.25 inch (0.6 cm). Use slow-release acidifying nitrogen fertilizers, such as sulfur coated urea. Acidification with ammonium-based N-sources such as ammonium sulfate also reduces disease severity over time. Most of the annual usage of nitrogen fertilizer should be confined to the autumn months. Core cultivation alleviates damage in compacted soils. On sunny days, it is not uncommon for the upper 2 inches (5.1 cm) of soil to have temperatures 5 to 10°F (2.8 to 5.6°C) above the air temperature. Irrigating during sunny periods will elevate soil temperature because water efficiently conducts heat. Hence, syringing to avoid wetting of soil is preferred to irrigation on hot and sunny days, particularly on greens.

Preventive applications of fenarimol (Rubigan), propiconazole (Banner), triadimefon (Bayleton), or curative applications of thiophanate (CL 3336, Fungo 50) drenches may provide a satisfactory level of control on close-cut Kentucky bluegrass lawns or fairways. To control the disease in chronically affected annual bluegrass on greens, propiconazole or triadimefon should be applied at 21- or 28-day intervals from mid-May (or about 2 to 3 weeks after crabgrass seed germinates) through August. Propiconazole, triadimefon, and other triazole fungicides do not require watering-in, but they tend to be more effective if applied in high water dilutions or when treated sites are syringed immediately after treatment. Thiophanates and fenarimol, however, provide improved control when applied in large volumes of water or when watered-in immediately after application. Fungicides are ineffective if turf is allowed to enter drought-induced dormancy.

Fusarium Blight

Pathogens: *Fusarium culmorum* (Sm.) Sacc., *F. poae* (Peck) Wollenweb.
Primary Hosts: Bentgrasses, bluegrasses and creeping red fescue
Predisposing Conditions: Hot and dry periods in summer

The name Fusarium blight is presently used to describe crown rot or root rot symptoms in situations where signs of *Fusarium* spp., such as spores and pink mycelial growth, are abundant. *Fusarium* crown and root rots occur in drought stressed or senescing tissues. They are most common in semi-arid regions. Affected turf thins out in an irregular pattern rather than circular patches. Summer patch and necrotic ring spot, described previously, are recognized as diseases with circular patch symptoms but where *Fusarium* spp. are not involved as primary pathogens. These *Fusarium* spp. also may cause leaf spotting when days are warm and nights are cool in late spring or summer. Leaf lesions are tan, oval to elongated in shape and may have a purple-colored border. These *Fusarium* species also are common damping-off pathogens.

Management

The cultural and chemical measures for managing Fusarium blight are the same as those described for summer patch. Fusarium leaf spot and damping-off can be arrested with a broad-spectrum fungicide such as chlorothalonil (Daconil 2787), iprodione (Chipco 26019) or vinclozolin (Curalan, Touche, Vorlan).

Southern Blight

Pathogen: *Sclerotium rolfsii* Sacc.
Primary Hosts: Bentgrasses, bluegrasses, and bermudagrasses
Predisposing Conditions: Warm to hot and humid weather

Southern blight is primarily a disease of annual bluegrass, creeping bentgrass, and bermudagrass grown on golf courses (Plate 2–55). Kentucky bluegrass and perennial ryegrass lawns, sports fields, or fairways, however, also may be injured. The disease is more prevalent in warm regions, particularly Southern California and southeastern states. The disease has been reported as far north as Washington D.C. in the east, and northern Illinois in the Midwest. In southern California, the disease may begin in spring and continue throughout the summer, whereas elsewhere it tends to be more prevalent during hot and humid periods of late spring and summer.

The disease initially appears as small, circular, yellow patches. Patches increase in size rapidly and may produce frog-eyes or crescents up to 3 feet (1 m) in diameter. On greens, patches are yellow or reddish-brown in color. Inspection of wet foliage or thatch during the morning hours may reveal the presence of grayish-white mycelium and/or sclerotia. Sclerotia are produced in abundance and are initially white. In time, sclerotia develop a mustard-brown color and are so large that they appear similar to sulfur coated urea granules in color and size.

Management

On greens and tees, the disease is managed by applying ammonium sulfate, reducing thatch through verticutting or core cultivation, and by applying fungicides such as flutolanil (ProStar) or triadimefon (Bayleton). For larger areas, renovation of damaged turf with less susceptible turfgrass species may be more cost-effective than the use of fungicides.

Helminthosporium Leaf Spot, Melting-Out and Red Leaf Spot

Pathogens:	*Bipolaris* spp. and *Drechslera* spp.
Primary Hosts:	Bentgrasses, bluegrasses, bermudagrasses, fescues, and perennial ryegrass
Predisposing Conditions:	Warm to hot temperatures and frequent cycles of wet and dry weather

In summer, Kentucky bluegrass, perennial ryegrass, fine leaf fescues, and other grasses may decline due to invasion by *B. sorokiniana*. This fungus also may cause leaf spot and melting-out phases. *Bipolaris sorokiniana* normally is most severe during wet summers when temperatures exceed 80°F (>26°C) and humidity is high. This disease generally is aggravated when infected stands are subsequently subjected to drought stress. The leaf spot and melting-out symptoms are the same as described previously for *Drechslera*-incited diseases.

During warm and wet periods, *B. cynodontis* may become a severe crown, stolon, and root rot pathogen of bermudagrasses. Dark-brown spots on leaves and sheaths, which are oval-shaped to elongated, are a common symptom. Lesions may have a tan center. Red leaf spot (*D. erythrosphila*) is a summer disease of bentgrasses, particularly colonial bentgrass and Toronto creeping bentgrass. Leaf lesions appear as small reddish spots. Leaves develop a brick-red color and often die back from the tip.

Management

See the "Helminthosporium Leaf Spot and Melting-Out" section for management of these diseases.

Curvularia Blight

Pathogen:	*Curvularia geniculata* (Tracy and Earle) Boedijn, *C. lunata* (Wakk.) Boedijn, others
Primary Hosts:	Bentgrasses and bluegrasses
Predisposing Conditions:	High temperatures and drought stress

Spores of *Curvularia* spp. are found in abundance on most grass tissues that have been damaged by heat and/or drought stress. These fungi can even be seen growing and sporulating on stems and roots. The ability of *Curvularia* spp. to be pathogenic is in some doubt, but most pathologists believe they can cause disease when plants have been stressed by a combination of one or more factors including heat, drought, soil compaction, low mowing, and scald. Affected plants die back from the tip, appear yellowish, and the stand thins out in an irregular pattern. Brownish leaf lesions similar to those caused by Helminthosporium diseases may or may not be evident.

Management

Little has been documented regarding chemical or cultural management of *Curvularia* blight. Cultural practices that alleviate stress such as increasing mowing height, avoiding drought stress, alleviating soil compaction, and proper fertility are highly recommended. Fungicides effective in controlling Helminthosporium diseases also may reduce the severity of *Curvularia* blight.

Nigrospora Blight

Pathogen: *Nigrospora sphaerica* (Sacc.) Mason
Primary Hosts: Kentucky bluegrass, creeping red fescue, perennial ryegrass, St. Augustinegrass
Predisposing Conditions: Warm, rainy periods in summer

Nigrospora blight develops during warm, humid and rainy weather in summer. In St. Augustinegrass, however, it is most likely to occur in spring and early summer. In cool-season grasses, the disease appears as 3 to 6 inch (7.6 to 15.2 cm) diameter patches and mimics dollar spot. When leaf surfaces remain wet for long periods, the pathogen produces an abundance of white, fluffy foliar mycelium that can be seen during early morning hours. Leaves die back from the tip, and necrotic tissue is tan in color. Leaf lesions in Kentucky bluegrass appear as tan bands. These lesions are similar to those of dollar spot; i.e., where tan and green tissues meet there is a brown stain. Succulent leaves of Kentucky bluegrass overstimulated by nitrogen may develop a water-soaked or oily appearance. Leaves and tillers collapse and plants may die rapidly in circular patches 3 to 8 inches (7.6 to 20.3 cm) in diameter. This patch symptom is very similar to that produced by summer patch.

Management

This disease tends to be a foliar blight and only occasionally is severely damaging. Apply balanced N + P + K fertilizer primarily in fall to cool-season grasses, and avoid more than 1.0 lb N/1000 ft^2 (50 kg N/ha) in late spring-early summer to ensure turf vigor. Avoid night irrigation and do not mow when foliage is wet. Delay applications of herbicides until fall or other periods when the disease is no longer active. Broad-spectrum fungicide sprays may be warranted where the disease is chronic and destructive.

Copper Spot

Pathogen:	*Gloeocercospora sorghi* Bain & Edgerton ex Deighton
Primary Host:	Velvet bentgrass
Predisposing Conditions:	Warm to hot and humid periods in summer

Copper spot is primarily a disease of velvet bentgrass in New England (Plate 2–56). The disease may develop on rare occasions in creeping bentgrass and bermudagrass. As the name implies, the disease appears as copper-colored patches 1 to 2 inches (2.5 to 5.1 cm) in diameter. This disease mimics dollar spot on greens, except infected foliage develops a reddish-brown or copper color. During wet periods, leaves are covered with small, slimy, salmon-pink to orange-colored masses of spores, which are borne on structures called sporodochia. When dry, sporodochia are bright-orange and resinous. Very small, black sclerotia are produced in large numbers in dead leaf tissue. Sporodochia and sclerotia can be seen with the aid of a hand lens. Unlike dollar spot, only leaves are blighted and plants seldom die. Turf slowly recovers from injury following the advent of cool and dry weather.

Management

Avoid mowing affected turf early in the morning when leaves are wet. Speed leaf drying by dragging or poling greens. Do not apply water-soluble nitrogen fertilizers when the disease is active. Manage velvet bentgrass with low to moderate amounts of nitrogen at all times, and apply limestone if soil pH is in the acid range. Broad-spectrum contact or local penetrants such as chlorothalonil (Daconil 2787) and iprodione (Chipco 26019) effectively control copper spot. Thiophanates (CL 3336 or Fungo 50) and triadimefon (Bayleton) provide a longer residual control.

Fairy Rings

Pathogen(s):	Many basidiomycetes
Primary Hosts:	All turfgrasses
Predisposing Conditions:	Warm and moist weather with intermittent periods of drought stress in summer

The turf diseases known as fairy rings may be caused by any one of 60 species of fungi. These fungi can cause the formation of rings or arcs of dead or unthrifty turf, or rings of dark green, luxuriantly growing grass (Plates 2–57, 2–58, and 2–60). Fairy ring fungi belong to a group known as the Basidiomycetes or "mushroom fungi." These fungi primarily colonize thatch or organic matter in soil and generally do not directly attack turfgrass plants; however, some are weakly parasitic.

Fairy rings are classified into three types according to their effects on turf:

Type 1: Those that kill grass or badly damage it.

Type 2: Those that stimulate grass by forming rings of dark green turf.

Type 3: Those that do not stimulate grass and cause no damage, but produce mushrooms in rings.

The most destructive rings are of the Type 1 variety (Figure 2.1). Type 1 rings are very common, especially in lawns and golf course fairways that previously had been pasture or where tree stumps or lumber had been buried. Type 1 rings normally appear as circles or arcs of dark green, fast growing grass. The most common fungus known to cause Type 1 fairy rings is *Marasmius oreades* (Bolt. ex Fr.) Fr. Type 1 rings are distinguished by three distinct zones: an inner lush zone where the grass is stimulated and grows luxuriantly; a middle zone where the grass may be dead; and an outer zone in which the grass is stimulated. The distance from the inside of the inner zone to the outside of the outer zone may range from a few inches (3 to 6 cm) to 3 feet (>90 cm) wide. The green stimulated zones are the result of the breakdown of organic matter, which releases nitrogen and causes the stimulated growth. The outer green zone is caused by the breakdown of thatch by the fairy ring fungus, which liberates nitrogen. The inner green zone develops in response to the release of nitrogen as bacteria degrade aging or dead fungal mycelium produced in previous years. The formation of the three zones is noticeable from early spring to winter.

Mushrooms of the fungus causing a Type 1 ring are produced in the bare zone or at the junction of the bare and outer zone. Rings, however, may not produce mushrooms for several years, especially on closely-cut greens. The

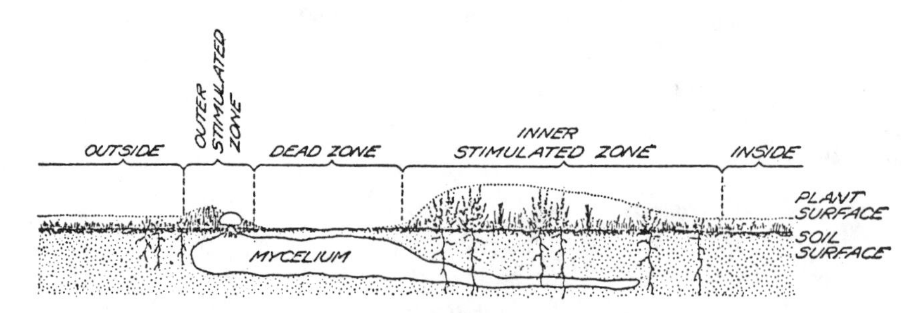

Figure 2.1. A bisect of the *Agaricus tabularis* ring. The vegetation on the inside and outside does not differ noticeably. The outer stimulated zone in which the fungus fruits are produced is separated from the inner stimulated zone by a bare zone in which plants are only occasionally found. The distribution of the mycelium in the soil is also indicated in the illustration. From H.L. Shantz and R.L. Piemeisel, "Fungus Fairy Rings in Eastern Colorado and Their Effects on Vegetation," *J. Agric. Res.* 11:191–245 (1917).

underside of the mushroom cap is composed of gills, upon which spores are produced. The importance of spores in the spread of fairy ring fungi is not well understood.

A fairy ring is broken when its mycelium encounters an obstacle such as a rock, pathway, or unfavorable soil condition. The ring may also disappear for no apparent reason. In general, two fairy rings will not cross one another; i.e., at the point of intersection the growth of each ring stops. This obliteration at the point of contact is believed to be caused by the production of self-inhibitory metabolites that also will antagonize other members of the same or different fungal species. On slopes, the bottom of the ring usually is open, giving the appearance of an arc rather than a ring. This may be due to the downward movement of self-inhibitory metabolites that prevent fungal development in turf on the lower side of the ring.

Rings vary in size from a few inches (15 cm) to 200 feet (60 m) or more in diameter, and become larger each year. The annual radial growth ranges from 3 inches (7.6 cm) to as much as 19 inches (48 cm). The rate of outward movement, as well as overall diameter of rings, is determined by soil and weather conditions. Growth of a fairy ring begins with the transport of fragments of fungal mycelium and possibly spores. The fungus initiates growth at a central point and continues outward in all directions at an equal rate. The fairy ring generally is first observed as a cluster of mushrooms. Rings fade in the fall or winter and the bare zone often is not visible. Loss of visibility is due to the general brownish appearance of dormant turf during winter and because the turf is not metabolizing nitrogen in large enough quantities to produce the lush green inner and outer zones.

Fairy rings have been observed in areas where soil pH has ranged from 5.1 to 7.9. It is likely that fairy rings will occur under any soil condition that will support turfgrass growth. Nearly all of the commonly cultivated turfgrasses are known to be affected by fairy ring fungi.

Type 1 fairy ring fungi kill vegetation primarily by rendering infested soil water repellent. Hence, the dead zone is due to mycelium of the fungus that accumulates in such large amounts in soil that it prevents entry of rain or irrigation water, and thus kills the plants by drought. Furthermore, soil in the dead zone normally is dry when compared to adjacent soil. It is quite characteristic for grass on the outer edge of the dead zone to display the blue-gray color of turf under drought stress. Generally, if a plug of soil is removed from the edge of an active fairy ring the dense, white, thread-like network of mycelium can be seen clinging to soil and to roots of grass plants. When environmental conditions are optimum for fungal growth, the white mycelium can easily be seen on the surface of the thatch layer. Fairy ring infested soils normally have a mushroom odor, even if fungal mycelium is not evident. Although some fairy ring fungi are known to parasitize roots and produce compounds toxic to roots, it is likely that most damage to turf can be attributed to the fungal mycelium rendering the soil impermeable to water.

Management

Control of fairy rings is made extremely difficult by the impermeable nature of the infested soil. Chemical control has been ineffective because the fungus grows deeply into the soil and lethal concentrations of fungicide do not come into contact with the entire fungal body. There are three approaches to combating fairy rings: (a) suppression, (b) antagonism, and (c) eradication.

Suppression is the most practical approach to combating fairy rings in most situations. The suppression approach is based upon the premise that fairy rings are less conspicuous and less numerous where turf is well watered and fertilized. This method of control involves a combination of core cultivation, deep watering, and proper fertilization. Core cultivation is beneficial since it aids in the penetration of air and water. The entire area occupied by the ring, to include a 2 foot periphery beyond the ring, should be plugged to remove soil cores on 2 to 4 inch (5.1 to 10.2 cm) centers. The area should then be irrigated to a depth of 4 to 6 inches (10.2 to 15.2 cm). Use of a wetting agent should help improve water infiltration. The ring area should be re-treated in a similar fashion at the earliest indication of drought stress; that is, repeat the process whenever the dark green grass turns blue-gray and begins to wilt. When a coring unit is not available, a deep root feeder with garden hose attachment may be useful to force water into the dry soil. Forcing water into dry soil with a tree root feeder is the best management approach where only a few fairy rings are present. Apply recommended amounts of nitrogen at the appropriate time of year to help mask fairy rings. Fairy rings, however, can be stimulated by excessive nitrogen or organic matter.

The antagonism approach is based upon the observation that rings exhibit mutual antagonism, i.e., elimination when they come into contact with one another. This method involves removal of the sod by striping or killing the turf with a nonselective herbicide. The soil must be rototilled repeatedly in several directions until the mycelium infested soil has been thoroughly mixed. The soil is then prepared in the usual manner for seeding or sodding. This method has been shown to be very promising, but has had only limited testing. Fairy rings, however, are known to recur in sod fields within 2 to 3 years after harvest and establishment of a new crop. This may be due to poor or inadequate mixing of soil prior to seeding.

There are two methods of eradication: fumigation and excavation. Both methods are laborious, costly and not always successful. Prior to fumigation, remove the sod from an area 2 feet (60 cm) to the inside and 2 feet (60 cm) to the outside of the green rings. It is essential not to spill any soil or sod onto the healthy grass. The soil is then loosened to a depth of 6 to 9 inches (15 to 23 cm). Then select a fumigant, the most common being methyl bromide, dazomet (Basamid) and metam-sodium (Metam, Vapam). Dazomet and metam-sodium are safer and easier to handle. Dazomet is applied in granular form. Metam-sodium is applied in water and should be poured uniformly on the loosened soil with a sprinkling can. The site must be tilled prior to application and then rolled immediately after treatment. Ensure that the solution or granules are kept off healthy grass as they will scorch or kill the leaves. The treated area can be covered with polyethylene to retain vapors. Rolling soil, however, provides a sufficient seal and covering is only required for methyl bromide. At the conclusion of the waiting period stated on the label (10 to 20 days or longer depending on soil temperature) remove the cover and allow the soil to air out for 2 weeks or until the chemical odor is no longer evident. For methyl bromide, treated areas only need to be covered for 1 to 2 days, and seeding can begin within a day after the cover is removed. Finally, add fresh soil, if needed, and reseed or sod. These chemicals should be used well beyond the drip line of trees and shrubs. Fumigation with methyl bromide should be carried out only by a specially licensed pesticide applicator. Extreme precautions must be taken to ensure that children or pets do not come into contact with these fumigants. Consult labels of fumigants for more precise information on handling and applying these chemicals.

The second alternative to fairy ring eradication is to carefully dig out and discard all infested soil in the ring. This would involve removal of soil to a 12 inch (30 cm) depth, and the excavation should be wide enough to extend at least 2 feet (60 cm) beyond the outermost evidence of the ring. The excavation must then be filled with fresh uncontaminated soil and the area reseeded or sodded.

Superficial Fairy Ring

Pathogens: *Coprinus kubickae* Pilat & Svrcek, *Trechispora* spp., others
Primary Hosts: Bentgrass greens and fairways
Predisposing Conditions: Warm, rainy periods in summer

In addition to fairy rings, several mostly unidentified basidiomycetous (mushroom) fungi cause three other diseases of turf known as superficial fairy rings, localized dry spots, and yellow ring. Superficial fairy rings sometimes are referred to as "white patch" since they appear primarily in bentgrass greens or fairways as white, circular patches that range from 3 inches (7.6 cm) to 3 feet (90 cm) in diameter (Plates 2-61 and 2-62). At the edge of these well defined, circular patches are 1 to 2 inch (2.5 to 5.1 cm) fringes of white mycelium. This disease appears during warm and rainy periods in summer, and is most commonly observed under conditions of low nitrogen fertility.

Upon close inspection, one can observe that the older leaves in the whitish fringe die prematurely, and have a bleached-white appearance. Although the fungus appears unsightly it does not infect leaves. However, it causes premature death of leaves and sheaths by blocking incoming sunlight, which results in the rapid breakdown of chlorophyll in these tissues. Once leaves and sheaths die, the fungus uses the necrotic tissues as a source of nutrition. Thatch degradation by the fungus may cause shrinkage, which disfigures or interferes with the trueness of a putting surface. In extreme cases, SFR fungi will reduce turf density. These fungi develop principally in thatch and do not penetrate more than 0.5 to 1.0 inches (1.3 to 2.5 cm) of underlying soil.

Management

Superficial fairy rings are more common in poorly nourished turf. Application of 0.5 lb nitrogen per 1000 ft^2 (25 kg N/ha) along with mechanical disruption of thatch and fungal mycelium by verticutting, spiking, or core cultivation helps to mask or minimize the adverse effects of these fungi. There are no fungicides known that control SFR fungi.

Localized Dry Spot

Pathogens: Unidentified basidiomycetes
Primary Hosts: All greens and bentgrass fairways
Predisposing Conditions: Hot and dry periods of summer

Localized dry spots are common on high-sand content greens or mineral soil greens that have been aggressively topdressed with sand. They normally develop in new golf course greens and bentgrass fairways within 2 to 3 years of seeding, and tend to decline in severity over time. Localized dry spots appear as solid patches of wilted or dried-out turf (Plates 2–60 and 2–63). Their appearance sometimes is preceded by fairy ring development or simply by the presence of numerous mushrooms. Patches can be circular and range from a few inches (6 to 8 cm) to several feet (0.5 to 1.0 m) in diameter, or they may appear as large irregularly-shaped areas of wilted or dead turf. Soil within the patches remains bone dry despite frequent irrigation. Water will penetrate the thatch, but not the thatch-soil interface, and will usually run off dry spot areas. Plants within affected patches develop a purplish color that is indicative of wilt, and eventually die as a result of drought stress.

The cause of localized dry spots has been attributed to unidentified basidiomycetous (mushroom) fungi. These fungi cannot be isolated from samples obtained from the hydrophobic (water repellent) soil. It is believed that water repellency is caused by the breakdown of older fungal mycelium, which releases substances that provide a coating of organic material around individual sand particles. This organic coating causes adjacent sand particles to pack or bind together, and renders the soil impervious to water infiltration. The water repellent, i.e., hydrophobic condition, normally is restricted to the upper few inches (3 to 6 cm) of soil. Removal of thatch alone will not significantly improve water infiltration.

Management

Verti-draining, core cultivation, or water injection cultivation in combination with frequent application of a wetting agent will help to alleviate this condition. Water injection cultivation is least destructive and quite effective; however, it does not cure the condition. Keeping turf alive in localized dry spots requires numerous daily syringes and weekly to bimonthly treatment by water injection during dry summer periods. Isolated spots can be individually treated by frequent probing with a tree deep-root feeder that injects water.

Fungicides are not likely to affect the severity of localized dry spots once they have appeared.

Yellow Ring

Pathogen:	*Trechispora alnicola* (Bourd. & Galzin) Liberta
Primary Host:	Kentucky bluegrass
Predisposing Conditions:	Warm and wet weather from spring to fall

This pathogen belongs to the class of mushroom fungi, which also are responsible for fairy rings. As its name implies, this disease causes characteristically yellow rings to develop in Kentucky bluegrass (Plate 2–59). There are no other known hosts, and the rings may come and go. Normally there are large amounts of white fungal mycelium present in the thatch layer, which produces a strong, mushroom odor.

Management

Given the ephemeral nature of this disease, and the fact that it causes no permanent damage to turf, control measures are seldom recommended. Verticutting or core cultivation disrupts fungal growth, reduces thatch, and may speed the disappearance of the yellow rings. Fungicides are ineffective.

Gray Leaf Spot

Pathogen:	*Pyricularia grisea* (Cooke) Sacc.
Primary Hosts:	St. Augustinegrass and perennial ryegrass
Predisposing Conditions:	Prolonged periods of warm and humid weather

Gray leaf spot is principally a disease of St. Augustinegrass, but may also injure bermudagrass, centipedegrass, and bahiagrass. It occasionally is a problem in perennial ryegrass, and in rare situations attacks bentgrasses and fescues. In warm-season grasses, leaf spots appear during prolonged periods of high humidity from spring to fall. In cool-season grasses, gray leaf spot occurs in late summer following extended periods of high humidity and heat stress.

On warm-season grasses, leaf lesions are initially tiny, but increase in size to form circular spots or oblong lesions. Lesions are gray, and bordered by a narrow band of purple or brown colored tissue. When dew is present, lesions have a grayish-felt appearance. The felted spot is comprised of large numbers of spores, which are easily seen with the aid of a microscope. The lesions can increase rapidly and kill leaves. Dying leaves turn yellow or grayish-tan, and the stand may have the appearance of having been subjected to severe drought stress.

Among cool-season grasses, perennial ryegrass grown on fairways in the transition zone is occasionally blighted. Symptoms in ryegrass can mimic those typical of late season, nonuniform blighting by brown patch; however, no foliar mycelium is produced. Leaf lesions are oblong, gray or brown in color, and bear large populations of spores. This disease can be very disfiguring and cause loss of turf density. The pathogen, however, blights leaves but usually does not kill perennial ryegrass or other grass species.

Management

This disease is enhanced by frequent applications of water-soluble nitrogen in summer. After the disease subsides, the application of a balanced N-P-K fertilizer, with half of the nitrogen from a slow-release source, will stimulate recovery. Deep and infrequent irrigation during daytime hours, and raising the mowing height helps to alleviate damage. Avoid using herbicides or plant growth regulators when the disease is active. Application of a broad-spectrum fungicide, such as chlorothalonil (Daconil 2787), will effectively control the disease.

Cercospora Leaf Spot

Pathogen: *Cercospora* spp.
Primary Host: St. Augustinegrass
Predisposing Conditions: Prolonged periods of humid and wet weather in summer

Cercospora leaf spot is primarily a disease of St. Augustinegrass (*Cercospora fusimaculans* Atk.), but it also may injure bentgrass, buffalograss, and bermudagrass. The disease is favored by warm, humid weather and long leaf surface wetness periods in summer. Leaf spot lesions are similar in appearance to those of gray leaf spot. Lesions may be linear to oval-shaped with tan to gray centers and a purple border. In severely infected stands, leaves turn yellow, wither, and the turf loses density. Stem tissues are not attacked and plants seldom die.

Management

Avoid night irrigation since this will increase leaf surface wetness period, and encourage spore germination and production. Water deeply but infrequently during early morning hours. In poorly nourished stands, apply 0.5 to 1.0 lb of nitrogen per 1000 ft^2 (25 to 50 kg N/ha) to stimulate turf growth. If appropriate, selectively prune trees and shrubs to improve air circulation and sunlight infiltration. A broad-spectrum fungicide may provide effective control.

White Blight

Pathogen: *Melanotus phillipsii* (Berk. & Broome) Singer
Primary Host: Tall fescue
Predisposing Conditions: Hot, humid weather

Tall fescue is the only reported host for this pathogen. This disease develops during very hot and humid periods, and primarily in sunny places. It rarely occurs north of Washington D.C. Circular patches range from 3 to 14 inches (7 to 36 cm) or greater in diameter. Leaves die back from the tip and become light tan to white in color. Leaf blades become matted and foliar mycelium may be evident. Small, tannish-white, stalkless mushrooms 1/8 to 1/4 inch (3 to 6 mm) wide develop on blighted leaves. The presence of these mushrooms developing on blighted foliage is the key diagnostic sign for white blight.

Management

Avoid drought stress by watering deeply but infrequently. Blighting is restricted to leaves; therefore, fall applications of a balanced N-P-K fertilizer will stimulate growth and rapid recovery of damaged turf. Overseeding severely thinned-out areas may be required. There are no known chemical control measures; however, fungicides that effectively control brown patch may reduce disease severity.

Slime Mold

Pathogens: *Physarum* spp., *Fuligo* spp., others
Primary Hosts: All turfgrasses
Predisposing Conditions: Warm, rainy periods in summer

Slime molds are primitive fungal-like organisms that move about as protoplasmic amoebae and feed on dead organic matter in thatch (Plate 2–64). During prolonged periods of warm, wet overcast weather in summer, they migrate onto turfgrass leaves to reproduce. Once on the leaves they differentiate into fruiting bodies called sporangia. These sporangia appear as small rounded, ball-like clusters about the size of a pin head, and are gray, black or purple in color. When the sporangia mature and dry they are typically gray, and appear similar to cigarette ashes. Spores are discharged and fall into thatch where they develop into tiny, shapeless masses of protoplasm. These slime molds are not parasitic, but if the turf is not routinely mowed they can block sunlight from leaf surfaces and on rare occasions these leaves may turn yellow.

Management

Slime mold protoplasm and sporangia are removed by mowing or irrigating the turf. If turf is not growing, and therefore not being mowed, the sporangia can be easily brushed off leaves. Fungicide use is seldom, if ever, warranted for slime mold control.

Rust

Pathogen: *Puccinia* spp.
Primary Hosts: Kentucky bluegrass, perennial ryegrass, and zoysiagrasses
Predisposing Conditions: Prolonged periods of overcast weather or shaded environments in late summer and fall

There are many species or forms (known as races) of rust fungi that attack nearly all turfgrasses to include bluegrasses, bermudagrass, buffalograss, ryegrasses, St. Augustinegrass, and zoysiagrass. Stem rust (*P. graminis* Pers.) of Kentucky bluegrass, crown rust (*P. coronata* Corda) of perennial ryegrass, and zoysiagrass rust (*P. zoysiae* Dietel) are the most common and important rust diseases of turf (Plates 2–65 and 2–66). These rust diseases most commonly are observed during cool, moist periods of late summer and fall. They are most damaging to poorly nourished turf and turfs grown under a low mowing height or in shade. In most regions of the U.S., rusts do not often cause serious turf damage. However, in some environments marked by long periods of wet and overcast weather, such as coastal areas from northern California to Canada, the rusts are chronic and debilitating diseases. They are particularly severe in turfgrass seed-producing areas in the Pacific Northwest.

Rust-affected turfs exhibit a yellowish or reddish-brown appearance from a distance. Close inspection of diseased leaves reveals the presence of conspicuous red, black, orange or yellow pustules. These powdery pustules are comprised of huge numbers of spores. Several types of spores are produced by rusts and these fungi have complicated life cycles. During most seasons, rust-affected plants generally appear healthy. Kentucky bluegrass turfs simultaneously infected with stripe smut and rust can be severely thinned out in late summer.

Management

In most regions, rust-affected turfs can be effectively maintained by employing sound cultural practices. A balanced N + P + K fertility program is most often preferred to fungicides in situations where rust is damaging poorly nourished turfs. Irrigate early in the day to ensure leaf dryness prior to nightfall; irrigate deeply but infrequently; increase mowing height; and increase mowing frequency. By increasing mowing frequency, leaves bearing immature spores are removed and this reduces the potential for more leaf infections.

Sterol-inhibiting fungicides such as propiconazole (Banner) and triadimefon (Bayleton) effectively control rust diseases in a single spring or fall application. Contact fungicides are not very effective and multiple applications are required to reduce rust injury.

Seedling Diseases/Damping-Off

Pathogen: *Pythium* spp., *Rhizoctonia solani*
Bipolaris spp., *Drechslera* spp., *Curvularia* spp., *Fusarium* spp., others
Primary Host: All seeded species
Predisposing Conditions: Seeding during warm, humid or wet periods

Under a well-defined set of environmental conditions, many common soil fungi can parasitize seeds and seedlings. Seed decay or damping-off of seedlings most often occurs when cool-season grasses are seeded in late spring, summer, or early fall when high temperatures and high humidity prevail. Conditions of high temperature and humidity, especially at night, slow seed germination and reduce seedling vigor. Furthermore, these are the same conditions that are conducive to growth of *Pythium* spp. and *Rhizoctonia solani. Fusarium* spp., *Bipolaris* spp., and *Curvularia* spp. also are destructive under conditions of high temperature and/or drought stress. During winter and spring following a fall seeding, young plants can be damaged by Fusarium patch (*Microdochium nivale*), or leaf spot and melting-out diseases (*Drechslera* spp.). Hence, temperature and moisture extremes that enhance fungal growth but reduce growth and vigor of seedlings are the major factors that predispose seeds and seedlings to damping-off diseases.

Damaged stands often appear patchy with severely thinned-out areas nonuniformly distributed throughout areas of good density. Stands also may appear spotty with small patches of yellowed plants distributed nonuniformly in a stand. Close inspection of yellowed plants may reveal the presence of brown leaf spots caused by *Fusarium, Curvularia, Bipolaris,* or *Drechslera* spp. *Pythium* spp. and *R. solani* progress rapidly (24 to 48 hours) during warm and humid weather, initially causing seedlings to appear darkened and water-soaked. These seedlings soon shrivel, collapse, and turn a necrotic brown color. Dead seedlings may be matted and have a greasy appearance.

Other factors that may enhance damping-off are as follows: poor seed to soil contact; seed planted too deeply; excessive nitrogen fertilization; excessive irrigation; poor soil water drainage; poor air circulation; old and slow to germinate seed; and seed contaminated with pathogenic, fungal spores. Some cultural approaches to minimize damping-off are listed in Table 2.5.

Fungicides may be used preventively or curatively to avert damping-off. A preventive approach would include use of fungicide-treated seed. Captan and thiram are good, protectant fungicides, but they may be ineffective against

Table 2.5. Cultural Approaches to Reduce Seed and Seedling Diseases

Delay renovation or seeding of cool-season grasses until cooler periods in fall.
Ensure proper seed to soil contact by rolling, and avoid planting seed too deeply. In older turfs, remove thatch prior to renovation or overseeding.
Use certified seed of regionally adapted cultivars.
Provide for surface water drainage by proper grading before seeding.
During establishment, use a balanced N-P-K fertilizer that has at least half of the nitrogen component in a slow-release form.
Avoid topdressing seedlings with water-soluble nitrogen fertilizers in spring and summer.
Syringe the seedbed frequently, but do not allow water to inundate or wash seeds or seedlings.
Avoid nighttime irrigation and time afternoon watering so the foliage is dry prior to nightfall.
Maintain a proper cutting height and remove clippings where practical.

Pythium damping-off. Metalaxyl- or ethazol-treated seed (trade name Apron and Koban, respectively) will protect seeds from *Pythium* spp., but not other pathogens. Fungicidal dusts can be prepared in the absence of commercially treated seed by mixing wettable powder formulations at a rate of 1% to 2% by weight with seed (e.g., 1 to 2 lb [454 to 908 grams] of fungicide per 100 lb [45.4 kg] of seed). Seed dusting should be performed in a sealed container, in a well vented room or outside, and workers should wear a respirator, gloves, and a spray suit.

Foliar sprays or soil drenches also are effective in controlling seedling diseases. A Pythium-targeted fungicide with contact activity [e.g., chloroneb (Terraneb SP.), or ethazole (Koban)] should be tank-mixed with a broad spectrum fungicide [e.g., chlorothalonil (Daconil 2787), iprodione (Chipco 26019), mancozeb (Fore) or vinclozolin (Curalan, Touche, Vorlan)] and applied to the seedbed and allowed to dry prior to the first irrigation. If hot and humid conditions prevail during the seedling emergence period, a foliar spray of the tank-mix combination should be reapplied as needed on a 14- to 21-day interval. In addition to the aforementioned Pythium-targeted fungicides, fosetyl-aluminum (Aliette), metalaxyl (Subdue), or propamocarb (Banol) could be used on emerged seedlings.

VIRUS DISEASES

Viruses that infect grasses are pathogenic entities that consist of nucleic acid (usually RNA) covered by a protective protein coat. Viruses are unable to grow, they are immobile, they exhibit no metabolic activity outside of living cells, and they have no means of penetrating cells. Despite these limitations, they are common inhabitants of grasses.

Viruses gain entry into grasses primarily by being vectored by insects or they may be seedborne. They also may be mechanically transmitted in sap, which is rare in nature yet it is a primary means of St. Augustine decline transmission. When a virus gains entry to a cell it biochemically triggers the plant to produce nucleic acid and protein identical to that of the invading virus. As huge amounts of virus particles are produced, they are spread throughout the plant and cause it to lose vigor, turn chlorotic, and in rare cases die. While viruses are common in many grasses, little is known about virus diseases of turfgrasses. St. Augustine decline and centipede mosaic are the only well-described virus diseases of turf.

St. Augustine Decline and Centipede Mosaic

Pathogen: Panicum mosaic virus
Primary Host: St. Augustinegrass and centipedegrass
Predisposing Conditions: During periods of active turf growth in spring and summer

This disease is primarily a problem in Gulf Coast and southeastern states from Texas to South Carolina. Panicum mosaic virus causes a decline in both St. Augustinegrass and centipedegrass (Plate 2–67). Early symptoms include a mild yellowing and leaf stippling that mimics mite damage. Leaves may develop a mottled yellow-green mosaic, which becomes more pronounced or severe the longer plants are infected. In advanced stages, plants are stunted, lack vigor, and exhibit very slow stolon growth. In extreme cases, turf is severely chlorotic, leaves become necrotic and large areas thin or die out completely. It is uncommon, however, for panicum mosiac virus to kill large areas of turf. This virus is mechanically transmitted in sap; therefore, mowing wet leaves aids in the spread of the virus.

Management

Virus diseases cannot be controlled with chemicals. Due to the debilitating effects of this virus, these diseases are primarily managed by alleviating environmental stresses or renovating with resistant cultivars such as 'Floratam,' 'Floralawn,' 'Raleigh,' or 'Seville' St. Augustinegrass. Cultural practices, such as increasing mowing height, avoiding mowing when leaves are wet, preventing drought stress, and applying proper rates of balanced N-P-K fertilizers assist in maintaining turf vigor and density.

BACTERIAL DISEASES

Bacterial diseases are uncommon in turfgrasses. Plant pathogenic bacteria are single celled, usually rod-shaped, and they have rigid cell walls. They reproduce by binary fission and they may or may not be mobile. Bacteria have no means of penetrating plant cells, so they must enter plants through natural openings such as stomates and hydathodes, or through wounds. Once inside plants they cause injury by producing toxins or they can plug vascular tissues. By occluding vessels, for example, they prevent the movement of water and nutrients, and plants die primarily due to lack of sufficient water. There is only one recorded bacterial disease of turfgrasses in the U.S.

Bacterial Wilt

Pathogen: *Xanthomonas campestris* (Pammel) Dowson pathovar *graminis* (proposed)
Primary Host: 'Toronto' creeping bentgrass
Predisposing Conditions: Warm or cool, rainy periods in spring and fall

Because the first recorded host for this disease was 'Toronto' (also known as C-15) creeping bentgrass, it was initially referred to as C-15 Decline (Plate 2–68). The disease also may attack Cohansey, Nimisilia, and Old South German creeping bentgrass cultivars. A closely related biotype, *X. campestris* pathovar *poannua* causes a similar wilt disease in annual bluegrass, and is being developed as a biological control agent for *P. annua* where it is considered a weed.

In 'Toronto' creeping bentgrass, the bacteria are primarily limited to xylem vessels in roots, but they may be detected in crown and leaf tissue (Plate 2–69). Once the xylem elements of a large number of roots become plugged with masses of bacterial cells, the plants begin to wilt. Initial symptoms therefore appear as wilt and leaves develop a blue-green color. This stage is short-lived, and leaves rapidly turn brown and shrivel. In annual bluegrass, individual plants turn yellow and die in spots about the size of a dime. When there is coalescence of numerous dead plants, the nonuniform browning can mimic anthracnose.

In the laboratory, a diagnostician will cut affected leaves with a razor blade and look for oozes of streaming bacterial cells on a microscope slide. Large areas are destroyed in a nonuniform pattern within a few days. Adjacent plants in higher-cut collars or fairways may display little or no injury. This disease is favored by periods of heavy rainfall followed by cool nights and warm days, and is therefore most likely to appear in spring and fall.

Management

Increasing mowing height reduces disease severity dramatically, but negates the playability of greens. Mowing turf when leaves are dry may slow progress of the disease. Antibiotics suppress bacterial wilt, but they are very expensive, difficult to handle, and generally do not provide an acceptable level of control. In situations where the disease is chronically severe, greens should be renovated with another cultivar of creeping bentgrass.

PLANT PARASITIC NEMATODES

Nematodes are known to cause extensive injury to turfgrasses in warm temperate and subtropical areas of the U.S. (Plate 2–70). In northern climates, the loss of turfgrass that can be attributed to nematodes is unknown. However, nematodes may be more troublesome in the transition zone and northern regions than is commonly believed. This would be particularly true following a mild winter. In general, parasitic nematodes most actively feed on turfgrass roots during environmental periods favorable for growth of the grass. Hence, feeding would be more active on cool-season species in spring and fall, while heaviest feeding on warm-season grasses would occur during summer. The injurious effects of this feeding, however, generally do not become noticeable until turf is subjected to environmental stresses in mid- to late summer.

Nematodes are very small, eel-like worms that range from 1/75 to 1/8 inch (< 0.2 to 3 mm) in length. Nematodes reproduce by eggs, which hatch to liberate larvae. Larvae molt 4 times before reaching adult size. Each female is capable of producing hundreds of eggs and the entire life cycle for most species is completed in 5 to 6 weeks under suitable conditions. Because plant pathogenic nematodes are obligate parasites, they must feed on living tissues in order to grow and reproduce. Most nematodes are capable of attacking a wide range of plant species, and can survive on weeds in the absence of turfgrasses. Most nematodes store large quantities of food, which enables them to survive long periods in soil in the absence of suitable plants. Many survive in frozen soils and may overwinter in living roots or in dead plant tissues.

Literally millions of nematodes can inhabit a few square feet (or a square meter) of soil, but most nematodes are nonpathogenic. Most nonpathogens actually perform a beneficial service in soil by helping to degrade organic matter. Although there are thousands of species, only about 50 species are known to parasitize turfgrasses. All plant parasitic nematodes bear a hollow, spear-like structure called a stylet. The stylet is similar to a hypodermic syringe, and is used to inject enzymes into plant cells. Simultaneously, partially digested food is withdrawn. Plant pathogenic nematodes are grouped according to their feeding habit. Endoparasitic nematodes partially or totally burrow into plant tissues and feed primarily within, whereas ectoparasitic nematodes feed from the plant surface, although a small portion of the body may be embedded. The ectoparasitic nematodes are more commonly injurious to turfgrasses than endoparasitic nematodes. Nematode activity is favored by warm and moist soil conditions. Nematode populations generally peak in June or July and again in late August or early September. Their activity is enhanced in light textured soils and reduced in compacted or fine textured soils where aeration becomes restricted. Nematodes are unable to move more than a few millimeters in soil, but they may be transported over longer distances by moving water, soil, equipment and man.

The symptoms of nematode injury include yellowing, stunting, wilting or early signs of drought stress, and thinning of the stand. These symptoms are related to the injury nematodes inflict upon the root system. Therefore, symptoms of injury may not become noticeable until water becomes limiting. Due to the similarities between environmental stress symptoms and nematode injury, the source of the problem is difficult to diagnose. Like many so-called weak or secondary fungal pathogens, nematodes may not cause much of a problem until environmental extremes reduce the vigor of a turf. There often is no pattern to nematode injury, but affected areas generally appear as streaks or oval-shaped areas. Severe infestations may result in a near total loss of grass plants, which are soon replaced by weeds. Inspection of roots may or may not reveal some indication of nematode feeding. Roots may exhibit one or more of the following symptoms: swellings, red or brown lesions, excessive root branching, necrotic root tips, and root rot. Some of the more common plant parasitic nematodes that are known to injure turfgrasses are listed in Table 2.6.

Turfgrass areas damaged by nematodes are especially prone to wilt and do not respond readily to an application of fertilizers or fungicides. This lack of response may be a good indicator of a nematode problem. In this situation, a soil sample should be sent to a nematologist for analysis. Soil should be

Table 2.6. Common Plant Pathogenic Nematodes Known To Be Injurious to Turfgrasses

Common Name/Genus	Feeding Group	Turfgrasses Injured/ Root Symptoms
Lesion (*Pratylenchus* spp.)	Endoparasitic	Cool- and warm-season grasses are injured. Root lesions initially minute and brown, but enlarge and may prune the root system.
Lance (*Hoplolaimus* spp.)	Ectoparasitic	Warm-season grasses, creeping bentgrass, and annual bluegrass are injured. Causes swelling of roots followed by necrosis and sloughing of cortical tissues. Can be endoparasitic and migratory.
Ring (*Macroposthonia* spp. = *Criconemoides* spp.)	Ectoparasitic	Especially important in centipedegrass; also injures Kentucky bluegrass, bentgrass, bermudagrass, and zoysiagrass. Roots are stunted; brown lesions on roots.
Spiral (*Helicotylenchus* spp.)	Ectoparasitic	Cool- and warm-season grasses injured. Roots are poorly developed with premature sloughing of cortical tissues.
Stunt or stylet (*Tylenchorhynchus* spp.)	Ectoparasitic	Warm- and cool-season grasses injured. Roots shortened, shriveled, brown lesions evident on roots.
Root-knot (*Meloidogyne* spp.)	Endoparasitic	Warm-season (especially zoysiagrass) and cool-season grasses (especially bentgrass) injured. Galls (i.e., swellings or knots) on roots. Galls may be small and difficult to see.
Stubby root (*Paratrichodorus* spp.)	Ectoparasitic	Warm-season grasses, Kentucky bluegrass and tall fescue injured. Large brown lesions on roots; swelling on root tips.
Sting (*Belonolaimus* spp.)	Ectoparasitic	Bermudagrass, zoysiagrass, and St. Augustinegrass are injured. Lesions evident, especially root tips.
Dagger (*Xiphinema* spp.)	Ectoparasitic	Warm-season grasses (especially zoysiagrass) and perennial ryegrass are injured. Root lesions are reddish brown to black and sunken.
Pin (*Paratylenchus* spp.)	Ectoparasitic	Kentucky bluegrass, and fine and tall fescues are injured. Tillering and rooting increased, but fewer lateral roots. Distinct lesions on roots.

collected from a dozen or more areas at the edge or interface between healthy and injured turf. Sampling from severely thinned areas may yield unreliable results because these obligate parasites will not survive in large populations in the absence of living plants. Samples should also be collected in the root zone region, normally the upper 3 to 6 inches (7 to 15 cm) of soil. A more complete discussion on soil sampling for nematodes, as well as other agents of disease, was provided in an earlier section of this chapter. Various methods for extraction are used in the laboratory. Unfortunately, there are no reliable data correlating nematode number per sample and expected degree of turf injury in the field. The nematologist, however, generally will be able to make a relatively good management recommendation based on species and number of nematodes present in a sample.

Management

An absolute determination of a nematode problem from visual symptoms and even soil analysis is difficult. Frequently, the best indication of a nematode problem is a positive response from a nematicide. It should be pointed out, however, that turf green-up frequently occurs following a nematicide application, presumably because the death and subsequent decay of nematodes, as well as other invertebrates, liberates nitrogen. Nematicides are highly toxic, organophosphate derivatives. They must be handled with extreme caution and used only according to the procedure and at the rates given on the label. Because the target of a nematicide is in soil, it is essential the chemical be thoroughly watered-in, and core cultivation prior to application will facilitate the downward movement of the chemical. Ethoprop (Mocap) and fenamiphos (Nemacur) are the only nematicides registered for use on turf, and they must be applied by a certified pesticide applicator. Both nematicides may only be applied to commercial turfgrass sites, such as sod farms or golf courses. Both are available in granular form and both can discolor turf when applied during hot weather.

Extensive reliance on nematicides can develop into an intractable problem in which nematode populations rebound rapidly following chemical treatment. In some situations, such as extremely high sting nematode population densities in bermudagrass, fumigation may ultimately become the only solution. Fumigants are of little practical value in most situations because they kill turf as well as most other living organisms they come into contact with. Obviously, fumigants would only be used prior to establishing a turf or if total renovation is desired. A brief discussion on fumigants and their proper use is provided in the section on managing fairy rings.

Because nematodes only may be injurious during summer stress periods, cultural practices that alleviate stress will help to minimize injury. Such practices would include judicious syringing, increasing the mowing height, and use of a balanced fertilizer. Application of water-soluble nitrogen fertilizer during periods of environmental stress in summer will place an additional (and perhaps lethal) stress on an already dysfunctioning root system under attack from nematodes. There are some research data indicating that use of organic forms of nitrogen fertilizer (e.g., sewage sludges) can discourage development of high populations of some parasitic nematodes, when compared to the use of inorganic forms. Granular chitin-protein materials, such as Clandosan, are reported to control nematodes in some crops. These materials, however, do not appear to have any activity against plant parasitic nematodes found in turfgrasses.

ALGAE AND BLACK-LAYER

Algae contain chlorophyll and are very simple plants. These plants do not parasitize grasses, but they are included here because they are highly invasive and often out-compete grasses for space in wet or shaded environments. Black layer is a soil physical malady that acts continuously to damage plants and develops under conditions similar to those that encourage algae.

Algae

Pathogen: *Nostoc* spp., *Oscillatoria* spp., *Chlamydomonas* spp. *Hantzschia* spp., others
Primary Hosts: All turfgrasses grown in wet shade, particularly on putting greens
Predisposing Conditions: Prolonged periods of warm, overcast and rainy weather

Algae contain chlorophyll and often are described as primitive plants because they lack roots, stems and leaves. The types of algae that proliferate in wet turf soils are primarily blue-green and filamentous (Plates 2–62 and 2–71). The filaments are long and are comprised of a single cell. These algae reproduce by fission, and individual cells may develop thick walls and behave like spores, which enable algae to survive unfavorable environmental conditions. These algae secrete a gelatinous substance and eventually form a black scum in thatch or on the surface of wet soils.

Algae can be a chronic problem on greens, especially those surrounded by trees and shrubs. Black algal scums are particularly common on greens that are compacted, low lying, adjacent to ponds and shrouded in trees. Their growth is encouraged by extended periods of rainy, overcast, and warm weather in summer. They are mainly a problem on greens because the very low mowing heights expose a much greater thatch surface. Algal scums slow water infiltration into soil and keep thatch wet for extended periods, further encouraging the proliferation of algae. Algal scums also impede oxygen and other gas diffusion into and out of soils, and may play a role in the formation of some black layers.

Management

Algae cannot be effectively controlled unless the conditions that predispose the turf to their growth are corrected. Control is therefore aimed at alleviating wet soil conditions by improving drainage, aeration, and by employing proper irrigation practices. This ultimately may require removal of trees and shrubs, installing drains, and possibly the rebuilding of greens using high-sand content soil mixes. Increasing mowing height is essential; persistent low mowing will only continue to aggravate the condition.

Application of ground agricultural limestone, hyrated lime or iron sulfate will provide some short-term alleviation by desiccating algae. Prior to application of any desiccant, the scum should be manually broken up by verticutting, spiking or core cultivation, and the debris raked off. Simultaneously, mowing height should be increased. A light application of nitrogen from ammonium sulfate (i.e., 0.10 to 0.25 lb N/1000 ft^2 or 5 to 12 kg N/ha) in combination with iron sulfate (1.0 to 2.0 oz/1000 ft^2 or 3 to 6 kg/ha) could be beneficial to putting greens that exhibit slow growth or poor vigor. During the fall, nitrogen fertility should be increased to stimulate tillering, which improves stand density. Fatty acid soaps (DeMoss) also reduce algal growth; however, they should only be applied when temperatures are below 80°F (27°C). When applied on hot and sunny days, these soaps dissolve the cuticle on grass blades, causing a severe burning of turf. Diluted bleach solutions and alkyl-ammonium chloride also may reduce algae growth. Most of the aforementioned materials should be used with caution during periods of high temperature stress, as they may discolor turf. Some fungicides, such as mancozeb (Dithane M-45, Fore) and chlorothalonil (Daconil 2787), help to arrest algae. To be effective, however, these fungicides need to be applied on two-week intervals, and applications should commence prior to the time algae blackening appears.

Black-Layer

Pathogen: *Abiotic*
Primary Host: High-sand content greens
Predisposing Conditions: Anaerobic conditions developing in response to a water saturated soil zone

Black-layer is a physical condition that is primarily associated with high-sand content greens or older greens modified with sand topdressing. Black-layer can develop in mineral soils and is commonplace in wetlands. Anaerobic conditions in soil are the cause of black-layer; however, the mechanism of their formation is not completely understood.

Symptoms first appear as turf develops a yellow or bronzed appearance, and eventually turf thins in irregular patterns. Loss of turf density generally is most severe in low lying, shaded areas where air circulation is poor. Inspection of the sand profile reveals the presence of a surface or subsurface black-layer.

Where layers are restricted to the surface, the thatch may have a black color. Subsurface layers are coal black in color and may develop several inches (3 to 12 cm) below the surface. Bands of blackened sand usually range from 0.25 to 1.0 inch (0.6 to 2.5 cm) in width, but bands greater than 1 inch (> 2.5 cm) in width are not uncommon. There is a foul, sulfurous, rotten-egg odor associated with affected soils due to the production of hydrogen sulfide gas by bacteria under anaerobic conditions.

Black-layer develops during periods of excessive rain or irrigation, especially on greens with poor internal soil drainage or with a high water table. The condition is frequently encountered where sand topdressing is layered over a heavier soil type or a thatch layer. Both situations can cause perched water tables to develop. While waterlogged soils of low oxygen content are invariably associated with black-layer, there are no other physical, chemical or cultural factors common to all turfs affected by this condition.

Management

The primary management tactics for black-layer control are to eliminate, avoid, or minimize perched water tables and other causes of soil waterlogging, such as excessive irrigation or plugged subsurface drainage systems. Leaks in irrigation lines can saturate soils for long periods before being discovered, and can predispose greens to black-layer. Control, however, begins with prevention. This is achieved by using proper sand particle sizes to construct greens or for topdressing mixes. Avoid topdressing materials that are not identical to the greens construction mix. Drainage systems should be properly designed and installed under new greens. Frequently check to ensure that subsurface drainage lines are clear and operating effectively. Irrigation practices that avoid ponding or waterlogging are essential. In situations where black-layers already have developed, greens should be syringed and otherwise judiciously irrigated to prevent waterlogging. Frequent core cultivation with wide diameter and deep tines will help to alleviate anaerobic conditions. Verti-drain and water injection cultivation also help to alleviate anaerobic conditions. Trees and brush that surround affected greens should be pruned or removed. Due to the potential for sulfur or iron to exacerbate black-layer, use of fertilizers and fungicides that contain large amounts of these elements, as well as gypsum, should be avoided. Water sources that are high in iron, magnesium, or manganese, or high in organic matter (i.e., pond or sewage effluent water) also should not be used.

Tables 2.7 and 2.8 list common diseases of major warm- and cool-season turfgrass species. Table 2.9 shows diseases arranged by tissues attacked and pathogen, and Table 2.10 presents the season of occurrence of common fungal turfgrass diseases.

Table 2.7. Summary of Common Diseases of Major Cool-Season Turfgrass Species

BENTGRASS	
Creeping, Colonial, Velvet	
Anthracnose	Red leaf spot
Bacterial wilt ('Toronto' creeping bent)	Red thread
Brown patch	Copper spot
Dollar spot	Snow molds
Fairy ring	pink, gray, snow blight
Leptosphaerulina blight	Southern blight
Localized dry spot	Superficial fairy ring
Nematodes	Take-all patch
Pink patch	Yellow patch
Pythium blight	Yellow tuft
Pythium root rot	
BLUEGRASSES	
Annual	
Anthracnose	Pythium blight
Ascochyta leaf blight	Pythium root rot
Bacterial wilt	Red thread
Brown patch	Snow molds
Dollar spot	pink, gray, snow blight
Fairy ring	Southern blight
Helminthosporium leaf spot/melting-out	Summer patch
Leptosphaerulina blight	Superficial fairy ring
Localized dry spot	Yellow patch
Necrotic ring spot	Yellow tuft
Nematodes	
Pink patch	
Kentucky	
Ascochyta leaf blight	Red thread
Brown patch	Rust
Dollar spot	Snow molds
Fairy ring	pink and gray
Helminthosporium leaf spot/melting-out	Smuts, flag or stripe
Leptosphaerulina blight	Southern blight
Necrotic ring spot	Yellow ring
Nematodes	Yellow tuft
Pink patch	
Powdery mildew	
FINE LEAF FESCUE	
Chewings and Creeping Red	
Anthracnose	Red thread
Ascochyta leaf blight	Snow molds
Dollar spot	pink and gray
Fairy ring	Summer patch
Helminthosporium leaf spot/melting-out	Southern blight
Pink patch	

Table 2.7. Continued

FINE LEAF FESCUE continued	
Hard and Blue Sheep Fescue	
Dollar spot	Pink patch
Fairy ring	Red thread
Helminthosporium leaf spot/melting-out	Snow mold, pink and gray
Leptosphaerulina blight	Summer patch
TALL FESCUE	
Ascochyta leaf blight	Pythium blight (southern U.S.)
Brown patch	Red thread
Dollar spot	Snow molds, pink and gray
Net-blotch	White blight (southern U.S.)
PERENNIAL RYEGRASS	
Brown patch	Leptosphaerulina blight
Dollar spot	Pink patch
Fairy ring	Pythium blight
Gray leaf spot	Red thread
Helminthosporium diseases	Rust
Brown blight	Snow molds
Leaf spot/melting-out	pink and gray
Net-blotch	

Table 2.8. Summary of Common Diseases of Major Warm-Season Turfgrass Species

BERMUDAGRASS	
Bermudagrass decline	Helminthosporium leaf spot/ melting-out
Brown patch[a]	Nematodes
Dollar spot	Pythium blight (deep south)
Fairy ring	Rust
Gray leaf spot	Spring dead spot
CENTIPEDEGRASS	
Brown patch[a]	Fairy ring
Centipede mosaic	Nematodes
Dollar spot	
ST. AUGUSTINEGRASS	
Brown patch[a]	Gray leaf spot
Cercospora leaf spot	Nematodes
Dollar spot	Nigrospora blight
Downy mildew	Rust
Fairy ring	St. Augustine decline
ZOYSIAGRASS	
Brown patch[a]	Nematodes
Dollar spot	Rust
Helminthosporium leaf spot/ melting-out	Large patch[a]
Fairy ring	Undescribed spring dead spot-like disease

[a] Brown patch is a spring, fall, and winter disease on these grasses in the deep south. In more northern ranges, brown patch may be important in summer in these grasses.

Table 2.9. Diseases Arranged by Tissues Attacked and Pathogen

Disease	Pathogen
I. PATCH DISEASES CAUSED BY ROOT INVADING PATHOGENS	
Summer Patch	*Magnaporthe poae*
Necrotic Ring Spot	*Leptosphaeria korrae*
Take-All Patch	*Gaeumannomyces graminis* var. *avenae*
Spring Dead Spot	*Leptosphaeria korrae, L. narmari,* or *Ophiosphaerella herpotricha*
II. FOLIAR DISEASES	
Dollar Spot	*Sclerotinia homoeocarpa*
Red Thread	*Laetisaria fuciformis*
Pink Patch	*Limonomyces roseipellis*
Brown Patch/Rhizoctonia Blight	*Rhizoctonia solani* and *R. zeae*
Leaf Spot (Helminthosporium)	*Drechslera* or *Bipolaris* spp.
Gray Snow Mold	*Typhula incarnata* and *T. ishikariensis*
Yellow Patch/Cool-Temperature Brown Patch	*Rhizoctonia cerealis*
Gray Leaf Spot	*Pyricularia grisea*
Zoysia Patch	*Rhizoctonia solani*
White Blight	*Melanotus phillipsii*
Nigrospora Blight	*Nigrospora sphaerica*
Leptosphaerulina Blight	*Leptosphaerulina trifolii*
Ascochyta Leaf Blight	*Ascochyta* spp.
Cercospora Leaf Blight	*Cercospora* spp.
III. FOLIAR, STEM, OR ROOT DISEASES	
Bermudagrass decline	*Gaeumannomyces graminis* var. *graminis*
Pythium Blight	*Pythium aphanidermatum, P. myriotylum, P. ultimum,* others
Pythium Root Rot	*Pythium graminicola, P. torulosum,* others
Pink Snow Mold	*Microdochium nivale*
Anthracnose	*Colletotrichum graminicola*
Nematodes	Many species
Melting-out	*Drechslera* or *Bipolaris* spp.
Fusarium Blight	*Fusarium culmorum* and *F. poae*
IV. DISEASES CAUSED BY OBLIGATE PARASITES	
Stripe Smut	*Ustilago striiformis*
Flag Smut	*Urocystis agropyri*
Powdery Mildew	*Erysiphe graminis*
Rust	*Puccinia* spp.
Yellow Tuft	*Sclerophthora macrospora*
V DISEASES THAT CAUSE HYDROPHOBIC THATCH OR SOIL, OR THATCH SHRINKAGE	
Fairy Ring	Many Basidiomycetes
Yellow Ring	*Trechispora alnicola*
Superficial Fairy Ring	*Coprinus kubickae,* others
Localized Dry Spot	Unknown Basidiomycetes

Table 2.9. Continued

Disease	Pathogen
VI. SEEDLING DISEASE	
Damping-Off	*Pythium* spp.; *Fusarium* spp.; *Rhizoctonia solani; Bipolaris* spp.; *Drechslera* spp.; *Curvularia* spp.; Others
VII. OTHERS	
Bacterial Wilt	*Xanthomonas campestris* pathovar *graminis*
St. Augustine Decline	Panicum mosaic virus
Centipede Mosaic	Panicum mosaic virus

FACTORS ASSOCIATED WITH FUNGICIDE USE

Arriving at the decision of whether to apply a fungicide to any turf area is often difficult and generally based on economic considerations. Aside from cost, the primary determinants in using a fungicide are the (a) prevailing environmental conditions, (b) host species and cultivars present, and (c) the pathogen(s). The environmental factor has unique implications in turfgrass pathology because the intensity and nature of turfgrass management greatly influences plant vigor and therefore the severity of diseases.

Promoting healthy growth through sound cultural practices is the first step in minimizing disease injury. Frequently, however, environmental stresses, traffic, and poor management weaken plants, predisposing them to invasion by fungal pathogens. When disease symptoms appear, it is imperative that a rapid and accurate diagnosis of the disorder be made. The prudent manager also attempts to determine those factors that have led to the development of the disease. The most common cause for extensive disease injury in turf can frequently be related to poor cultural practices. Such practices include frequent and close mowing, light and frequent irrigations, and inadequate or excessive nitrogen fertility. The development of excessive thatch, shade, poor air or soil water drainage, and traffic also may contribute significantly to disease problems. A good example is Helminthosporium diseases, which are particularly damaging when turf is mown too closely, given light and frequent irrigations, and thatch is thick as a result of excessive nitrogen fertilization. Despite hard work and adherence to sound cultural practices, diseases may become serious problems. This normally occurs when environmental conditions favor disease development, but not plant growth and vigor. For example, summer patch and brown patch are most damaging in cool-season grasses when high summer temperatures stress plants and impair their growth and recuperative capacity. In this situation, fungicides may be recommended in conjunction with cultural practices that promote turf vigor.

Fungicides may be applied preventively (i.e., before anticipated disease symptoms appear) or curatively (i.e., when disease symptoms first become evident). Applying a fungicide after the turf has been damaged significantly

Table 2.10. Season of Occurrence of the Most Common Fungal Turfgrass Diseases

Disease	Primary Season(s) of Occurrence	Primary Grasses Damaged[a]
Snow Molds	Late Fall to Spring	All species
Yellow Patch	Late Fall to Spring	ABG, CBG greens
Pythium Root Rot or Snow Blight	Early Winter to Spring	ABG, CBG greens ABG, CBG greens
Red Thread and Pink Patch	All year, especially Spring and Fall	ABG, CBG, FLF, KBG, PRG
Helminthosporium Diseases	All year, especially Spring and Fall	Most species
Take-All Patch	Spring to Early Winter	Bentgrasses
Necrotic Ring Spot	Early Spring to Early Winter	KBG
Spring Dead Spot	Spring	Bermudagrass
Bermudagrass Decline	Spring to Summer	Bermudagrass
Dollar Spot	Early Spring to Early Winter	Most species
Leptosphaerulina Blight	Spring to Fall	ABG, CBG, FLF, KBG, PRG
Gray Leaf Spot	Spring and Summer	PRG, St. Augustinegrass
Brown Patch	Summer to Early Fall	Most species
Summer Patch	Summer	ABG, KBG, FLF
Anthracnose	All year, especially Summer	ABG, CBG, FLF
Pythium Blight	Summer	ABG, Bermudagrass (deep south), CBG, PRG, TF (deep south)
Stripe and Flag Smut	Spring and Fall	KBG
Rusts	Late Summer and Fall	KBG, PRG, Zoysiagrass
Powdery Mildew	All year, especially Late Summer	KBG
Fairy Ring	Spring to Early Winter	All species
Yellow Tuft	All year, especially Spring and Fall	All species

[a] Hosts: ABG = annual bluegrass; CBG = creeping or colonial bentgrass; FLF = fine leaf fescues, including creeping red, Chewings, hard and blue sheep; KBG = Kentucky bluegrass; PRG = perennial ryegrass; and TF = tall fescue.

generally is a waste of time, money, and effort. Preventive fungicide treatment is recommended for chronically damaging diseases. This is particularly true on golf course putting greens in regions where snow molds, Pythium blight, brown patch, summer patch, and anthracnose are common. Curative applications are more economical and environmentally wise, but only if the disease can be treated before damage is excessive. The key to a successful curative fungicide program is vigilant scouting. In general, a single or possibly two, properly timed applications will provide effective control of most diseases encountered on lawns and athletic fields. Contact fungicides generally are less expensive and provide good control. Contact fungicides, however, may only provide 7 to 14 days of control under conditions of high disease pressure. Where sudden and severe, or chronic disease problems occur, a systemic (penetrant) alone, or a systemic plus a contact fungicide may be needed. Systemic or local systemic (local penetrant) fungicides provide 14 to 21 days protection during high pressure disease periods. Tank mixing a systemic plus a contact fungicide provides a slightly longer residual effect and a wider spectrum of disease control. Frequently, a fungicide only may be needed to help the turf better survive a high pressure disease period. Favorable changes in weather such as alternating hot-humid and cooler periods, however, provide the most effective means of reducing or eliminating disease problems in the summer.

Fungicide Use in Lawn Care

Proper use and selection of fungicides is too difficult and complicated for the vast majority of homeowners. Because of this, only well trained employees of lawn care companies can provide the most reliable lawn disease service. Fungicides, however, should not become a part of a normal application schedule. As a general rule, use of fungicides is not recommended in most home lawn situations because: (a) proper diagnosis and proper fungicide selection are difficult, (b) generally it is too late to achieve the economic and aesthetic benefits of a fungicide once extensive injury has occurred, (c) lawn care companies capable of only dry or granular applications do not have the proper spray equipment or they cannot obtain the desired fungicide(s) in granular form, and (d) it may be less expensive and better in the long run to overseed or renovate a damaged turf area with disease-resistant cultivars.

There are, however, several disease situations of lawn turf that are best controlled through a preventive fungicide application. Some notable situations include: (1) Kentucky bluegrass lawns injured in previous years by summer patch, necrotic ring spot, leaf smuts, and perhaps dollar spot, (2) perennial ryegrass lawns injured in previous years by Pythium blight, brown patch, dollar spot, or red thread, and (3) tall fescue and St. Augustinegrass lawns chronically damaged by brown patch. Many diseases, however, are effectively controlled with curative fungicide applications when disease symptoms first appear. For example, leaf spot in lawns of common-type Kentucky bluegrasses (e.g., Kenblue, Newport, Park, South Dakota, etc.) and creeping red or Chewings fescues (e.g., Pennlawn and Jamestown), red thread in perennial ryegrass, or gray leaf spot in St. Augustinegrass can be controlled effectively with 1 or 2 curative fungicide applications. Dollar spot disease is extremely common and if allowed to go unchecked, may cause extensive injury to Kentucky bluegrass, perennial ryegrass, creeping red fescue, and zoysiagrass lawns. When diagnosed in its early stages, however, dollar spot also is controlled effectively by fungicides. Given these situations, it becomes obvious that effective fungicide programs hinge upon: (a) knowledge of past disease problems in a particular lawn or neighborhood, (b) ability to distinguish between turfgrass species and sometimes cultivars within a species, and (c) ability to diagnose turfgrass diseases at an early stage in their development. Hence, in addition to the expense of fungicides and logistical problems associated with sending trucks to specialized fungicide accounts, the lawn care company also must educate its employees to scout and effectively diagnose diseases. This educational process is best achieved by in-house training programs and attending turf workshops and conferences.

Remember, once a disease has severely reduced stand density, overseeding or renovating with resistant cultivars is normally suggested. Fact sheets describing different diseases and a list of cultural practices that help minimize disease injury should be provided by the lawn care company to homeowners. Fact sheets on turf diseases may be available through Cooperative Extension Service offices.

Fungicide Use on Golf Courses

Where extremely high quality turf is required, fungicides will be needed in most years, and in nearly all areas of the U.S. Fungicide use is less common, however, in arid or semi-arid regions where diseases are not chronically severe. The indiscriminate use of fungicides or employment of numerous, preventive applications of fungicides for many diseases should be discouraged. Other than economic restraints, reasons why repeated fungicide applications may not be desirable include:

1. Development of fungicide resistant pathogens, which is most likely to occur with those fungi causing dollar spot and Pythium blight.
2. Disease resurgence, i.e., a phenomenon in which a disease recurs more rapidly and causes more injury in turfs previously treated with fungicides when compared to non-fungicide-treated sites.
3. A fungicide may control one disease, but encourage other diseases.
4. Fungicides may reduce the activity of beneficial microorganisms in the soil or enhance stem tissue production, which could lead to thatch buildups.
5. Phytotoxicity or objectionable plant growth regulator effects.

The development of fungal biotypes resistant to fungicides has been well documented. Resistant biotypes of the dollar spot fungus first developed as a result of repeated usage of cadmium-based fungicides and benzimidazoles (CL 3336, Fungo 50, Tersan 1991) on golf courses. Biotypes of the dollar spot fungus resistant to anilazine (Dyrene), iprodione (Chipco 26019), and sterol-inhibiting fungicides (Banner, Bayleton) also have been reported. The buildup of resistant biotypes of fungi occurs in response to a selection process that eventually enables a small, but naturally-occurring subpopulation of resistant biotypes to dominate in the fungicide-treated turfgrass microenvironment. Resistance problems can be delayed or averted by rotating fungicides with different modes of action, by tank mixing contact and systemic fungicides, or by tank mixing known synergists.

Turf managers have observed that some diseases may recur more rapidly and severely in turfs previously treated with fungicides, as compared to adjacent untreated areas. Dollar spot and brown patch are probably the most common diseases to exhibit this phenomenon. Disease resurgence is attributed to fungicides being toxic to beneficial microorganisms, which naturally antagonize and keep disease-causing fungi in abeyance.

Fungicides applied to control one disease may encourage other diseases. Benomyl (Tersan 1991) has been reported to enhance red thread, Helminthosporium leaf spot and Pythium blight. Thiophanates (CL 3336, Fungo 50) may increase crown rust in perennial ryegrass; iprodione (Chipco 26019) can increase yellow tuft; mancozeb (Fore) and flutolanil (ProStar) may enhance dollar spot; and chlorothalonil (Daconil 2787) can increase summer patch and stripe smut in Kentucky bluegrass. Encouragement of disease in these situations again may be attributed to offsetting the delicate balance between antagonistic and pathogenic microorganisms in the ecosystem.

When used repeatedly, certain fungicides have been shown to enhance thatch accumulation. Most studies have demonstrated only modest increases in thatch with fungicides. Benzimidazole fungicides, such as benomyl (Tersan 1991) and the thiophanates (CL 3336, Fungo 50), and sulfur-containing fungicides such as mancozeb (Dithane M-45, Fore) and thiram (Spotrete, Thiramad) can cause thatch to accumulate by acidifying soil. The effect of these fungicides is indirect; that is, they inhibit the thatch decomposition capacity of beneficial microorganisms by lowering soil pH. Some fungicides can enhance stem tissue production and also indirectly cause thatch to accumulate. Fungicides also may contribute to thatch buildup by being toxic to earthworms. Earthworms help reduce thatch by mixing soil with organic matter. Benomyl (Tersan 1991), mancozeb (Fore), anilazine (Dyrene), chlorothalonil (Daconil 2787), and various insecticides and nematicides have been shown to be toxic to earthworms.

The phytotoxicity that accompanies the usage of some fungicides generally is not severe. Most phytotoxicity problems occur when fungicides are applied to bentgrasses, particularly during periods of high temperature stress. Fungicides formulated as emulsifiable concentrates are most likely to cause a foliar burn when applied during hot weather. Mercury (Calo Clor) and PCNB (Terraclor) are likely to yellow turf when applied during warm weather. Repeated applications of sterol-inhibiting fungicides such as fenarimol (Rubigan), propiconazole (Banner), or triadimefon (Bayleton) may elicit a blue-green color in foliage of creeping bentgrass and other turfgrass species. Tank-mixing sterol-inhibiting fungicides with some plant growth regulators also may elicit an objectional level of discoloration.

It should be noted that many of the harmful side effects just described were either isolated events or occurred only after repeated use of one fungicide over the course of several years. Experienced turfgrass managers have long recognized that tank mixing known synergists and rotating fungicides with different modes of action greatly minimizes these potential problems. The importance of rapid and accurate disease diagnosis, and the judicious use of fungicides are integral in management programs where fungicides are commonly employed.

Types of Fungicides

Fungicides are somewhat mysterious chemicals because their behavior on or inside of plants, and how they physiologically affect microorganisms, is not clear. Fungicides often are divided into two groups: contact or systemic. This classification, and some other terms associated with fungicides, is misleading. For example, most fungicides that penetrate plant tissues (penetrants) are not fungicidal, but can provide both contact and systemic activity. Penetrants are actually fungistatic. That is, they only prevent growth or development of fungi and do not actually kill them. Several contact fungicides can kill fungal spores as they germinate, but even most contacts tend to be fungistatic. Another contradiction centers on the use of the word systemic. A systemic chemical by definition is capable of moving throughout a plant from leaves to stems to roots. In fact, the only truly systemic turf fungicide is fosetyl-aluminum (Aliette). Most other penetrants either remain localized or primarily move upward in the xylem with the transpiration stream. Several fungicides, such as the triazoles, move primarily upward from the point of tissue contact, but also are capable of lateral diffusion and limited downward movement. Downward or basipetal movement of triazoles is only a few millimeters. Hence, a triazole that contacts leaf tissue is highly unlikely to translocate to roots. However, chemical that contacts basal leaf sheaths, or runs down between leaf sheaths, may be transported into axillary buds and possibly stems.

In addition to contacts and so-called systemics, there is a third fungicide group known as site absorption or localized penetrants. These fungicides are absorbed into tissues, but they do not move beyond the site of uptake. Choroneb (Teremec SP), iprodione (Chipco 26019), and vinclozolin (Curalan, Touche, Vorlan) are localized penetrants. Their mode of action is non-site specific; that is, they interfere with two or more biochemical processes in susceptible fungi.

Contact fungicides provide activity outside of plants and protect only those tissues they contact. Because contact fungicides are subjected to removal from tissues by mowing and the forces of the environment (i.e., wash off, light degradation, microbial breakdown, etc.), they tend to provide a relatively short period of protection. The number of contact fungicides available for use on turfgrass is dwindling and currently includes: chlorothalonil (Daconil 2787), ethazole (Koban), mancozeb (Dithane M-45, Fore), thiram (Spotrete,

Thiramad), and zineb. Contact fungicides are extremely important in chemical disease management programs because they generally are free of resistance problems and some make good synergists in tank mix combination with selected penetrants. Contact fungicides and localized penetrants normally interfere with several biochemical or physiological processes in susceptible fungi. In more technical terms, contact fungicides and localized penetrants are non-site specific, meaning that there are several genetic barriers that must be overcome in order for resistant biotypes to develop. Conversely, the systemic-type fungicides are generally single-site specific. That is, they only disrupt one specific biochemical or physiological process in a susceptible fungus. This process often is controlled by a single gene and therefore the probability of a resistant biotype developing is greatly increased. For example, triazole fungicides interrupt the production of ergosterol in sensitive fungi by blocking a single reaction. Ergosterol is used in minute quantities by fungi to produce membranes. In the absence of this sterol, membrane form and function is impaired and the fungus cannot grow. Because these fungicides interrupt only one of thousands of reactions occurring in the fungus, the probability of the existence of a small population of tolerant biotypes in the ecosystem is likely. Continuous use of fungicides with the same mode of action can selectively remove susceptible biotypes, which allows the resistant biotypes to proliferate and eventually dominate in a turf. Indeed, nearly all documented resistance problems are associated with systemics, while extremely few have been related to the use of contact fungicides.

Table 2.11 describes systemic behavior or movement of fungicides that penetrate plant tissues. Table 2.12 shows chemical class or properties of turfgrass fungicides, and Table 2.13 lists fungicides that are commonly used for controlling turfgrass diseases.

Fungicide Application

Most fungicides are diluted in water and sprayed onto turfgrasses. Granular forms of fungicides often are more expensive and contact fungicides applied on granules provide a level of residual control that is inferior to their sprayable counterpart. Systemic granular fungicides provide effective control of foliar blighting pathogens, but often have reduced activity against root pathogens. Granulars can move in surface water if a heavy rain occurs soon after application. This may leave turf in surface water drainage patterns unprotected. Granulars, however, have an important place in disease management programs. They can be used rapidly without the logistical problems associated with spraying. They are particularly useful in small units where diseases are localized and spraying is impractical. For example, if only a portion of 1 or 2 greens is showing disease symptoms it is more prudent to quickly spot treat with a granular fungicide rather than to prepare a tank for broadcast spraying.

Aside from improper sprayer calibration, perhaps the single greatest error in using fungicides is applying them in insufficient amounts of water to provide good plant coverage. Sprayable fungicides should be applied in a minimum of

Table 2.11. The Systemic Behavior or Movement of Fungicides that Penetrate Plant Tissues

Common Name	Trade Name(s)	Mode(s) of Action
	SYSTEMIC	
	Upward	
Flutolanil	ProStar	Inhibits a respiratory enzyme[a]
Propamocarb	Banol	Inhibits membrane function[a]
Thiophanate	CL 3336, Fungo	Inhibits mitosis
	Upward/Limited Downward	
Cyproconazole	Sentinel	Ergosterol inhibitor
Fenarimol	Rubigan	Ergosterol inhibitor
Metalaxyl	Subdue	Blocks RNA synthesis[a]
Myclobutanil	Eagle	Ergosterol inhibitor
Propiconazole	Banner	Ergosterol inhibitor
Terbuconazole	Lynx	Ergosterol inhibitor
Triadimefon	Bayleton	Ergosterol inhibitor
	True Systemic/Upward and Downward	
Fosetyl-Aluminum[b]	Aliette	Triggers plants' resistance mechanisms, blocks mycelial development, inhibits spore germination
	SITE ABSORPTION[b]	
	(*Taken-up In Tissue But No Significant Movement*)	
Chloroneb	Teremec SP, Terraneb SP	Interferes with enzymes[a]
Iprodione	Chipco 26019, Rovral	Membrane damage, inhibits cell division[a]
Vinclozolin	Curalan, Touche, Vorlan	Membrane damage, inhibits cell division[a]

[a] The precise mode of action of most fungicides other than the benzimidazoles, pyrimidine, and triazoles is unknown.
[b] Site absorption fungicides and fosetyl aluminum provide non-site specific modes of action.

1.0 to 2.0 gallons of water per 1000 ft^2 (420 to 840 liters/ha) or 45 to 90 gallons of water per acre. However, a higher water dilution of 3.0 to 5.0 gallons per 1000 ft^2 (1260 to 2100 liters/ha) is recommended by most manufacturers. Increasing water improves coverage and performance, which usually equates to longer residual effectiveness. For most diseases, fungicides must be allowed to dry on leaves prior to irrigating to be effective. Contact fungicides can lose most of their effectiveness if a rainstorm occurs prior to the fungicide drying on leaves. Even fungicides with systemic activity can exhibit reduced effectiveness if rain or irrigation occurs before the chemical completely dries on leaves. There are a few exceptions to this no post-application watering principle, and they largely apply to fungicides used to control root diseases. For example, thiophanates (CL 3336, Fungo 50) and some triazoles provide better summer patch control if watered-in before they have time to dry on leaf surfaces. Fungicides also should be sprayed through nozzles that atomize the spray droplet. Flat-fan, hollow cone, and raindrop nozzles generally are more effi-

Table 2.12. Common Chemical Name, Trade Names, and Chemical Class or Properties of Turfgrass Fungicides

Common Name	Some Trade Name(s)	Class/Type	Contact/ Penetrant[c]
Benomyl[a]	Tersan 1991	Benzimidazole	P
Chloroneb	Teremec SP, Terraneb SP	Substituted aromatic hydrocarbon	P
Chlorothalonil	Daconil 2787, Thal-o-nil	Substituted aromatic hydrocarbon	C
Cyproconazole	Sentinel	Triazole	P
Ethazole/Etridiazole	Koban, Terrazole	Substituted aromatic hydrocarbon	C
Fenarimol	Rubigan	Pyrimidine	P
Fosetyl-Aluminum	Aliette	Ethyl phosphonate	P
Flutolanil	ProStar	Benzamide	P
Iprodione	Chipco 26019, Rovral	Dicarboximide	P
Mancozeb	Dithane M-45, Fore, Lesco 4	Ethylenebis-dithiocarbamate	C
Metalaxyl	Subdue	Acylalanine	P
Myclobutanil[b]	Eagle	Triazole	P
Propamocarb	Banol	Carbamate	P
Propiconazole	Banner	Triazole	P
Quintozene	PCNB, Terraclor	Substituted aromatic hydrocarbon	C
Terbuconazole[b]	Lynx	Triazole	P
Thiophanate-ethyl	Cleary's 3336	Benzimidazole	P
Thiophanate-methyl	Fungo 50	Benzimidazole	P
Thiram	Spotrete, Thiramad	Dialkl dithiocarbamate	C
Triadimefon	Bayleton	Triazole	P
Vinclozolin	Curalan, Touche, Vorlan	Dicarboximide	P

[a] Voluntarily withdrawn from market; future status unknown.
[b] Names proposed or pending U.S. EPA registration.
[c] Contact = Fungicide is only active on leaf and sheath surfaces.
Penetrant = Fungicide is absorbed and can provide activity both on the outside and inside of plant tissues.

cient than nozzles that deliver a large droplet, such as flood jet nozzles. Overall, flat-fan nozzles are most often used for delivering fungicides as well as herbicides and plant growth regulators. Low pressure produces larger droplets, and can be another cause of reduced effectiveness. Pressure in the spray boom at delivery should be in the range of 30 to 60 psi (207 to 414 kpa).

Sprayers need to be accurately calibrated prior to mixing fungicides. Recheck calibration after every 3 days of use or more often. Screens and nozzles should be visually checked prior to each spray to ensure uniform delivery of the fungicide. Turn on the agitation system before adding fungicides, and allow it to run continuously. Spray tanks should be filled halfway with water before adding any fungicides. When tank-mixing products, always place water-insoluble materials, which are formulated as wettable powders,

Table 2.13. Some Commonly Used Fungicides for Controlling Turfgrass Diseases and Algae

Disease	Fungicide	Schedule and Comments
Anthracnose (*Colletotrichum graminicola*)	Banner[a] Bayleton[a] Daconil 2787 Dithane M-45 Fore Rubigan[a]	Apply systemics[a] preventively on 21-day intervals. Best to tank mix a systemic[a] with a contact[b] for curative treatment. Apply curative sprays as needed on 14-day intervals. Spring coring and use of low levels of N-fertilizer with fungicide may be beneficial.
Brown Patch (*Rhizoctonia solani*)	Banner Chipco 26019 Cleary's 3336 Curalan Daconil 2787[b] Dithane M-45[b] Fore Fungo 50 ProStar Spotrete[b] Terraclor[b,c] Touche	Apply on 10- to 14-day intervals during hot humid weather especially when night temperatures exceed 68°F (>20°C). Contacts,[b] however, may provide only 7–10 days control during stress periods. [c]Apply only to tall fescue, perennial ryegrass, or warm-season lawn turfs; will yellow close-cut turf in summer.
Cool Temperature Brown Patch (*Rhizoctonia cerealis*)	Chipco 26019 Daconil 2787 Dithane M-45 Fore Terraclor	Apply as needed; normally once or twice on a 21-day interval. Terraclor may yellow turf if applied when air temperature is above 65°F (>18°C).
Copper Spot (*Gloeocercospora sorghi*)	Bayleton Cleary's 3336 Daconil 2787 Rubigan	Apply once or twice on a 14- to 21-day interval.
Dollar Spot (*Sclerotinia homoeocarpa*)	Banner[a] Bayleton[a] Chipco 26019[a] Cleary's 3336[a] Curalan[a] Daconil 2787[b] Fungo 50[a] Rubigan[a] Touche[a] Vorlan[a]	Contacts[b] provide 10- to 14-days control; systemics[a] provide 14- to 28-days control. Alternate systemics[a] with contacts[b] to delay development of tolerant strains. Do not alternate systemics with the same mode of action. Maintain adequate N levels and increase mowing height when disease is active.
Fairy Rings (*Marasmius oreades* and others)	No chemical control available	Frequent coring, judicious watering and maintaining adequate N fertility will alleviate symptoms.
Gray leaf spot (*Pyricularia grisea*)	Banner Daconil 2787	Apply as needed on a 10- to 14-day interval.

Table 2.13. Continued

Disease	Fungicide	Schedule and Comments
Helminthosporium Leaf Spot, Melting-Out, and Net-Blotch (*Drechslera* spp., *Bipolaris* spp.)	Chipco 26019 Curalan Daconil 2787 Dithane M-45 Fore Terraclor Touche Vorlan	Sufficient control usually is provided by 1 or 2 applications made on a 14- to 21-day interval. Avoid drought stress, light and frequent watering, excessive thatch, and high N fertility in spring. Overseed with improved cultivars.
Powdery mildew (*Erysiphe graminis*)	Banner Bayleton Rubigan	A single application generally is sufficient. Reduce shade and improve air movement. Avoid excessive N fertility, drought stress, and increase height of mowing.
Pythium blight (*Pythium* spp.)	Aliette[a] Banol[a] Koban[b] Subdue[a] Terraneb SP[b]	Apply contacts[b] on 5- to 10-day intervals, particularly during hot, humid weather when night temperatures exceed 65°F (18°C). Systemics[a] provide 14 to 21 days of control or longer. Aliette and Banol should be applied preventively for best results. Water early in the day, and improve drainage and air circulation. Avoid water soluble N.
Red Thread (*Laetisaria fuciformis*) and Pink Patch (*Limonomyces* spp.)	Banner Bayleton Curalan Chipco 26019 Daconil 2787 ProStar Touche Vorlan	Generally, only 1 or 2 applications on 14- to 21-day intervals are necessary. Apply 0.5 to 1.0 lb N/1000ft^2 if turf has poor vigor.
Rust (*Puccinia graminis* and other *Puccinia* spp.)	Banner[a] Bayleton[a] Daconil 2787[b] Dithane M-45[b] Fore[b]	Apply contacts[b] on a 7- to 10-day interval as needed or apply a systemic[a] once when symptoms appear. Improve turf vigor by maintaining adequate N levels.
Stripe Smut (*Ustilago striiformis*) and Flag Smut (*Urocystis agropyri*)	Banner Bayleton Rubigan	A single application in May or during mid- to late October provides good control. Avoid drought stress and increase mowing height.
Slime Molds (*Physarum cinereum*) and other Slime Molds	No fungicide necessary	Poling, hosing down of leaves with water, and mowing removes fruiting structures.

Table 2.13. Continued

Disease	Fungicide	Schedule and Comments
Gray Snow Mold or Typhula Blight (*Typhula incarnata,* and *T. ishikariensis*)	Banner Bayleton Chipco 26019 Curalan ProStar Terraclor Terraneb SP Touche Vorlan	Apply before the first heavy snow or onset of cold, rainy weather. Repeat as needed during mid-winter and early spring snow melts.
Pink Snow Mold or Fusarium Patch (*Microdochium nivale*)	Banner Bayleton Chipco 26019 Curalan Fungo 50 Terraclor Touche Vorlan	Both snow molds may produce symptoms during cool, wet weather in the absence of snow cover. Where chronic, apply a fungicide just prior to cold and wet weather.
Spring Dead Spot (*Leptosphaeria korrae, Ophiosphaerella herpotricha,* others)	Banner Rubigan	Apply a fungicide in mid-September or about 30 days prior to expected bermudagrass dormancy. Avoid applying N after mid-August. Expect only 30–60% reduction in disease.
Southern Blight (*Sclerotium rolfsii*)	Bayleton ProStar	Spray or spot apply a fungicide as needed.
Summer Patch (*Magnaporthe poae*)	Banner Bayleton Cleary's 3336[c] Fungo 50[c] Rubigan[c]	For lawns or fairways apply in early June and again in early July. For greens, begin treatment in mid-May and apply 2 to 4 times on 28-day intervals. Increase mowing height, and irrigate deeply at onset of drought. Avoid excessive thatch, compaction, and high N fertility in spring. Use slow-release and acidifying N fertilizers. Water-in fungicide before it dries on leaves.[c]
Take-All Patch (formerly known as Ophiobolus Patch) (*Gaeumannomyces graminis* var. *avenae*)	Rubigan	Apply 2 to 3 times on a 21-day interval when symptoms first become evident. Acidify soil with ammonium-based nitrogen fertilizers.
Yellow Tuft (*Sclerophthora macrospora*)	Aliette Subdue	One or 2 annual applications in spring or fall provides excellent control.

Table 2.13. Continued

Disease	Fungicide	Schedule and Comments
Nematodes	Mocap Nemacur	See label restrictions. Nematicides may only be used on commercial sites to include sod farms and golf courses.
Necrotic Ring Spot (*Leptosphaeria korrae*)	CL 3336 Fungo 50 Rubigan	Apply a fungicide once in April or May.
Algae	Daconil 2787 Dithane M-45 Fore Iron sulfate	Apply on 10- to 14-day intervals. Core or verticut algal scums before treatment. Increase mowing height and ensure proper N-P-K nutrition.

Note: References to trade or brand names do not constitute an endorsement, guarantee, or warranty. Consult the label for more information on spray frequency, tank-mixing, and other important facts. It is important to read the label because some states have not approved all fungicides or the specific labeling of all diseases noted. Furthermore, not all labels have been rewritten to distinguish recent changes in fungal taxonomy. No discrimination is intended against products not mentioned.

dry dispersable granules or flowables, into the tank first. Soluble materials are added to the tank after insolubles, which includes formulations such as emulsifiable concentrates, liquids, or soluble powders. Do not tank-mix more than one emulsifiable concentrate, as turf burning may occur, particularly when treating greens. Never tank-mix fungicides with insecticides formulated as emulsifiable concentrates. Low water dilutions also increase the possibility of phytotoxicity when applying emulsifiable concentrates. In general, fungicides should not be tank-mixed with insecticides or herbicides unless otherwise stated on labels. For example, insecticides targeted for white grubs and some preemergence herbicides targeted for annual grass weeds should be watered-in immediately, and this practice would likely negate any benefits of a fungicide. Whenever in doubt, apply materials separately rather than in tank-mix combination. Thoroughly clean the spray tank, hose, boom, and nozzles after each use. Many disasters have occurred when a fungicide was applied through an improperly cleaned sprayer that was previously used for a nonselective herbicide application.

Little information exists regarding the chemical interactions of tank mixes. Most well-known chemical incompatibilities are noted on pesticide labels. There are two general types of incompatibilities: chemical and physical. Chemical incompatibilities generally occur when the pH of the final solution, or the presence of one of the compounds, reduces the efficacy or increases the phytotoxicity of a pesticide. Some examples of chemical incompatibilities are as follows: mixing lime or an alkaline reacting fertilizer with a benzimidazole or an ethylenebis-dithiocarbamate fungicide can reduce their effectiveness; tank-mixing iron sulfate with an emulsifiable concentrate may cause phytotoxicity; and tank-mixing a triazole fungicide with some plant growth regulators may discolor or damage annual bluegrass or creeping bentgrass greens. The pH of the final tank-mixture should be between 6.5 and 7.0. Additives are available for adjusting pH of spray solutions. A pH meter should be purchased by managers who spray pesticides more than a few times per year. These meters require frequent calibration, and stock buffer solutions should be purchased for the purpose of recalibration.

Physical incompatibility normally is associated with excessive foaming or settling-out of particles. Mixing prepackaged mixtures of 2,4-D + MCPP + dicamba with some wettable powder fungicides may cause formation of a precipitate (i.e., solid particles that separate out of the suspension or solution to form a solid material at the bottom of the tank). Mixing chlorothalonil (Daconil 2787) flowable with fosetyl-aluminum (Aliette) also may form a precipitate. Physical incompatibility can indicate that there is an equipment problem. For example, wettable powders mixed without sufficient agitation or without a sufficient amount of water will clog screens. Prewetting and creating a slurry is helpful in getting wettable powders into suspension, especially when spraying with a small quantity of water. It is important to always keep the agitation system running, even during breaks or when in transit.

It is important to mix only enough material that can be sprayed in one day. Chemicals will interact in the tank and if enough time elapses the effectiveness of pesticides will diminish. Temperature also influences pesticide effectiveness. As temperature in the tank is increased, the reaction rate of chemicals will increase and the likelihood of reduced efficacy is enhanced. Time and temperature, however, affect the performance of insecticides and fertilizers more significantly than fungicides.

As previously noted, many incompatible combinations are listed on pesticide labels. Frequently, however, compatibility questions arise, especially when dealing with new formulations of pesticides or when unusual pesticide combinations are being considered. It therefore becomes necessary to test the compatibility of a mix yourself. This is best achieved through a simple, two-step test. Step 1 involves placing a mixture of the precise dosage of pesticide plus the appropriate amount of water in a quart (0.5 to 1.0 liter) jar for 30 minutes. If separation of chemicals occurs or if materials settle out it is probably unwise to use the mixture. Step 2 should be performed regardless of results acquired in Step 1. In Step 2 the mixture is applied to turf. Preferably, the mixture should be applied during adverse environmental conditions such as hot, dry weather and intentionally overlapped to ensure that phytotoxicity does not occur. A minimum of 48 hours should elapse before the response can be properly evaluated.

ACKNOWLEDGMENTS

I wish to express my sincere gratitude to colleagues for reviewing and sharing valuable information contained in the manuscript: Drs. H.B. Couch,

K. Kackley-Dutt, G.E. Holcomb, N. Jackson, R.T. Kane, N.A. Tisserat and J.M. Vargas, Jr.

REFERENCES

Sanders, P.L. *The Microscope in Turfgrass Disease Diagnosis*. The Pennsylvania State University, University Park, PA, 1993. 28 p.

Shurtleff, M.C., T.W. Fermanian, and R. Randell. *Controlling Turfgrass Pests*. Prentice-Hall, Englewood Cliffs, NJ, 1987. 449 p.

Smiley, R.W., P.H. Dernoeden, and B.B. Clarke. *1992 Compendium of Turfgrass Diseases*. 2nd Edition. American Phytopathological Society, St. Paul, MN, 1992. 98 p.

Smith, J.D., N. Jackson, and A.R. Woolhouse. *Fungal Diseases of Amenity Turf Grasses*. E. & F.N. Spon, London, 1989. 401 p.

Turfgrass Patch Diseases Caused By Ectotrophic Root-Infecting Fungi. B.B. Clarke and A.B. Gould (Eds.). American Phytopathological Society. St. Paul, MN, 1993. 161 p.

Vargas, J.M., Jr. *Management of Turfgrass Diseases*. Lewis Publishers, Chelsea, MI, 1993. 294 p.

CHAPTER 3

Turfgrass Insect and Mite Management

David J. Shetlar

GOAL OF INSECT AND MITE MANAGEMENT IN TURF

Home lawns, sports turf, golf courses, and sod farms have different needs and requirements for their usage of turf. Therefore, each of these turf usage areas would require different levels of pest management. Generally, the goal of insect and mite management is the one of integrated pest management (IPM); that is, keep pest damage to an acceptable aesthetic or damage level while using monitoring and appropriate control methods. In order to achieve this goal, several important principles must be understood:

I. **Pest Identification:** The turf habitat provides suitable living space for a multitude of animals. These animals can be beneficial, damaging, or of no real importance. Turf managers must be able to identify each of these animals and determine if any control action is necessary. March fly larvae feeding on dead turf killed by a disease may be mistaken for the cause of the dead turf. Obviously, disease management is needed, not insect control.

II. **Pest Life Cycles:** Each insect and mite found in turf has a unique life cycle. Turf managers must become familiar with these life cycles because certain stages of these life cycles are often resistant to controls. Other stages are quite vulnerable to control, and these should be the target of control actions.

III. **Turf, a Unique Environment:** Turfgrass is a special environment with unique attributes. Thinking of turf as a regular field crop will result in pest management failures. Turf is a perennial plant cover with distinctive zones (Figure 3.1). Each zone is utilized by pest insects and mites. Pests in the upper, **foliar/stem zone** of turf are often conspicuous and fairly easy to control because of their exposure. Pests in the **stem/thatch zone** can evade detection for some time, until their damage begins to show as discolored foliage. Pests located in the **thatch/soil zone** also evade detection until their damage to turf roots results in significant turf loss. Pests located in the stem/thatch zone and thatch/soil zone are often the most difficult to manage with pesticides unless irrigation is available to wash the pesticides into the areas occupied by the pests.

Thatch and the high organic matter formed in the upper inch of soil bind and restrict chemical controls. Pesticides that work in agricultural soils with 1% to 3% organic matter simply do not move well in the 3% to 10% organic matter soils under turf. Finally, there are nuisance pests associated with turf. Though these bees, ants, ear-

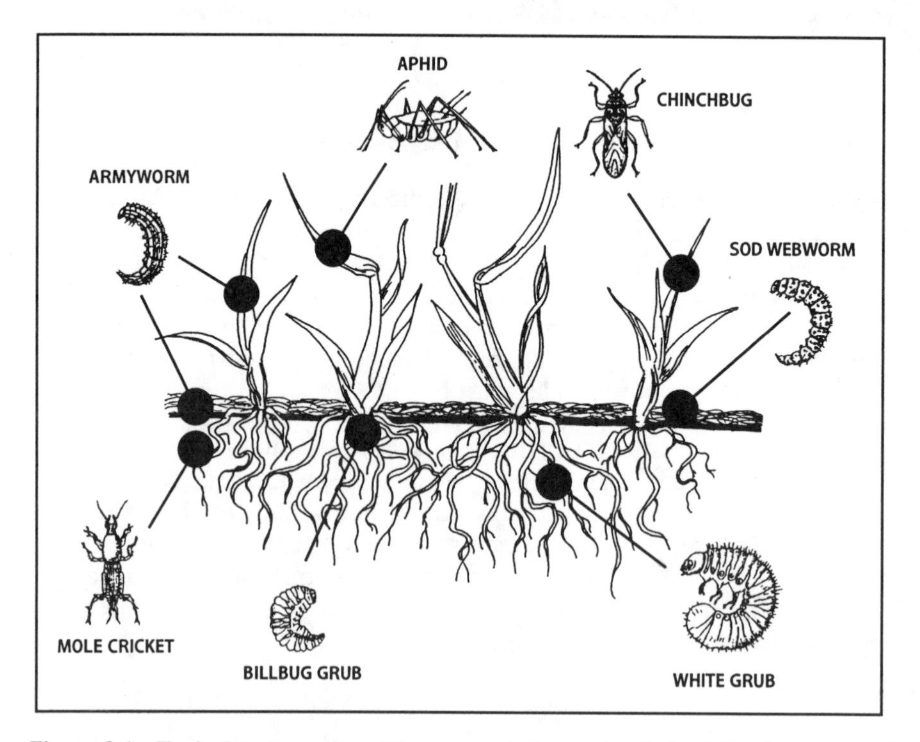

Figure 3.1 Typical turf profile with representative pests of the foliar/stem, stem/thatch, and thatch/soil zones.

wigs, and others are "people problems" which rarely damage the turf directly, they have to be managed.

IV. Monitoring: Pest monitoring is one of the most important principles in pest management. By using regular monitoring, the turf manager can detect new pests and determine when, or if, existing pests will reach damaging numbers. Monitoring is also used to determine if a pest is in a vulnerable stage of its life cycle.

V. Select Appropriate Controls: Pesticides (chemical controls) are powerful pest management tools, but using pesticides as the only tool will eventually fail. IPM approaches pest control as a system of decisions and control tactics. In pest management, we must constantly use cultural and biological controls to their fullest potential. This can help reduce the unwarranted usage of pesticides, preserve their usefulness, and develop a more diversified and resilient turf habitat.

I. PEST IDENTIFICATION

Formal training in entomology, the study of insects, can certainly help the turf manager make correct identifications of the various insects, mites, and other invertebrates encountered in the turf environment. The pictures and illustrations provided in this book can help. However, even seasoned managers can make errors in identification. Therefore, managers should learn to make contact with experts located at Land Grant universities (through their Cooperative Extension Services), consultants associated with professional turf organizations (e.g., golf course associations, lawn care associations, etc.), and other managers.

No single book can answer all pest identification problems. Turf managers are encouraged to build a library of information. See the general references located at the end of this chapter for additional information on insects, mites, and other animals associated with turf.

Insects and Mites Associated With Turf: An Introduction

Many invertebrate and vertebrate animals can be associated with turf. The turfgrass environment provides a sizeable source of food and living space for a variety of plant feeders, predators, scavengers, and decomposers. The vast majority of invertebrate species associated with the turf habitat can be considered beneficial or inconsequential in habits. However, a limited number of herbivores feed on grass leaves, stems, and roots. When in sufficient numbers, these herbivores become significant pests. Others, such as fleas, spiders, ants and wasps can cause concern because of their nuisance ability to sting, bite, or simply get in the way.

The invertebrates most commonly encountered by the turf manager are:

1. Phylum Nematoda. Nematodes or roundworms are micro- to macroscopic in size and include numerous plant pests, decomposers, and animal parasites. Those that attack turf roots are generally too small to see without the aid of a microscope. Larger nematodes appear to be worms without segments, are pointed at each end and are constantly coiling and twisting (Figure 3.2). These larger species are usually beneficial decomposers or parasites of other animals.

2. Phylum Mollusca. Snails and slugs are small to large sized animals without segmentation of the body (Figures 3.3 and 3.4). The slugs are merely snails which have secondarily lost or reduced the shell. Most are plant feeders or decomposers. They are usually considered a nuisance in turf.

3. Phylum Annelida. Segmented worms include the earthworm (Figure 3.5) as well as numerous smaller, closely related worms. Though considered beneficial because of their soil aeration habits and ability to aid in the decomposition of organic matter, they can become a nuisance pest when they create "lumpy" turf in lawns or leave castings on golf course greens and tees.

4. Phylum Arthropoda. The Arthropoda is the largest group of animals. They are characterized by having segmented bodies with segmented legs. The group includes the crustaceans (sowbugs and pillbugs as well as shrimps and crabs), arachnids (spiders and mites), chilopods (centipedes), diplopods (milli-

Figure 3.2 A free-living nematode, *Xiphinema* sp. (redrawn from Cobb).

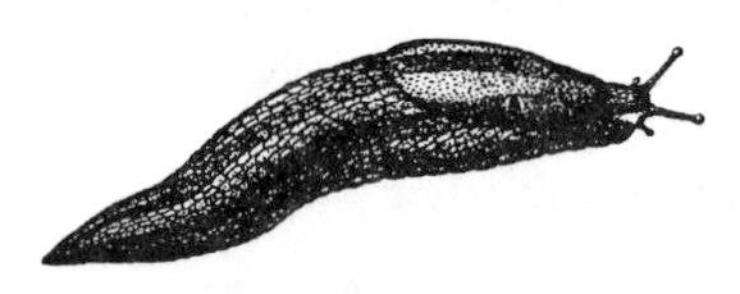

Figure 3.3 Typical garden slug. (USDA)

Figure 3.4 Typical terrestrial snail. (N. Carolina Agr. Ext. Serv.)

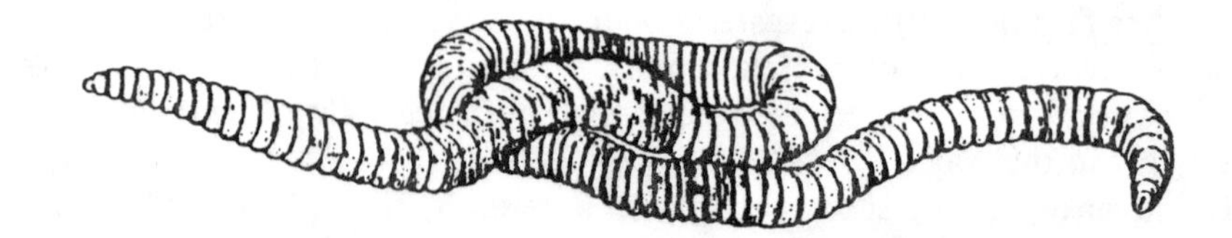

Figure 3.5 An earthworm. (redrawn from Storer & Usinger).

Plate 2-1.
Pink and gray snow mold in creeping bentgrass.

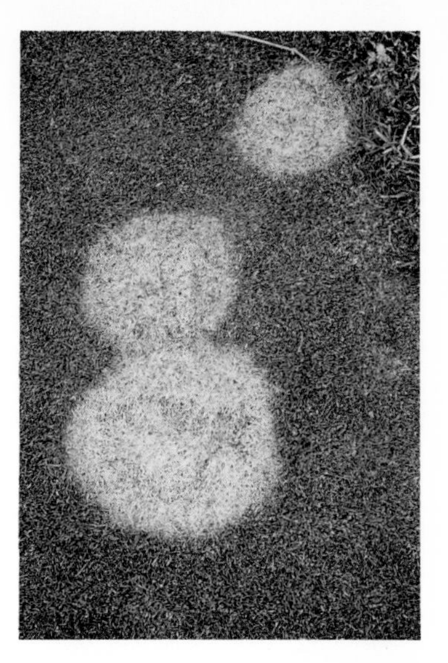

Plate 2-2.
Pink snow mold in creeping bentgrass.

Plate 2-3.
Fusarium patch (*M. nivale*) in creeping bentgrass.

Plate 2-4.
Gray snow mold in tall fescue.

Plate 2-5.
Yellow patch (aka cool temperature brown patch, *R. cerealis*) in bentgrass.

Plate 2-6.
Yellow patch (*R. cerealis*) in bentgrass.

Plate 2-9.
Pink patch mycelium causing water soaking of foliar tissues. Courtesy of N. R. O'Neill.

Plate 2-7.
Red thread patches in perennial ryegrass.

Plate 2-8.
Red threads or sclerotia of *L. fuciformis*.

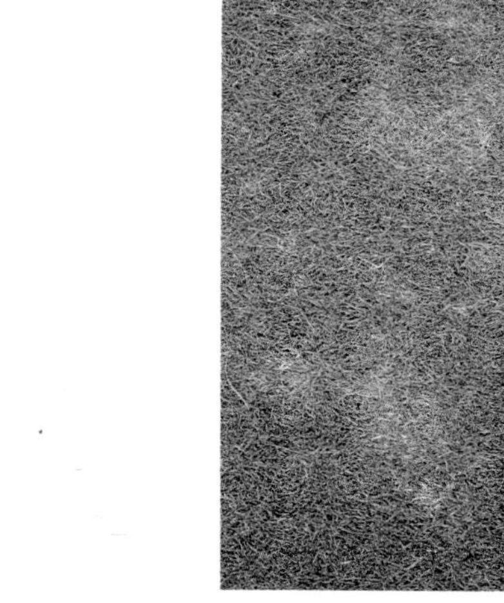

Plate 2-10.
Pink patch in bentgrass.

Plate 2-11.
Melting-out (D. poae) in common Kentucky bluegrass (right) contrasted with a resistant cultivar (left).

Plate 2-12.
Melting-out (B. cynodontis) in bermudagrass.

Plate 2-13.
Leaf spot lesions (D. poae) on Kentucky bluegrass. Courtesy of N.R. O'Neill.

Plate 2-14.
Leaf spot (D. poae) lesions and blighting of a bluegrass leaf.

Plate 2-15.
Take-all on a creeping bentgrass green.

Plate 2-16.
Take-all on a bentgrass tee.

Plate 2-17.
Reddish-brown-colored take-all patches in bentgrass.

Plate 2-18.
Yellow-colored take-all patches in bentgrass.

Plate 2-20.
Necrotic ring spot in Kentucky bluegrass.

Plate 2-19.
Necrotic ring spot in a Kentucky bluegrass lawn. Courtesy of N. Jackson.

Plate 2-21.
Spring dead spot patches in bermudagrass.

Plate 2-22.
Spring dead spot patches in bermudagrass.

Plate 2-23.
Coalescence of patches and nonuniform loss of bermudagrass due to spring dead spot.

Plate 2-24.
Large patch (*R. solani*) disease of zoysiagrass. Courtesy of N.A. Tisserat.

Plate 2-25.
Pythium root rot in creeping bentgrass.

Plate 2-26.
Dollar spot in Kentucky bluegrass.

Plate 2-27.
Mycelium of the dollar spot fungus in perennial ryegrass.

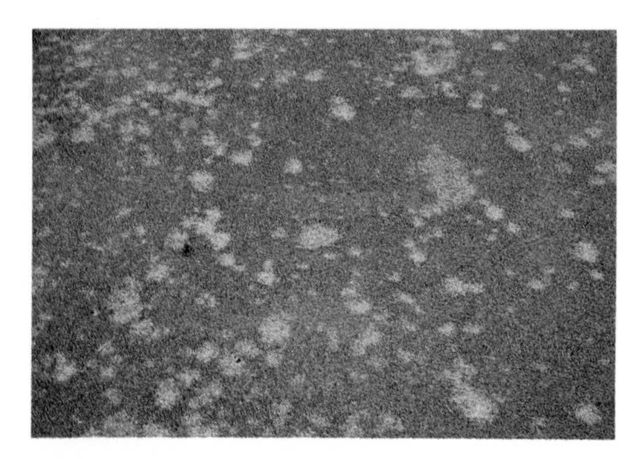

Plate 2-29.
Dollar spot and brown patch in creeping bentgrass.

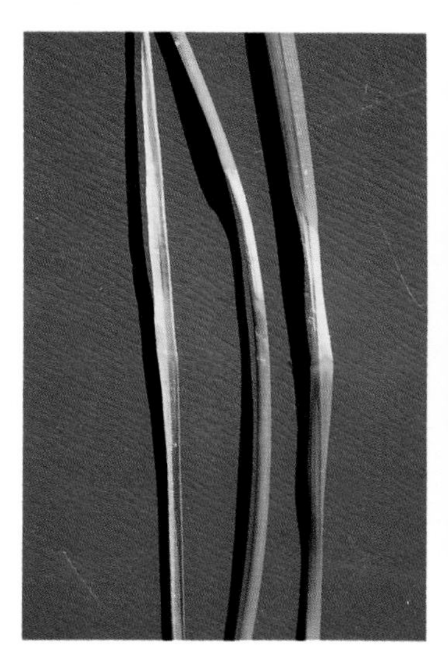

Plate 2-28.
Hour-glass shaped lesion and tip-dieback of Kentucky bluegrass leaves affected with dollar spot.

Plate 2-30.
Kentucky bluegrass leaves shredding and curling due to stripe smut.
Courtesy of R.W. Smiley.

Plate 2-31.
Small, orange speckling of anthracnose in annual bluegrass plants.

Plate 2-32.
Bentgrass thinning due to anthracnose basal rot.

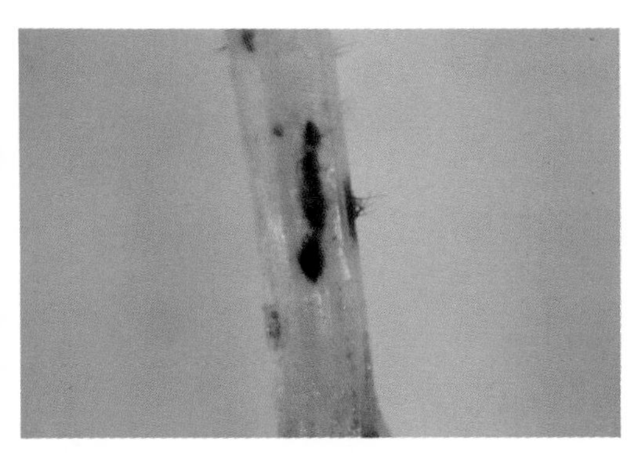

Plate 2-33.
Acervulus (black hairs) and black infection mats of the anthracnose fungus on a bentgrass stolon.

Plate 2-35.
Leptosphaerulina blight in annual bluegrass.

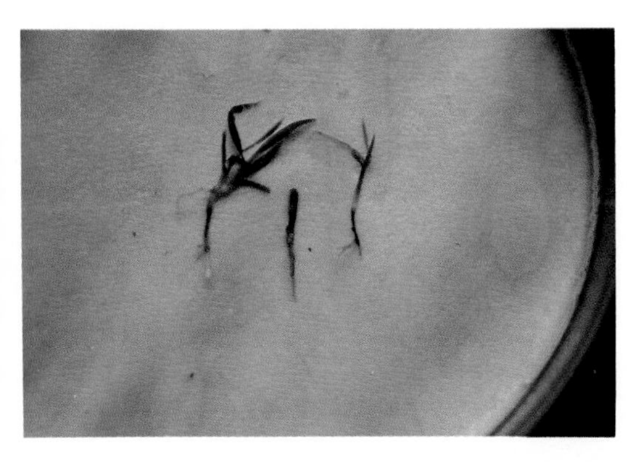

Plate 2-34.
Black stem bases of bentgrass plants with anthracnose basal rot.

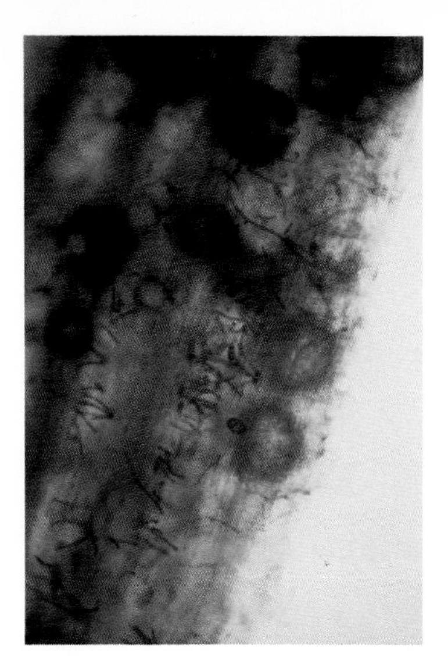

Plate 2-36.
Fruiting bodies of Leptosphaerulina spp. in a bentgrass leaf.

Plate 2-37.
Yellow tufted bentgrass plants.

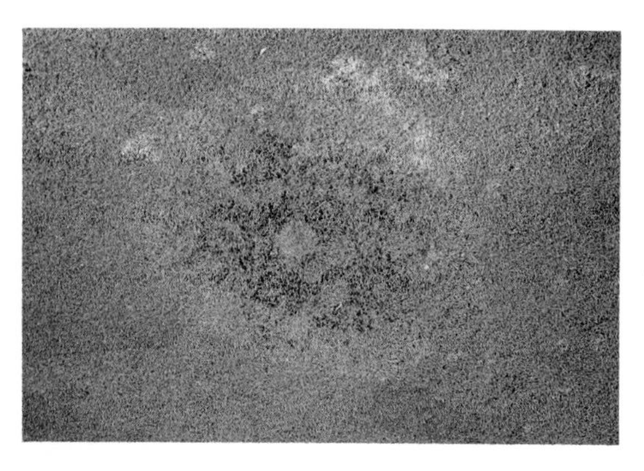

Plate 2-38.
Brown patch, blackish blue-green algae and dollar spot in bentgrass.

Plate 2-39.
Brown patch smoke rings in bentgrass.

Plate 2-40.
Brown patch in colonial bentgrass.

Plate 2-41.
Brown patch in tall fescue.

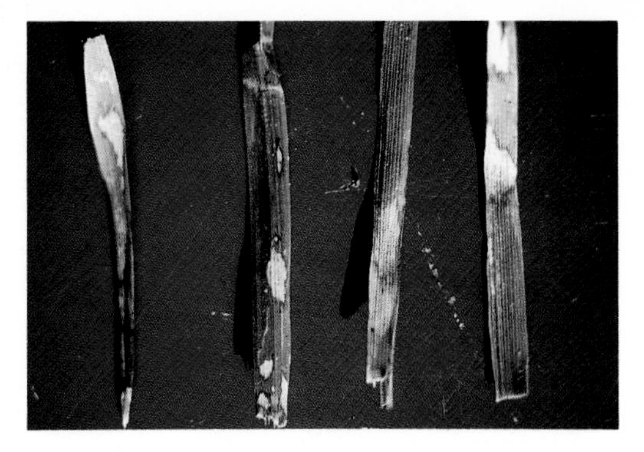

Plate 2-42.
R. solani lesions on tall fescue leaves.

Plate 2-43.
R. solani foliar mycelium in perennial ryegrass.

Plate 2-45.
Copper-colored Pythium blight patches in bentgrass.

Plate 2-44.
Pythium blight in bentgrass.

Plate 2-46.
Pythium blight patches with smoke rings in perennial ryegrass.

Plate 2-47.
Pythium blight in water drainage areas on a perennial ryegrass fairway.

Plate 2-48.
Summer patch in a Kentucky bluegrass fairway.

Plate 2-49.
Summer patch "crater pits" with bronzed peripheries in a Kentucky bluegrass lawn.

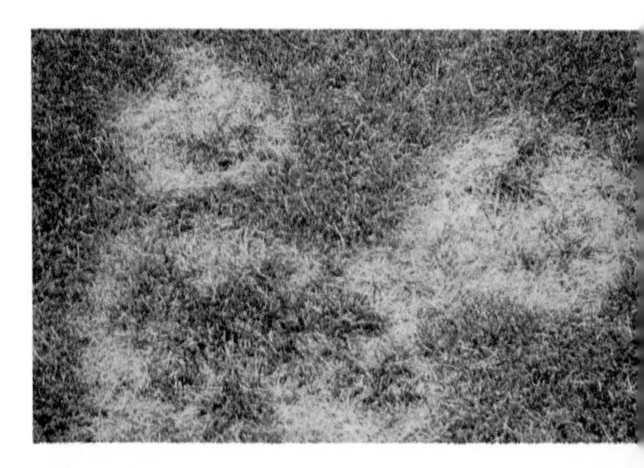

Plate 2-50.
Circular patches of summer patch in Kentucky bluegrass.

Plate 2-51.
Summer patch in annual bluegrass.

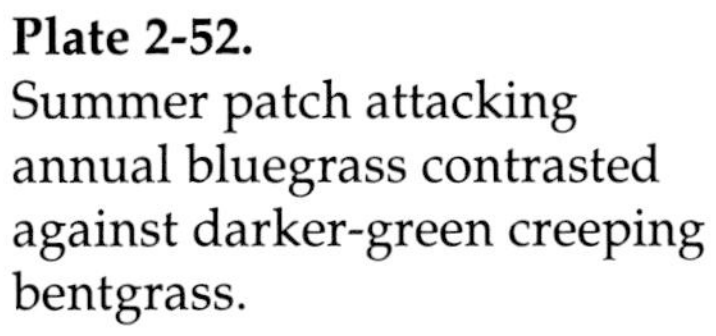

Plate 2-52.
Summer patch attacking annual bluegrass contrasted against darker-green creeping bentgrass.

Plate 2-53.
Summer patch in annual bluegrass.

Plate 2-54.
Summer patch in creeping red fescue.

Plate 2-55.
Sclerotia of S. *rolfsii*.

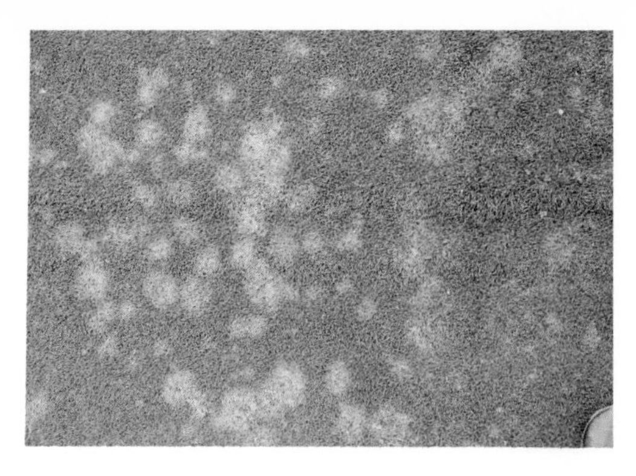

Plate 2-56.
Cooper spot in velvet bentgrass.

Plate 2-57.
Two converging fairy rings in Kentucky bluegrass.

Plate 2-58.
Type I fairy ring in perennial ryegrass.

Plate 2-59.
White wefts of mycelium in thatch produced by the yellow ring fungus.

'late 2-60.
'airy rings and localized dry spots in a
›utting green.

Plate 2-61.
Superficial fairy rings in
bentgrass.

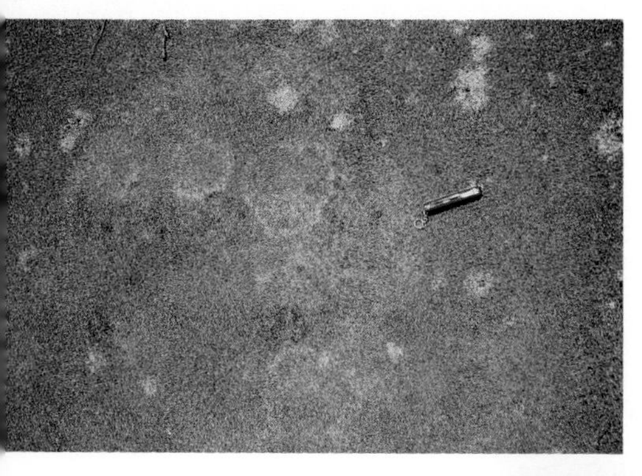

'late 2-62.
›rown patch, dollar spot,
›uperficial fairy rings, and
›lackish blue-green algae
ı bentgrass.

Plate 2-63.
Localized dry spots in a putting
green.

Plate 2-64.
Fruiting bodies of a slime mold
fungus.

Plate 2-65.
Rust in a Kentucky bluegrass lawn.

Plate 2-66.
Rust pustules on zoysiagrass leaves.

Plate 2-67.
Chlorosis and mosaic of St. Augustine decline. Courtesy of N. Jackson.

Plate 2-68.
Bacterial wilt in annual bluegrass.

Plate 2-69.
Bacterial wilt in Toronto creeping bentgrass. Courtesy of D.J. Wehner.

Plate 2-70.
Plant parasitic nematode damage in a putting green. Courtesy of T.R. Turner.

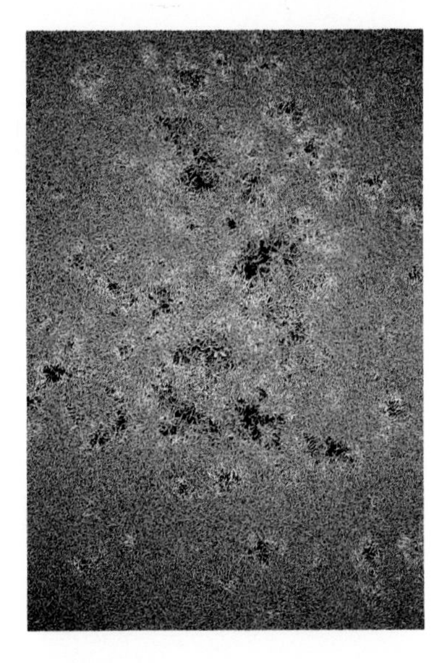

Plate 2-71.
"Blackish" blue-green algae colonizing bareground areas created by dollar spot disease in bentgrass.

Plate 2-72.
Moss growing in a dense turf.

Plate 3-1.
"Witchesbrooming" of bermudagrass caused by bermudagrass mites, *Eriophyes cynodoniensis* Sayed.

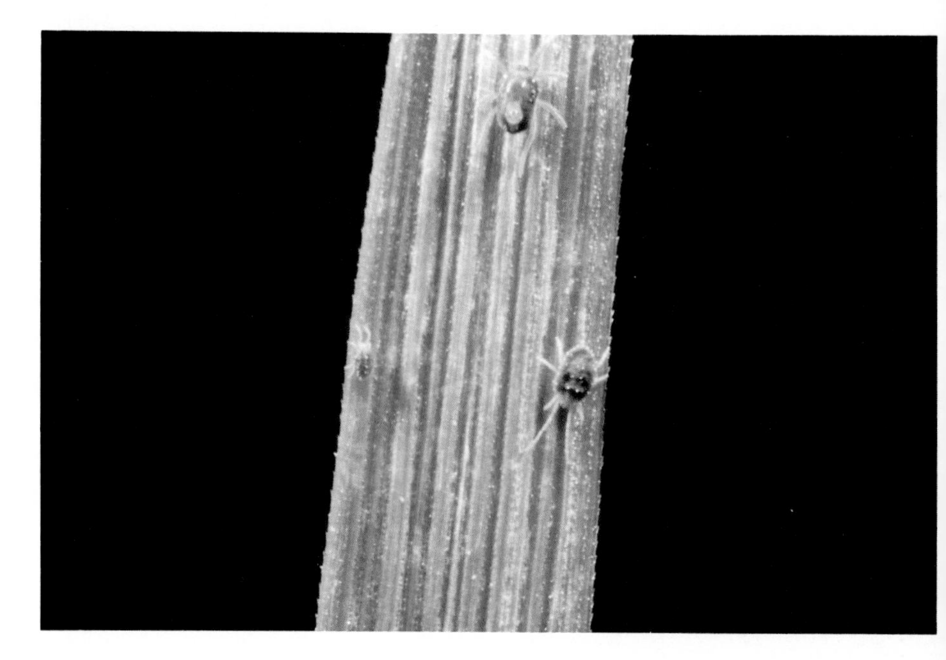

Plate 3-2.
Winter grain mite, *Penthaleus major* (Duges), adult (above) and clover mites, *Bryobia praetiosa* Koch, adult (lower right) and nymph (lower left).

Plate 3-3.
Tawny mole cricket, *Scapteriscus vicinus* Scudder, adult female.

Plate 3-4.
Southern chinch bug, *Blissus insularis* Barber, adult.

Plate 3-5.
Twolined spittlebug, *Prosapia bicincta* (Say), adult.

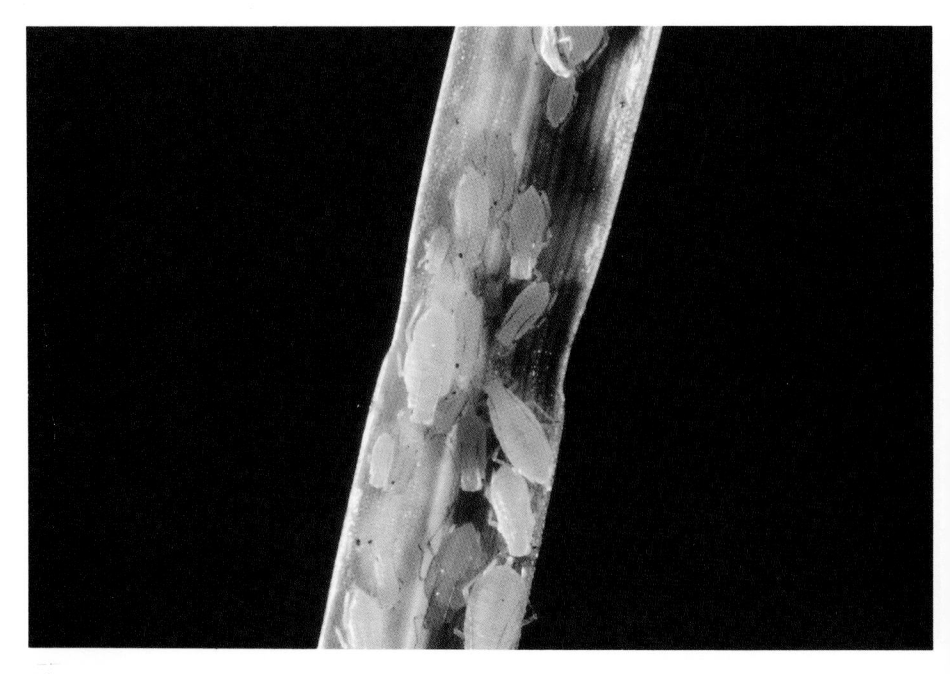

Plate 3-6.
Greenbugs, *Schizaphis graminum* (Rondani).

Plate 3-7.
Rhodesgrass mealybugs, *Antonina graminis* (Maskell), on bermudagrass node. Note long excretory wax tubes.

Plate 3-8.
Bermudagrass scales, *Odonaspis ruthae* Kotinsky.

Plate 3-9.
Ground pearls, *Margarodes meridionalis* Morrison, collected from roots of centipedegrass.

Plate 3-10.
Adults of the common North American white grubs which infest turfgrasses. (top row, left to right) May/June beetle, *Phyllophaga* spp.; black turfgrass ataenius, *Ataenius spretulus* (Haldeman); Asiatic garden beetle, *Maladera castanea* (Arrow); Oriental beetle, *Exomala orientalis* (Waterhouse); (middle) Japanese beetle, *Popillia japonica* Newman; (bottom row, left to right) European chafer, *Rhizotrogus majalis* (Razoumowsky); southern masked chafer, *Cyclocephala lurida* Bland; northern masked chafer, *C. borealis* Arrow; and, green June beetle, *Cotinis nitida* (Linnaeus).

Plate 3-11.
Annual bluegrass weevil, *Listronotus* (=*Hyperodes*) *anthracinus* (Dietz), adult.

Plate 3-12.
Bluegrass billbug, *Sphenophorus parvulus* Gyllenhal, larva in crown.
Note sawdust-like frass above larva.

Plate 3-13.
Bluegrass billbug, *Sphenophorus parvulus* Gyllenhal, adult and larva.

Plate 3-14.
Larger sod webworm, *Pediasia trisecta* (Walker), adult in resting position.

Plate 3-15. A sod webworm larva in thatch with green frass pellets.

Plate 3-16. Tropical sod webworm, *Herpetogramma phaeopteralis* Guenee, adult.

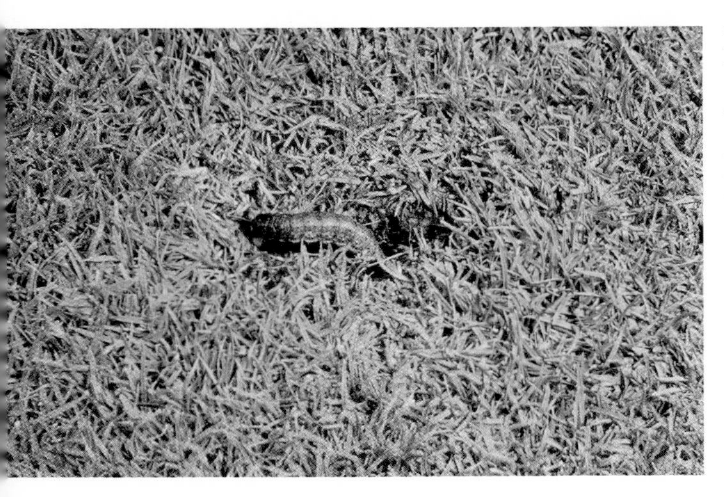

Plate 3-17. Black cutworm, *Agrotis ipsilon* (Hufnagel), larva and typical feeding spot on short cut bentgrass.

Plate 3-18. Armyworm, *Pseudaletia unipuncta* (Haworth), larva on grass stem.

Plate 3-19. March fly, *Bibio* sp., larvae and pupa (left).

Plate 3-20. Cicada killer, *Sphecius speciosus* (Drury), adult.

pedes), symphylans (garden centipedes), and the insects. Each of these groups have a special set of characters which are used in identification. However, all arthropods have: an exoskeleton broken down into segments and body regions (head, thorax, and abdomen) which must be shed at periodic intervals to accommodate growth; paired, jointed appendages (legs, mouthparts, and occasional abdominal appendages), a ventral nerve cord and dorsal pumping heart.

Classes of Arthropods

The six major groups (classes) of Arthropods are the Arachnida, Crustacea, Chilopoda, Diplopoda, Symphyla and Insecta.

The Arachnida includes spiders, mites and ticks, scorpions, daddy-long-legs, and other less common groups (Figures 3.6 to 3.10). All normally have four pairs of legs, a single pair of oral sensory appendages called pedipalps (either leglike or pincherlike), a pair of pincherlike structures near the head called chelicerae, no antennae, and two body regions (the cephalothorax and abdomen). Spiders have the two body regions separated by a constriction, while most of the other groups have the two regions broadly joined. Since most arachnids are terrestrial, they use tracheal tubes, book lungs, or simple diffusion (small mites) in order to respire. Though arachnids are generally feared, most are considered beneficial predators and scavengers. However, some mites attack turf plants, and ticks (large, blood sucking mites) are commonly found in turf where their host animals reside.

The Crustacea are usually aquatic (crabs, shrimp and crayfish) though a few

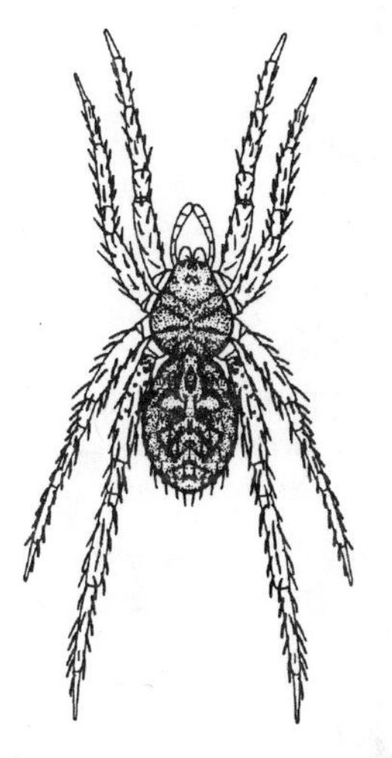

Figure 3.6 A wolf spider (Lycosidae). (redrawn from Eddy & Hodson)

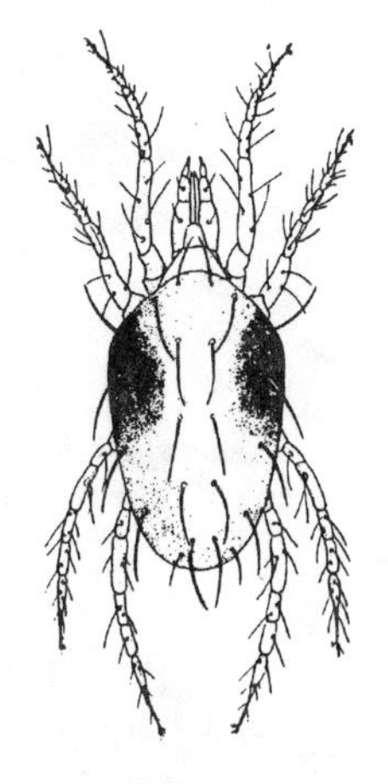

Figure 3.7 Twospotted spider mite, *Tetranychus urticae* Koch, adult female. (USDA)

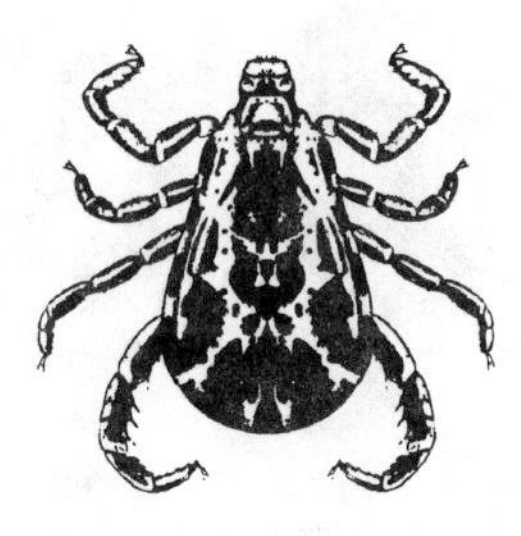

Figure 3.8 American dog tick, *Dermacentor variabilis* (Say), adult male. (CDC)

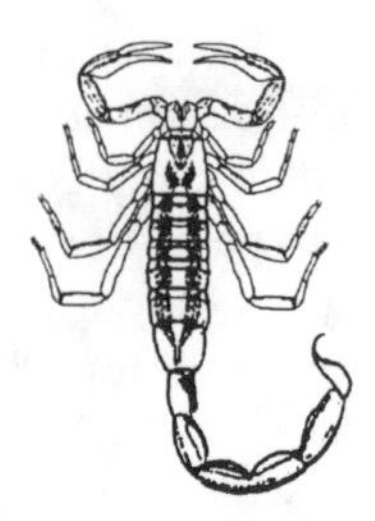

Figure 3.9 A scorpion, *Centruroides vittatus*. (CDC)

Figure 3.10 A phalangid or "daddy-long-legs." (CDC)

groups have adapted to terrestrial life (sowbugs and pillbugs, the order Isopoda) (Figure 3.11). Crustaceans have two pairs of antennae and the body divided into two regions—a head and trunk or cephalothorax and abdomen. The terrestrial isopods have the second pair of antennae reduced to a small stub and have the head/trunk body design. Isopods are considered beneficial decomposers though they may feed on grass seedlings.

The Chilopoda or centipedes are terrestrial and have a head and trunk region with a single pair of antennae (Figure 3.12). Each trunk segment has only one pair of legs attached to the side of the body. The first pair of legs have been modified into pincherlike fangs located just below the mouth. Most centipedes are beneficial predators, though some are large enough to bite humans if disturbed.

The Diplopoda or millipedes are terrestrial and also have a head and trunk region with a single pair of antennae (Figure 3.13). Each trunk segment has been joined with another (diplosegments) so that it appears that each visible

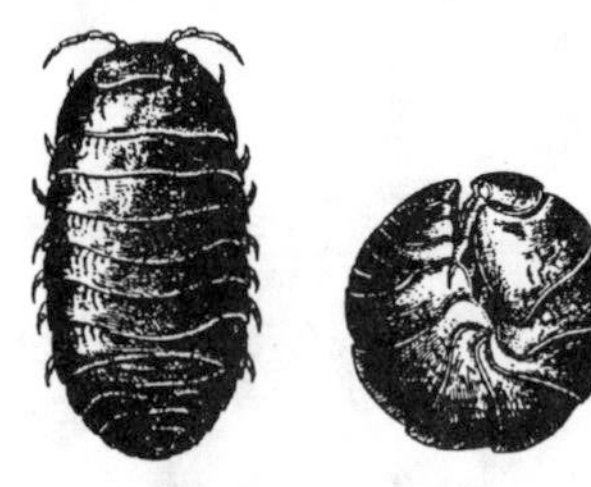

Figure 3.11 A common pillbug, *Porcelleo* sp., dorsal view (left) and disturbed individual rolled into ball (right). (USDA)

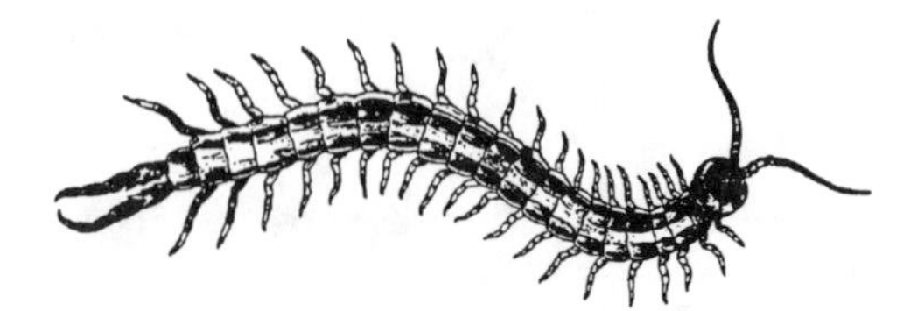

Figure 3.12 A common centipede, *Scolopendra obscura* Newport. (USDA)

Figure 3.13 A typical polydesmid millipede with flattened upper segmental plates. (CDC)

segment has two pairs of legs. Millipedes are considered beneficial scavengers, though many feed on living plant tissues, especially of seedlings.

The Symphyla or garden centipedes are terrestrial, have a head and trunk with a single pair of antennae but no eyes or body pigmentation (Figure 3.14). They look like tiny centipedes, but they do not have fang-legs and most of the species feed on root hairs of plants. Occasionally, they reach large numbers in the soil and their feeding may cause stunting of the plants.

The Insecta is the largest group of animals and the members are characterized by having three distinct body regions (head, thorax, abdomen), a single pair of antennae, three pairs of legs, and adults usually have two pairs of wings. They respire through a set of air tubes, trachea, which run throughout the body. There are numerous orders of insects which contain members that attack or are associated with the turfgrass environment. Probably of greater importance is whether the insect belongs to one of two categories according to life cycle or metamorphosis. The more primitive insects have a gradual (simple or incomplete) life cycle, while the vast majority have a complete (complex) life cycle.

II. PEST LIFE CYCLES

All pests of turf, whether weeds, diseases, or insects have unique life cycles. By understanding these life cycles, the turf manager can approach management by emphasizing controls targeted to the most vulnerable stage of a pest.

Insects and mites can be divided into two general groups with similar life

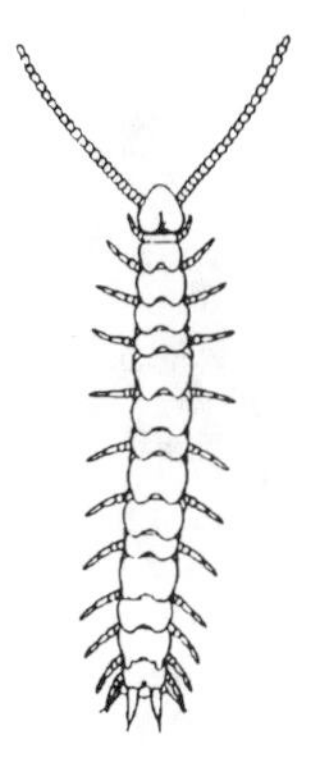

Figure 3.14 Garden symphylan, *Scutigerella immaculata* (Newport). (USDA)

cycles—gradual and complete metamorphosis. Those insects and mites with a gradual metamorphosis have immatures and adults that look much alike and generally live and feed in the same habitat. Chinch bug nymphs and adults have the same general body shape and feed on plant juices with their piercing, sucking mouthparts. On the other hand, insects with a complete metamorphosis have immatures that are very different in form than the adult, and these two stages usually feed on different foods and live in different habitats. Cutworm larvae live in the turf and have chewing mouthparts, while the adults fly to flowering plants and have sucking mouthparts. These adults return to the turf to lay eggs.

Insects and mites with gradual life cycles are often easier to control because all stages are present and the only decision to be made is whether the population is large enough to cause problems. However, pests with complete life cycles are usually vulnerable to management during some critical phase of their cycle. Understanding when these periods occur and sampling just prior to the event, turf managers can make the most use of their control options.

Insect Metamorphosis

Insects with a gradual metamorphosis (Figure 3.15) have three life stages: egg, nymphs and adult. Eggs hatch into the first nymph (called the first instar), which can add body mass but must shed its exoskeleton in order to continue growth. The second nymph (called the second instar) continues the process. As the nymphs molt and mature, they develop external wing pads (if they have wings as adults), and eventually molt into the adult stage (last or adult instar). Most insects with a gradual metamorphosis have egg, nymphs and adults located in the same habitat and feeding in the same manner. Though the nymphs may have different coloration than the adult, they have the same general body shape and form.

Insects with a complete metamorphosis have four life stages: egg, larva, pupa and adult. The eggs hatch into the first larva (called the first instar), which can add body mass but must shed the exoskeleton in order to continue to grow. The second larva (called the second instar) continues the process, usually for a predetermined number of times. Most cutworm and sod

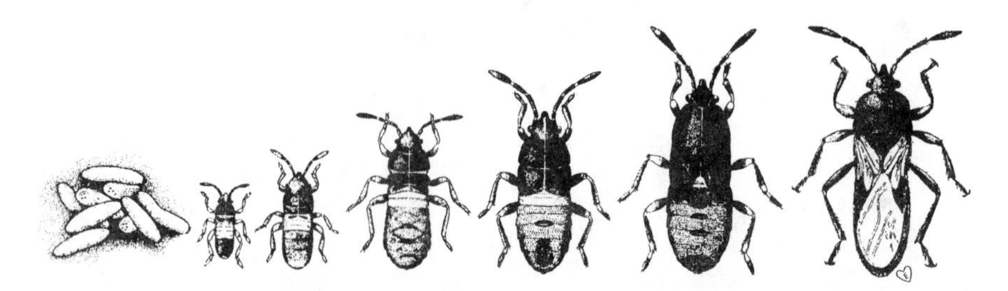

Figure 3.15 Hairy chinch bug, *Blissus leucopterus hirtus* Montandon, life stages. (left to right) eggs, 1st, 2nd, 3rd, 4th, 5th, instar nymphs, winged adult. (USDA)

webworm larvae have 5–6 instars while white grubs only have three instars (Figure 3.16). The last larval instar then molts into a transformation stage, the pupa. Within the pupal exoskeleton, most of the larval body is rearranged and reproductive organs are formed. The pupa usually displays the eventual adult appendages of wings and legs. The winged adult stage emerges from the pupal case, expands the wing pads, and is ready to continue the life cycle. Insects with complete life cycles usually have larvae which live in different habitats than the adult. White grubs feed in the soil, while the adult beetles often feed on the leaves of trees and shrubs. Cutworms feed on the leaves and stems of turf, while the adults must obtain sustenance from the nectar of flowers. Pests with complete life cycles can cause special problems for turf management because the eggs and pupae are usually resistant to controls. Likewise, if the larvae or adults are susceptible to controls, they are usually in different habitats.

The larvae of insects with a complete metamorphosis are usually given special names. Scarab beetle larvae are called white grubs; moth and butterfly larvae are caterpillars; most fly larvae are maggots (Figure 3.17); and, larvae of bees, wasps, and ants may be grublike, caterpillarlike or maggotlike.

Mite Life Cycles

Plant feeding mites associated with turf can be treated as if they have a simple metamorphosis. All mites have four stages: the egg, larva (an immature

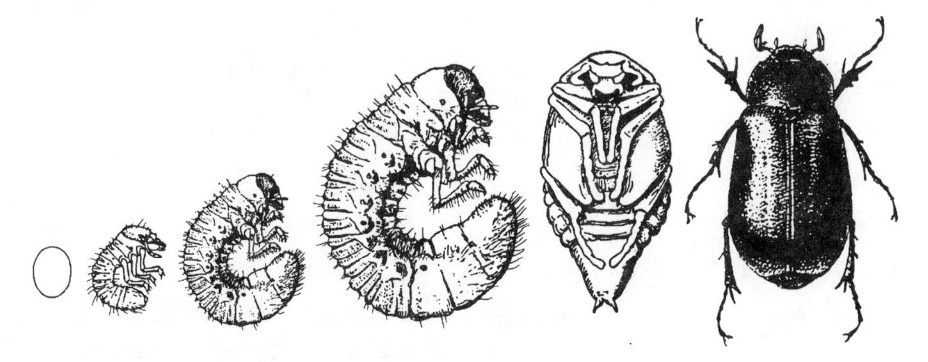

Figure 3.16 A June beetle, *Phyllophaga* sp., life cycle showing relative sizes of stages. (left to right) egg, 1st instar, 2nd instar, 3rd instar larva, pupa, adult. (USDA)

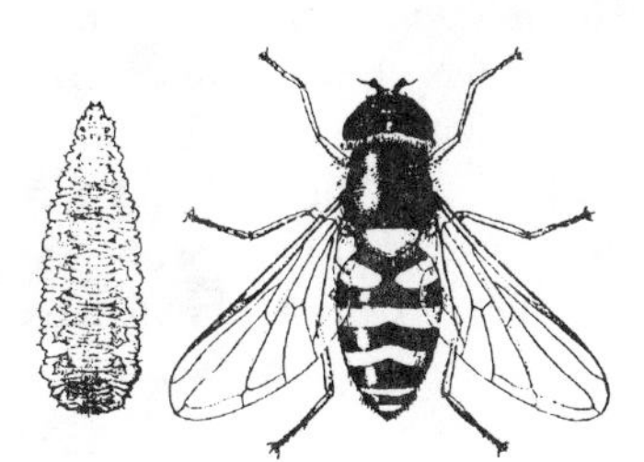

Figure 3.17 Typical syrphid fly larva (left) and adult (right). (USDA)

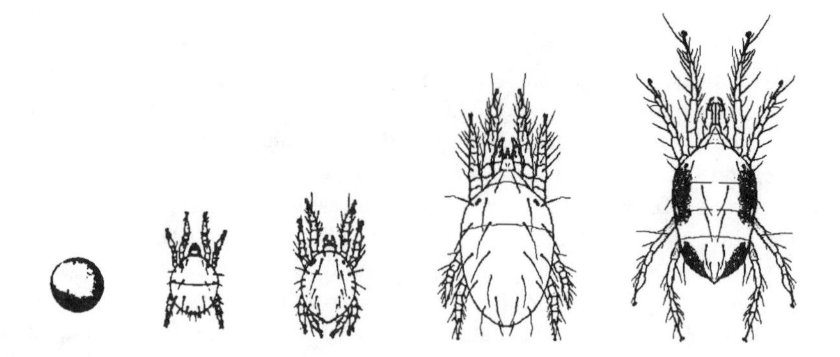

Figure 3.18 Twospotted spider mite, *Tetranychus urticae* Koch, life stages. (left to right) egg, larva, nymph I or protonymph, nymph II or deutonymph, and adult. (USDA)

with only three pairs of legs), two nymphal instars (protonymph and deutonymph), and the adult (Figure 3.18). Usually, all stages are found on or near the host. The adults of some mites migrate out of their normal habitat in order to lay eggs in protected areas. The eggs of most mites are resistant to chemical controls, which makes mites difficult to manage if the eggs enter periods of dormancy. The eggs of the winter grain mite go dormant for the entire summer, while the eggs of the Banks grass mite may be either summer or winter dormant.

Ticks are parasitic mites with considerably more complicated life cycles. Their larvae (called seed ticks), nymphal instars and adults may require different hosts and can travel considerable distances while feeding on their hosts.

III. TURF, A UNIQUE HABITAT

Most grasses used as turf are perennials which have great ability to recover from environmental stress and pest damage. Many species of grasses react to environmental stresses by going dormant during winter cold or summer drought. When managers try to force grasses to maintain color and growth during these normal dormancy periods, additional stress can be caused, and pests often take advantage of these situations. Home owners often comment that the neighbor who does nothing to the turf never has chinch bug or grub problems, while their highly managed and watered lawn seems to be under constant attack. In this case, the pests are merely avoiding the dormant turf and utilizing the resources of the actively growing turf. In other cases, golf course managers cut turf so short that it cannot develop deep and extensive root systems. This stressed turf may be less tolerant of white grubs feeding on the roots.

Grasses come in infinite varieties with a wide range of attributes. Wise turf managers must learn how to select the best varieties with resistance and tolerance to the environmental stresses and pests located in the area. Renovation with resistant, tolerant cultivars should be the first line of defense against insects and mites attacking turfs.

Most turf can tolerate low to moderate levels of pests. By simply watering or fertilizing, much insect and mite damage can be masked. On the other hand, over reliance on watering and fertilizing can result in disaster if the irrigation is suddenly cut off, or the pest populations build beyond the turf's ability to mask damage and regrow.

Most turfgrasses produce a layer of dead and living organic matter—thatch. This thatch serves as a natural insulating layer for mediating soil temperatures and maintaining moistures. However, if this thatch layer becomes too thick, many insects and mites also take advantage of this mediated habitat. Thatch is very high in organic matter, which can serve as a substrate for adsorption of pesticides. Therefore, if a thatch layer is too thick, insecticides applied for grub or mole cricket control will never reach their targets.

Turf constantly renews its roots and stolons. This renewal, along with the activity of earthworms and other invertebrates, continues to increase the organic matter content of the soil immediately below the thatch layer. This organic matter increase is considered a major benefit of turf culture and can help explain why turfgrass serves as a living filter to reduce groundwater contamination. On the other hand, this increase may also help build up levels of "organic feeders" such as white grubs and mole crickets.

IV. MONITORING

Pest monitoring is simply using those techniques and tools which allow the turf manager to determine **when** and **if** control action is needed.

Many turfgrass managers apply pesticides for control of anticipated turfgrass insect and mite pests through regularly scheduled "programs." These applications may be called "preventive-", "round-" or "calendar-" timed applications. On the other hand, some insect pests are not expected, and applications are made after some damage has been detected.

Preventive Applications are made to prohibit or eliminate a perceived, potential pest problem. These applications are often identified as "grub proofing," "guaranteed insect control," and in the weed area—"crabgrass prevention." The most common reasons given for using preventive applications are:

- The number of lawns or area to be treated or the golf course may be considered too large to time applications, using sampling and monitoring.
- This application is "insurance" against damage which is important for customer relations (including a greens committee), and some of the

pesticides are inexpensive enough to make the application whether insects are a problem or not.

- Less training is needed by the turf manager/applicator to apply a predetermined control product at a specified time.
- Lawn care customers "expect" something to be done, and golf course visitors don't want to see anything but perfect turf.

Though the reasons for making preventive applications seem appropriate, there are certain problems associated with this strategy:

- A damaging pest population may not have occurred, and therefore the pesticide application was not needed. This leads to questions of environmental concern.
- Applications of pesticides that are not necessary may encourage development of pest resistance or accelerated pesticide degradation. In either case, the usefulness of the pesticide may be lost.
- Merely having the pesticide in the tank or on the fertilizer granule increases the chance of pesticide misapplication.
- Making a pesticide application, whether needed or not, reduces the professional status of the manager/applicator. This is especially important in a period of increasing public concerns about the environment and safety.

Occasionally, preventive pesticide applications are warranted. Where pests are certain to occur (because of previous experience or predictive models indicate that a major pest outbreak will occur), certain insecticides are more effective when used as preventives. Insecticides with sustained residual ability or those with insect growth regulator action are often more effective when used as preventives.

Of course, annual weedy grass prevention (through preemergent herbicides) is the tactic of choice because postemergent control of these pests is much more difficult and may require numerous pesticide applications. However, in thick turf without a history of crabgrass infestation or turf growing in heavy shade, indiscriminate use of preemergent herbicides may come under question. Likewise, some turf diseases must be prevented or significant turf damage will result. When predictive models indicate that a disease outbreak is bound to occur, a preventive (in this case, a protectant) fungicide application is in order.

Calendar Date Application is simply another form of preventive application. In this case, the application is made according to a calendar schedule. This type of application assumes that an "average year" occurs every year and the pests will reach damaging levels. These spray calendars are often developed at state universities and distributed through the Cooperative Extension Service. Spray calendars are provided as a **reference** for when things **normally** happen. Remember that "average years" are developed by meteorologists by averaging temperatures and rainfall amount over the previous 30-year period.

The reasons given for using calendar date applications are the same as those given for preventive applications, and the associated problems are the same. Unfortunately, the **yearly** weather can vary considerably from the "average." It is this yearly weather which affects the pest development. Warm weather causes insect pests to develop faster than the "calendar," and cool weather delays development.

Reactive Applications of Pest Controls are usually made because a damaging pest population was missed. The most common reasons given for using reactive applications are:

- The damaging pest population was unexpected, or the proper control window was missed.
- For some reason, the preventive or calendar application did not work.
- The pesticide(s) used to control this pest were too expensive to apply, so a gamble was taken on the chance that a damaging population would not occur.

Problems associated with making reactive applications are:

- Since poor sampling or monitoring was used, damage has occurred and people are upset.
- Damaging or noticeable pest populations may be more difficult to control. White grubs may be too large for effective control, the insects may not be in a susceptible stage, or the weather may be poor for assisting control product performance.
- If the preventive application didn't work, additional pesticides, through secondary applications, may be necessary.

Alternative Tools and Strategies for Timing of Controls

Useful alternative strategies for timing of controls are available and should be used in order to reduce the problems associated with preventive and reactive pesticide applications. These alternatives include active monitoring and sampling of pest populations, using weather-mediated models and pest mapping.

Active Monitoring and Sampling of pest populations is at the heart of all integrated pest management programs. Before proper controls can be applied, one needs to know if a pest is present and if its population or potential population will cause significant damage to the turf. In addition to the traditional **visual inspection** of the turf, several trapping and sampling tools are useful for monitoring turf insects and mites. The most common ones are:

- **Pitfall Traps** are simply cups or cans sunk into the turf to capture crawling insects such as billbug adults (Figure 3.19). Obviously, a pitfall trap is not appropriate where children may twist an ankle, but they

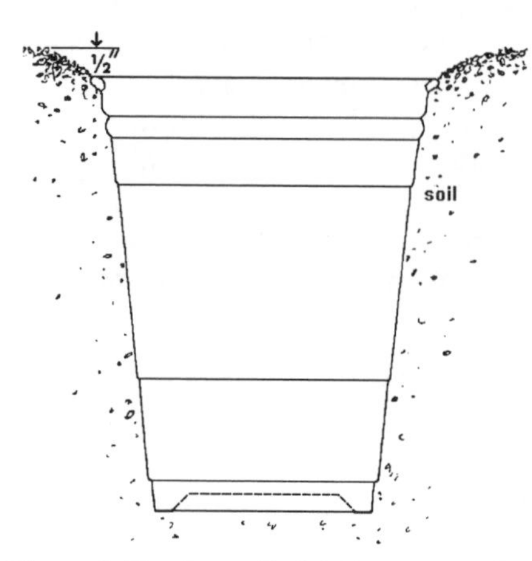

Figure 3.19 A small pitfall trap constructed with a 16–20 oz. plastic cup.

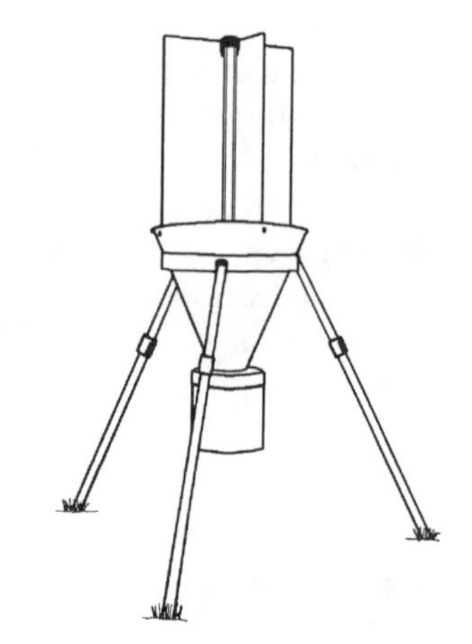

Figure 3.20 A typical light trap. (Ill. Coop. Ext. Serv.)

can be used next to flower beds or under a tree to check for insect activity.

- **Light Traps** which use "black lights" are very attractive to sod webworm, cutworm, June beetle, and masked chafer adults (Figure 3.20). Not every lawn needs to be monitored. However, a light trap in a neighborhood or section of town can assist you in determining whether insects are early or late according to normal "calendars."
- **Pheromone Traps** contain the sex and/or attractant chemicals used by sod webworms, cutworms, and some scarab beetles (Figure 3.21). These can be used, like light traps, to determine insect activity periods.
- **Cup Changer Samples** merely use a cup changer to pull plugs in a line across the turf in order to sample for billbug larvae or white grubs. You can quickly determine if a potential problem exists when the majority of samples contain a billbug or white grub larva. This sampling method is also referred to as "square foot samples" in other guides. In this case a knife is used to cut through the turf on three sides of a square foot area and the flap is lifted back to expose the insects below.
- **Disclosing Solution** (Soap Drench) uses one tablespoon of liquid dish-washing detergent per gallon of water for sod webworm and cutworm

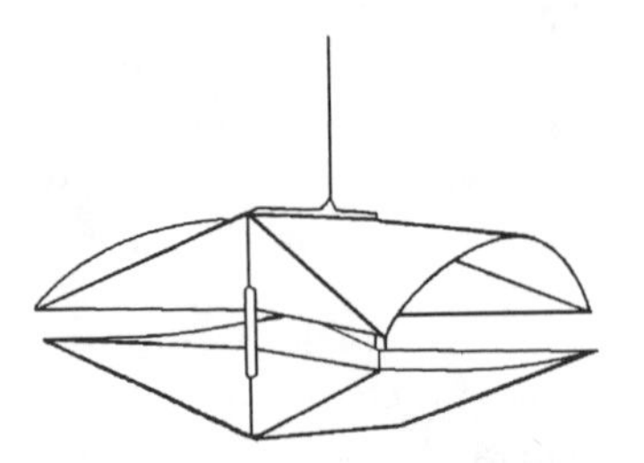

Figure 3.21 A "wing trap" type of pheromone trap.

monitoring. Simply spread two gallons of the mix over a square yard of turf and the caterpillars will quickly come to the surface. This is also useful in disclosing mole cricket nymphs. Adult mole crickets often need two flushings in order to force them to the surface.

- **Damage (Grid) Ratings** use a wood, metal, or plastic frame with cord or wire strung between the sides to form square grids (Figure 3.22). The frame is dropped onto the turf and presence or absence of pest activity is recorded for each grid; or, a rating is used for each grid and the average is taken.

Weather-Mediated Predictive (Degree-Day) Models are developed by monitoring weather parameters and comparing these to insect or mite activity. Though these models help determine better timing of controls, they still do not answer the question of whether the pests are present in sufficient numbers to cause damage or warrant controls. Models have been developed and published for chinch bugs, bluegrass billbugs, masked chafers, Japanese beetles and sod webworms. Most of these models have not achieved widespread usage because the models are too complex, or use base temperatures not normally used.

Pest Mapping is simply good record keeping. Since insect and mite pests generally require specific habitats in order to build up damaging populations, turfgrass areas which have had problems in the recent past are the most likely places in need of attention. In short, if a damaging pest population occurred last year in an area, the probability is much higher that the same thing will occur again. Keeping a useful record of pest occurrence is pest mapping.

Pest mapping in lawn care firms can be easily accomplished on a customer or neighborhood basis. Relatively young neighborhoods often seem to have very similar conditions from lawn to lawn. The developer may have purchased the sod or obtained seed from a single source. Therefore, most of the lawns will have the same varieties of grasses or mixes. If the "contractor's blend" was used, most of the lawns will have common perennial ryegrasses and fine fescues predominating. These will be very susceptible to sod webworms, chinch

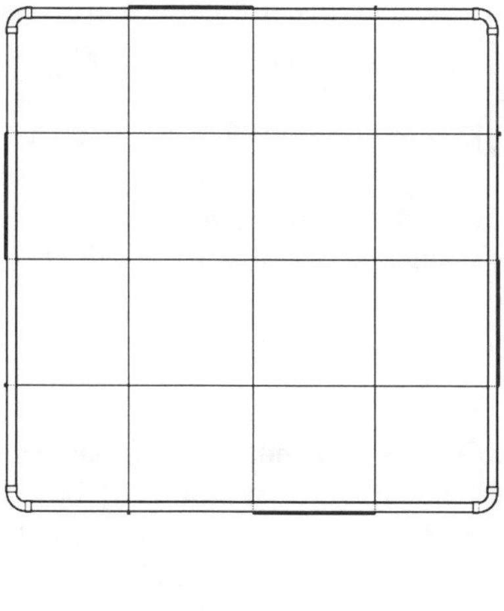

Figure 3.22 Frame strung into grids for sampling turf insects or their damage.

bugs, and billbugs. If pure Kentucky bluegrass sod was used, billbugs and white grubs will be common. By "mapping" neighborhoods with common characteristics, the wise lawn care manager can route better timing of controls into those neighborhoods that need them. In short, there is no sense in applying chinch bug insecticides in a billbug neighborhood.

Likewise, golf course superintendents commonly admit that certain putting greens or fairways have cutworms or white grubs, while others never have problems. Mapping and treating these high risk areas is wiser than "going wall to wall" with an application.

In order to accomplish lawn care pest mapping, train the staff to monitor for pests and record presence or absence. Obtain a large area map and "code" neighborhoods with high numbers of the same types of problem lawns. Use color-coded pins to mark different lawns in a neighborhood according to the type of pests found. Mark customer records with some codes which identify pest activity. Map neighborhoods with different aged lawns and try to identify the predominant turfgrasses present.

Pest mapping on golf courses and other larger facilities should be a simple matter. Insects and mites do not occur uniformly over a given area. Certain greens and tees will be attacked by cutworms, and only a few fairways will be damaged by white grubs. Keep in mind which areas have traditionally been problem spots. Use a "sampling crew" (two to three people who are trained to use proper sampling techniques) to scout the turf and determine if damaging populations are present, before the visual damage occurs. A crew of three can cover an 18-hole golf course in one day.

V. SELECTING APPROPRIATE CONTROLS

Pest Management versus Pest Eradication

Managing insects and mites which attack our turfgrass has generally relied on the use of pesticides. Whether this is good or bad is beyond the scope of this discussion, but we must ask whether alternative controls are available and appropriate. Before we can consider the alternatives, we should review our current concept of insect and mite pest management. Pest management as opposed to "eradication" implies that some pests will always be around. It is the goal of pest management to keep the pest populations down to a level where damage is not overly evident. In field crops, this has generally been termed an economic threshold level. In turfgrass management, the aesthetic threshold level (the population of a pest which causes noticeable, unacceptable, visual damage) is the term to be used.

Integrated Pest Management (IPM)

Integrated pest management or IPM is the selection, integration, and implementation of pest control (biological, chemical, and/or cultural) based on predicted economic, ecological, and sociological consequences (Figure 3.23). In other words, when a pest control action is used, we must consider the cost both to the ecosystem and human society. Using the IPM approach, three important concepts must be adopted:

A. No **single** pest control method will be successful. All of the control options—biological, chemical, and cultural must be used.

B. **Monitoring** (sampling) of the pest is constantly needed in order to evaluate the status (not present, present but not causing aesthetic damage, present and causing aesthetic damage, etc.) of a pest population.

C. **Mere presence** of a pest is not a reason to justify action for control.

There has been considerable misunderstanding about the IPM approach, IPM control options, and the underlying concepts. IPM is often called a "program," though it is really a method or way of approaching pest management. IPM is not a biological control or "organic" program, though biological controls are useful and organic materials can be used. It is not the "goal" of IPM to reduce or eliminate pesticide use, though pesticide usage is often dramatically reduced through pest monitoring and increased usage of biological and cultural controls. IPM is not the easiest nor least expensive management technique. Turf managers using IPM methods need the knowledge and training to make decisions, because IPM is a decision making process. The initial costs of investing time in monitoring and biological/cultural controls are eventually returned as improved turf and using fewer pesticides.

Monitoring in IPM

Monitoring techniques and tools have been discussed above, but the position of monitoring in IPM needs to be reemphasized. Monitoring pest activity and population levels is the key to successful IPM. Unfortunately, most feel that monitoring must be a complicated and time consuming process where someone must constantly watch each and every turf area. This is simply not true. Monitoring of pests in vast neighborhoods or even over large golf courses can be done in a multitude of ways—from visual inspection to using temperature-dependent (degree-day) developmental models. Another method

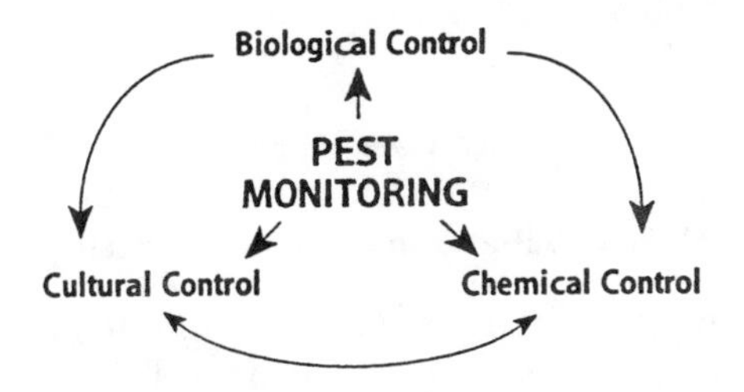

Figure 3.23 Integrated Pest Management (IPM) control options are selected after performing pest monitoring.

of solving the seemingly impossible task of monitoring pests in complex settings is the concept of **KEY PLANTS** and **KEY PESTS**:

- **Key Plants** are trees, shrubs, and turfgrasses that are known to have perennial pest problems. As an example, birch trees always get leafminers, aphids, and borers while red oaks rarely get significant pests. Tall fescues and bermudagrass can withstand most surface pests, while bentgrass and fine fescues readily succumb.

- **Key Pests** are those that cause significant damage or may kill trees, shrubs, or turf. These key pests often have special times (windows of opportunity) when they are susceptible to controls. Aphids or galls in oaks are rarely significant enough to warrant controls, while peach tree borers in ornamental plums need special attention. Clover mites may discolor turf foliage, but they rarely kill it. On the other hand, white grub and mole cricket populations have to be constantly monitored because they can kill vast areas of turf.

The Control Options

As mentioned above, IPM uses three general control options—biological, chemical, and cultural controls. These are our alternatives, and we must understand the benefits and limitations of each option. Since we are dealing with turf culture, many of the pest problems can be traced back to the direct result of poor maintenance. In other words, turf placed in urban habitats, pushed to perform in stressful conditions (e.g., mowed too closely, allowed to develop thick thatch layers, etc.), or not suitably adapted (e.g., fine fescue in the sun) are the ones most likely to be severely attacked by pests. Therefore, let us look at the Cultural Control option first.

Cultural Controls

The cultural control option should be our **first** consideration as an alternative in turfgrass IPM. Cultural controls in field crops have generally included: sanitation, crop rotation, tillage, host plant resistance/tolerance, mechanical/physical destruction, and quarantine. If we look at these techniques, we may wonder how they relate to turf culture. Though we use different terms, these techniques are commonly used and need to be emphasized more.

A. **Sanitation** helps remove inoculum or hiding areas of pests. Pruning, raking of leaves and destruction of heavily infested plant stock are sanitation techniques useful on our urban landscapes and nurseries. In turf, thatch removal and management are similar operations.

B. **Crop Rotation** is generally used in field crops (e.g., corn rotated with soybean) but should be considered for turf. Though we have little scientific knowledge on how a turf stand changes over time, anecdotal evidence suggests that original blends of various grass species are not maintained over time. Patches of light colored perennial rye will compete with darker bluegrass. Improved bermudagrasses may not compete well with coarse types, and uneven growth and color results. Likewise, many turf areas were developed without the benefit of modern, resistance factors. These older areas should be renovated and replaced when possible. Therefore, renovation is the equivalent of crop rotation.

C. **Tillage** in field crops exposes resting pests, and breaks up the soil for better air and water movement. In turf, core aeration, verticutting, and top dressing are similar processes.

D. **Host Resistance** uses plants which are less susceptible to pest attack (tolerance) or produce actual toxins (antibiosis) which kill or stop pest growth. This tactic can be one of the most important in turf culture. In fact, most insects and diseases which are currently problems can be greatly reduced with the use of resistant turf cultivars. For people concerned with the overuse of pesticides, this is the major option to be considered.

E. **Mechanical/Physical** techniques are as simple as crushing the pest under foot to using large industrial vacuum sweepers to suck up pests. In our turf plantings, we need to constantly remind ourselves that simple crushing of pests can be an effective management tool! Recent evidence has indicated that core aeration can reduce grub populations by half. Adapting tools and equipment to fully use this technique are still in need of development.

F. **Quarantine** is a legal method of restricting movement of contaminated plant material. Unfortunately, this technique is rarely effective even though we know that many pest problems arrive on infested plant material. Therefore, we should pay special attention to avoid use of sod which may be infested with billbugs or chinch bugs.

G. **Good Turf Management** is one of the simple but commonly ignored methods of pest management. In other words, a “healthy” plant can generally fend for itself against insects, mites, and diseases. Therefore, one of the most important control alternatives that we can use is tending to the proper needs of turf—water, fertilizer, and mowing.

Biological Controls

Biological control is using **parasites, predators** and **pathogens** (diseases) to control pests. We have to realize that in urban landscapes and turf covered areas, there is a multitude of beneficial insects and mites which can prey on pests. In many cases, these naturally occurring beneficials will do a good job of controlling the pests if we do not disturb the system too much. As stated above, we usually disrupt this system by overusing pesticides which kill the beneficials better than the pests. On the other hand, there are occasions where we can actually increase these biological controls. The classical way to use the biological control option is through introductions, conservation, and augmentation.

- **Introductions** of exotic parasites, predators, or diseases are made when foreign pests become established. This is an attempt to create some of the checks and balances found where these pests are naturally controlled. Occasionally, foreign biological controls are found which may better control native pests.
- **Conservation** is using other control tactics, usually pesticides, so that they have the least adverse effect on predators and parasites. It can also be the providing of habitat or food needed by biological controls to improve their survival. In turf, we can use targeted sprays to those specific areas where pests are getting the upper hand. We can also plant flowers around turf areas which provide nectar and pollen to feed the adults of many of the parasitic insects.
- **Augmentation** is usually the rearing and release of biological control agents. Unfortunately, this technique is usually expensive and we must use those biological controls that fit into the definition of a "good" biological control.

A "good" biological control is probably not the correct term. We should use the term useful. Useful parasites, predators, and pathogens have the following characteristics:

A. **High reproductive potential** allows a biological control to keep up with the pest populations which also have high reproduction abilities.
B. **Good mobility** of a biological control means that it will be able to search out the pests or come into contact with the pests.
C. **Host-specific** biological controls will have reduced chances of adversely affecting nontarget organisms.
D. **Persistent** organisms will remain when pest populations become low and will carry over from season to season.
E. **Easily reared or encouraged** biological controls are generally less expensive and can be economically competitive with other controls.
F. **Tolerance of other controls** is a requirement of all our control tactics. In order to fit into true IPM, biocontrols need to be tolerant of cultural and chemical controls, if used.

In order to illustrate these concepts, let's look at a preying mantis versus a lady beetle. The preying mantis has one generation per year, eats anything in sight (including each other and other beneficials), usually ignores the small insects such as aphids, mites, and scales, often doesn't survive the summer to lay another egg case and is very sensitive to any pesticide. Therefore, preying mantids **do not** qualify as a useful biological control. On the other hand, lady beetles have many generations per year, they only eat a narrow range of pests (usually they are aphid, mite, or scale specialists), usually overwinter well, and can often withstand some of the softer pesticides, especially soaps and oils. Therefore, lady beetles easily qualify as a useful biological control.

Unfortunately, we often think that we have to actively introduce predators, parasites, and pathogens in our urban landscapes. Since most of these animals already exist, we merely have to be able to recognize them and avoid using cover sprays of pesticides.

Important **predators** commonly found in turf are:

A. **Lady beetles** are commonly sold as adults and are useful control agents in ornamental plants and crops, if properly handled (Figures 3.24 and 3.25). Lady beetles found in turf often indicate that aphids or mealybugs are present. These species are often not the same ones sold on the commercial market.
B. **Green lacewings** are also found in turf where aphids, mealybugs or mites are active (Figure 3.26). The predaceous larvae feed on these pests and can be effective in reducing populations. Eggs can be purchased and sprinkled over the area. Many of the reared lacewings prefer crop habitats and are of dubious benefit in turf.
C. **Ground beetles** and **rove beetles** are some of the most active and common predators present in most soil/turf habitats (Figure 3.27). Both the adults and larvae feed on a wide variety of pests. Unfortunately, most of these beetles are highly intolerant of pesticides. Target pesticide applications only to the places where pests are reaching damaging levels in order to conserve these beetles.
D. **Bigeyed bugs** are very common gray to black bugs which are commonly

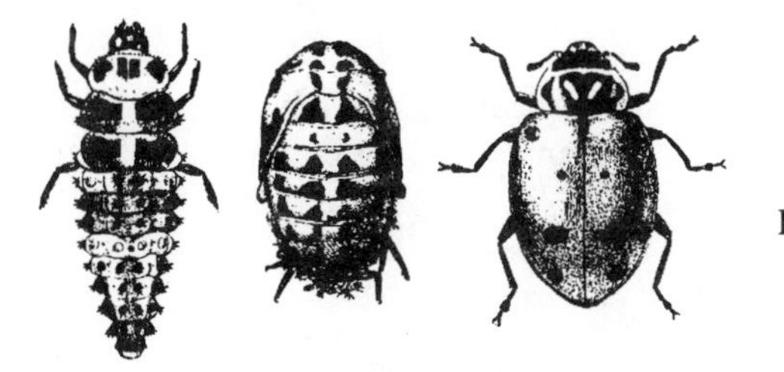

Figure 3.24 Convergent lady beetle, *Hippodamia convergens* Guerin-Meneville. (left to right) larva, pupa, adult (USDA)

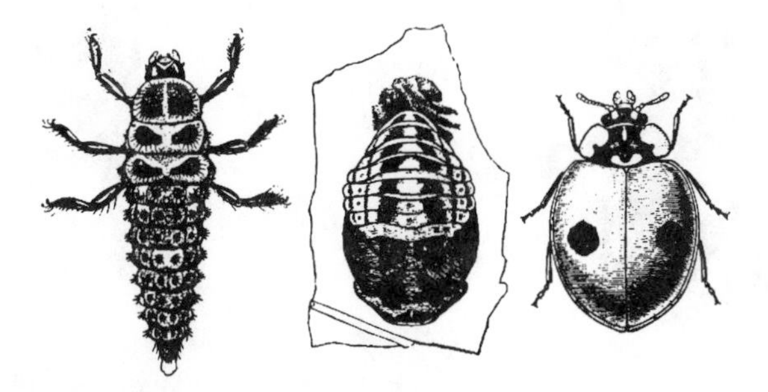

Figure 3.25 Twospotted lady beetle, *Adalia bipunctata* (Linnaeus). (left to right) larva, pupa attached to leaf, adult. (USDA)

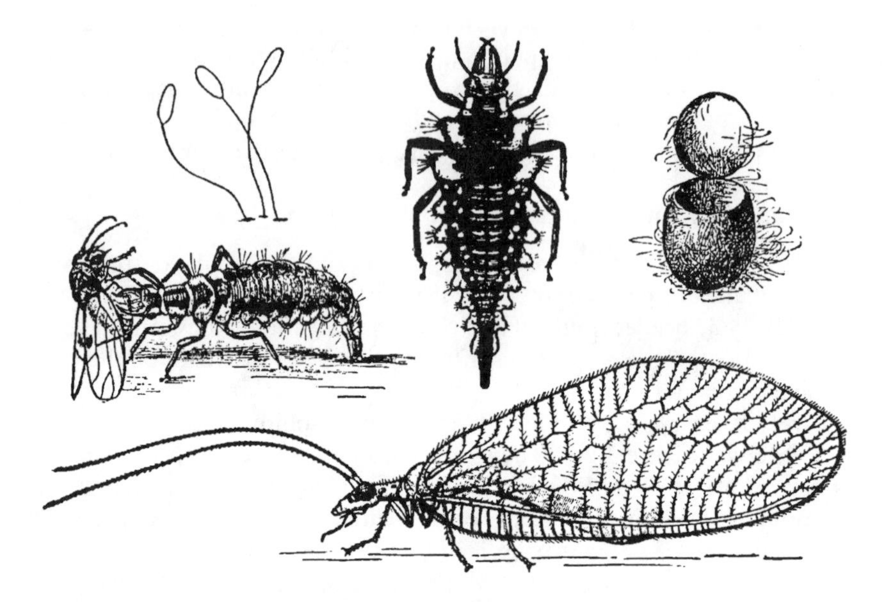

Figure 3.26 Green lacewing, *Chrysoperla oculata* (Say), stalked eggs, larva, opened cocoon, larva feeding on psyllid, and adult. (USDA)

confused with chinch bugs (Figure 3.28). The adults and nymphs prefer to frequent open or sparse turf areas where they suck dry any small insect unfortunate enough to wander by. Bigeyed bugs are common predators of chinch bugs.

Parasites are insects (often called parasitoids) with larvae which feed on the inside of their host, usually killing or sterilizing it. Some common parasites which frequent turf habitats are:

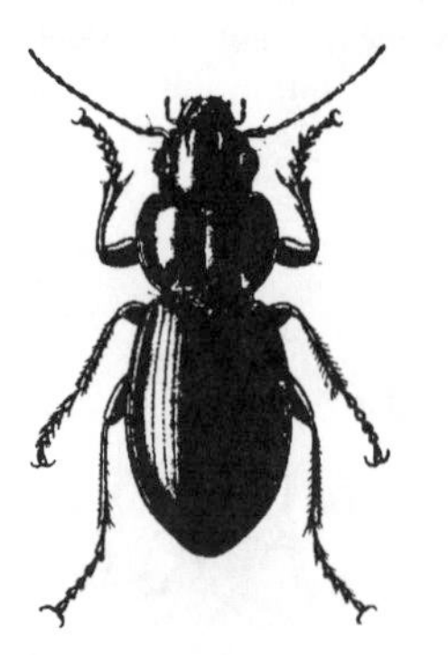

Figure 3.27 A ground beetle, *Evarthrus sodalis* LeConte, adult. (USDA)

Figure 3.28 A common bigeyed bug, *Geocoris bullatus* (Say). (after Rept. Ill. State Entomol.)

A. **Chalcid wasps** are usually very small to microscopic wasps which lay their eggs in the eggs, larvae, pupae or adults of other insects (Figure 3.29). Though not well studied or confirmed in turf insect pests, this group has been occasionally recovered from sod webworms, cutworms, billbugs, and mealybugs. In fact, an introduced species for control of the Rhodesgrass mealybug has been credited with keeping this pest under control in much of Texas. Because of their small size, these parasites go largely unnoticed.
B. **Tiphiid** and **scoliid wasps** are fairly large insects which attack white grubs but usually cause more alarm because of people's fear of being stung (Figures 3.30 and 3.31). Native species often keep masked chafer and green June beetle grub populations in check. They are usually hairy wasps with bright yellow or orange markings which hover over the turf during the summer months.
C. **Ichneumonid** and **brachonid wasps** are small to medium sized wasps which commonly attack caterpillars in turf (Figure 3.32). The larvae usually emerge from the dying host and spin small white or yellow cocoons. Small masses of these cocoons are occasionally found in the sparse turf left behind after cutworm or armyworm attacks.
D. **Tachinid flies** are medium sized, hairy flies which attack a variety of insects (Figure 3.33). In turf, species are known which attack May/June beetles, various caterpillars and mole crickets. The species attacking mole crickets was introduced from South America.

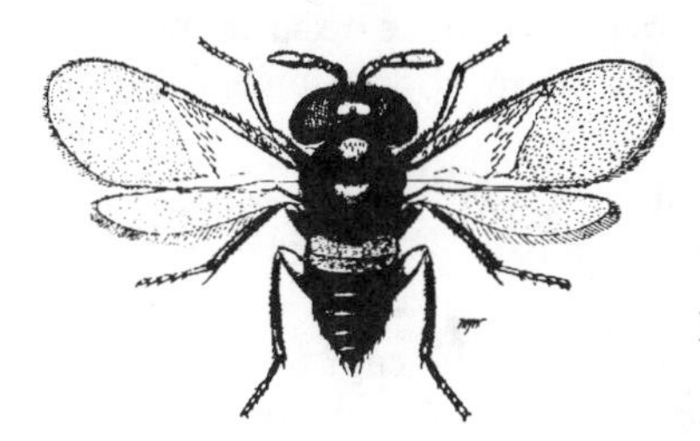

Figure 3.29 A chalcid wasp, *Aphelinus mali* (Haldeman), which parasitizes the greenbug. (USDA)

Figure 3.30 An imported tiphiid wasp, *Tiphia popilliavora* Rohwer, which attacks Japanese beetle white grubs. (USDA)

Figure 3.31 A scoliid wasp, *Scolia dubia* Say, which commonly attacks green June beetle larvae. (N. Carolina Agr. Exp. Sta.)

Figure 3.32 An aphid wasp, *Aphidius testaceipes* (Cresson), female laying egg in greenbug. (USDA)

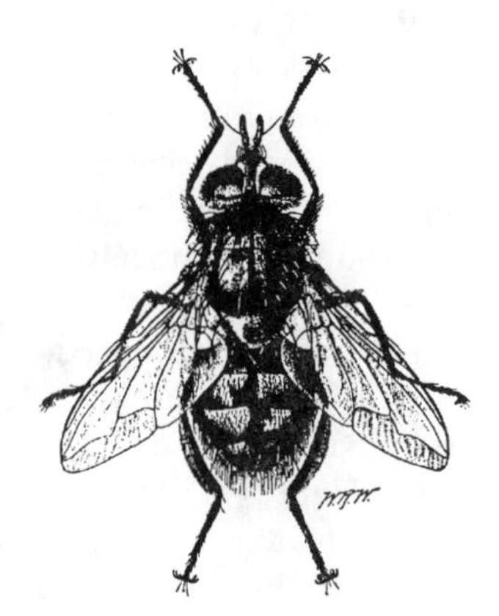

Figure 3.33 A typical tachinid fly, *Winthemia quadripustulata* (Fabricius), which commonly attacks armyworm larvae. (USDA)

Pathogens are simply a variety of disease-causing organisms which kill insects. They are usually bacteria, virus, fungi, and protozoa. Insect pathogens are nearly ideal in that they are very host specific. They are also very noninfective to vertebrates. Examples are:

A. **Bacteria** have been the easiest of the pathogens to utilize because they can often be reared "in vitro" (in artificial culture) and form spores fairly resistant to adverse environments. Examples are:

1. *Bacillus thuringiensis* (Bt)—has several strains which produce toxins lethal to various insect groups (and are thus technically a chemical control). The most common types are:

a. Bt '*kurstaki*' affects only young, leaf feeding caterpillars. Some strains are effective against armyworms and tropical sod webworms. Black cutworms and many of the standard sod webworms appear to be unaffected by these Bt strains, especially if they have reached detectable size in the turf.

b. Bt '*israelensis*' (= BTI) affects aquatic fly larvae such as mosquitos and black flies. These strains do not seem to be active against the turf infesting flies such as crane flies, Australian sod fly, and march flies. However, the BTI strains are useful around golf courses that have water hazards where mosquitos may breed.

c. Bt '*tenebrionis*' (= '*san diego*') affects certain leaf feeding beetles such as the elm leaf beetle. These strains do not seem to be active against white grubs.

2. *Bacillus popilliae* (= white grub milky disease) has one strain available which kills Japanese beetle grubs. Other strains have been identified which kill other species of grubs but these strains are not commercially available. Milky diseases seem to be "weak" pathogens which typically kill 30% to 50% of a

grub population. This can be significant in moderate grub infestations, but probably inadequate in heavy infestations.

3. *Serratia* sp. (= bacterial honey disease) has been recovered from various grubs but the only commercial preparations are currently used in management of the New Zealand grass grub. These are not available in the United States.

B. **Fungi** have been identified which attack a variety of turf infesting insects, but they are difficult to utilize because the spores are easily dried out or need high moisture and/or water to germinate. Examples are:

1. *Beauveria* spp.—have been identified infecting a wide variety of insects including bugs and beetles. A commercial strain is available in Europe for Colorado potato beetle control, and numerous companies have tried to develop commercial preparations for chinch bugs. Problems with formulating and maintaining viability have kept commercial preparations off the market.

2. *Metarhizium* spp.—have been identified infecting numerous soil insects including white grubs. No commercial strains are available in the United States.

C. **Viruses** are common pathogens of insects but are one of the most difficult to use because they require living insects to grow. Recent development of insect tissue culture has allowed for rearing of some of the virus strains but the only commercial product is nuclearpolyhedrosis virus (NPV) for gypsy moth control. Nothing is available for turf insect management.

D. **Entomopathogenic nematodes** are a group of tiny parasitic roundworms which carry a bacterium lethal to insects. Once a juvenile nematode gains entry into an insect it regurgitates the bacterium which paralyzes the insect. The nematode then feeds on the reproducing bacteria. Commercial products contain the infective juvenile stage (J3) of various species. Each species and strain of nematode seems to be most active against rather narrow groups of insects. These infective juveniles can be applied through conventional spray systems, but since they are living organisms, they need to be irrigated into the turf/thatch/soil before the application dries. The most commonly mentioned species are:

1. *Steinernema carpocapsae*, which has several strains good at attacking insects which live in the upper soil or on the soil surface. Biosafe®, Exhibit®, and Scanmask® are some of the commercial preparations.

2. *Heterorhabditis* spp. are better at attacking insects which live deeper in the soil. This group can also bore through the insect cuticle.

Chemical Controls

Chemical control has been considered to be the most successful pest management tool in turf maintenance. Unfortunately, we have overused and misused this option, so that many citizens are beginning to cast a weary eye on its use. Chemical control to most people means pesticides, though other chemicals such as attractants and pheromones are increasingly important in our IPM process. Even if pesticides are our principal weapon, we need to understand that not all pesticides are created equal. In IPM, we want to use the ideal pesticide—a material that only kills the target pest and has no other effect. Unfortunately, we don't have these "silver bullets." Most of the modern pesticides currently in use have short residual life spans (this reduces accumulation in the environment), are more selective (this reduces the chance of killing nontarget animals), and are used at lower rates (this reduces the total chemical "load" used). Because of these characteristics, we need to be able to better target our applications in order to control pests.

Another general public misconception about pesticides is that "natural" pesticides are better than "synthetic" pesticides. IPM principles do not make this distinction. Using pesticides in IPM is evaluated on economic, ecological, and sociological impacts together. In other words there are "natural" botanical insecticides (e.g., nicotine sulfate with an $LD_{50} = 55$ and a known carcinogen) which are much more toxic and have more adverse effects than some "synthetic" organic insecticides (i.e., acephate with an $LD_{50} = 866$). In short, chemical controls used in an IPM program should be selected on their total attributes.

By knowing that we do not have "ideal" pesticides, whether natural or synthetic, we must use great caution to limit their adverse effects. Generally, this means that we should only do **target applications** to those areas that need it—**not cover sprays**. General cover sprays (spraying everything, whether needed or not) tend to cause several problems. Cover sprays tend to tip the balance of control in favor of the pest. As incredible as this seems, cover sprays usually kill beneficial insects and mites (predators and parasites) better than they kill pests! Since pests usually have good reproductive ability, they "rebound" faster than their natural controls. This causes what we call **pest resurgence** and **secondary pest outbreak**. Cover sprays tend to cause development of resistance. Pests and potential pests often develop resistance to pesticides when they are under constant pressure from a specific pesticide. In other words, a few insects on a plant may not be causing significant damage, but if we constantly spray these insects, we are forcing them to develop resistance. Then, when they reach damaging levels, our pesticide is no longer effective. A more recently identified problem with general cover sprays of pesticides has been identified to be **enhanced** (accelerated) **degradation**. Since most of our current pesticides are organic compounds (i.e., containing carbon, hydrogen, and oxygen), microbes are able to use the chemicals as foods or nutrients. Generally these microbes are beneficial in aiding in the removal of these pesticides from the environment. However, when constantly "fed" through general cover sprays, these microbes "learn" to "eat" these pesticides more rapidly than normal. In summary, if we are going to use the chemical control option, we need to **use target sprays only when needed**.

In order to use the chemical control option to best manage turf-attacking insects and mites, the turf manager must have knowledge of the major chemical groups, and of the specific problems associated with using these compounds in the turf environment.

Chemical Control Groups

The chemical control option contains pesticides as well as repellents, attractants, or pheromones and desiccants.

A. **Pesticides** are chemicals which directly (i.e., acute toxicants) or indirectly (e.g., growth regulators) kill the target pest.

1. **Inorganics** are pesticides without carbon. They can be natural earth minerals or man-made compounds.

a. Boric acid is used for cockroach control inside houses or buildings, but is not registered for turfgrass usage.

b. Diatomaceous earth is the glass-like remains of single celled organisms, diatoms, which scratch insect cuticle or puncture gut cells. This material acts mainly as a desiccant, and is rarely useful in turf unless combined with an insecticide. This is probably because of the humid microclimate located within the turf canopy.

c. Elemental sulfur is an ancient control for insects and mites. Because of the large quantities needed for control, no products are currently available for usage against turf insects.

d. Heavy metal salts (e.g., mercury, lead, and arsenates) have been used extensively in the past, and until recently, were available for fungal and weed pest problems. These products are now generally considered too dangerous to use because of adverse effects on nontarget animals, as well as accumulation in the environment.

2. **Oils** are petroleum or plant based hydrocarbon chains that have insecticidal/miticidal activity. Toxic action appears to be accomplished by suffocation and/or cellular membrane disruption.

a. Petroleum oils are highly refined mineral oils used on dormant trees and shrubs (dormant oil) or even green plants (summer or horticultural oil). Petroleum oils have not been developed for turf usage because of the phytotoxicity problems.

b. Citrus oils (i.e., d-Limonene) have been shown to have insecticidal properties at low dosages. These are usually combined with soaps or other botanical insecticides. Some of these products are registered for surface insect control.

3. **Fatty acid salts** or **soaps** are generally man-made hydrocarbons which use an ion, usually potassium or sodium, to join together fatty acid chains. Fatty acid chains containing 6–10 carbons have insecticidal/miticidal properties. These soaps apparently kill by disrupting cell membranes. Soaps are useful for control of soft bodied insects such as aphids, mealybugs, and caterpillars, as well as spider mites. Soaps can cause phytotoxicity, so only registered products should be used for control of turfgrass pests.

4. **Microbial toxins** are molecules produced by bacteria, fungi, protozoa, and other microbes. Toxins like the Bt endotoxin are relatively low in toxicity to mammals, while botulism toxin is one of the most toxic molecules known to science. The toxins may be used by extracting the microbes from culture or by

using the whole organism. In either case, the chemicals produced by microbes are used, rather than expecting an infection from the microbes.

a. *Bacillus thuringiensis* (Bt) is a bacterium species that has numerous varieties which produce protein crystals toxic to a variety of insects and nematodes. Since these are considered biological controls by most, they have been more fully discussed above.

b. Avermectin-B (=Abamectin) is a powerful toxin derived from *Streptomyces* fermentation. This product is currently used in some miticides.

c. Chitin (=Clandosan®) is the chemical that makes up the exoskeleton of arthropods and nematodes. By adding chitin to the soil, microbes produce toxins (ammonia) and/or produce digestive enzymes that destroy the cuticle of insect and nematode pests. Field results in turf have not been consistent in efficacy.

5. **Botanicals** are plant extracts, usually alkaloid chemicals, that have insecticidal properties. Many people believe that since these are "natural" products, they are "safer" than other pesticides. Many of these chemicals can have strikingly adverse affects on mammals. Many cause severe allergic reactions (e.g., pyrethrin and sabadilla), have high toxicity (nicotine), or are even confirmed carcinogens (nicotine).

a. Pyrethrin is derived from a specific species of chrysanthemum originally grown in Iran. The natural product is an irritant to insect nervous systems and can cause quick knockdown, but many insects recover. To combat this phenomenon, pyrethrin is usually mixed with a synergist such as piperonyl butoxide (PBO) or rotenone to provide better kill of insects. Some people are very allergic to the material. Pyrethrin-containing products are available for turf, but little testing on efficacy has been performed.

b. Rotenone (=Cube, Derris) is derived from the roots of a tropical legume plant. The material is highly toxic to fish and was originally used by South American indians to collect fish for food. Pigs also have a strong reaction to the material. Rotenone products are generally registered for vegetable and flower garden pest control.

c. Sabadilla is an alkaloid derived from a South American lily seed. Though it has low dermal toxicity, it is a powerful irritant which, if inhaled, can cause severe circulatory and respiratory failure. This product is registered for certain vegetable plants only.

d. Nicotine is a highly toxic alkaloid ($LD_{50} = 55$) derived from tobacco and is a known carcinogen. Nicotine sulfate is still available for some garden plant pest control.

e. Azadirachtin or neem oil is an interesting botanical derived from an Asian tree grown in India and other countries. Neem oil is extracted from many parts of the plant, but the seeds provide the highest concentration of active material. Neem is used as a general cleaning chemical and is found in Asian toothpaste. It seems to act as a systemic with repellent and insect growth regulator effects. Several products containing azadirachtin are registered for turf caterpillar control.

f. Ryania is another alkaloid extracted from a tropical tree. The product has rather high oral toxicity, especially to dogs. It is registered for some vegetable crops only.

6. **Synthetic organics** are man-made compounds containing carbon, oxygen, and hydrogen, and are usually synthesized from petroleum-based compounds. This is the group most people refer to when they mention pesticide. Because of the diversity and number of materials in this group, no major attempt will be made to cover all of the products and compounds in this category. However, the reader is encouraged to consult other books on pesticides in order to become familiar with the actions and usage of these useful pest management tools.

a. Organochlorines (=Chlorinated hydrocarbons) are organics that usually have long residual life spans in the environment. This quality has caused most to be banned because they eventually end up in the food chain or cause damage to nontarget organisms. Chlordane and dieldrin were commonly used in the 1960s for grub control, which lasted several seasons.

b. Organophosphates usually have shorter residual life spans in the environment. Compounds in this category range from category 1 to 3 in toxicity and are generally neurotoxins. Most of the current insecticides/miticides registered for turf usage are in this category.

c. Carbamates usually have short or moderate residual life spans and range in category 1 to 3 in toxicity. Most are neurotoxic and several products are registered for turf usage.

d. Pyrethroids are synthetics that look and act like the botanical, pyrethrin. Pyrethroids range from category 1 to 3 in toxicity, though most are in categories 2 and 3. Where standard insecticides are used in pounds of active ingredient per acre, pyrethroids are used in fractions of pounds. Several pyrethroids are registered for surface insect and mite control in turf.

e. Insect growth regulators (IGRs) are man-made chemicals that look and act like natural insect hormones. For this reason, they generally have very low toxicity to mammals or other nontarget animals. IGRs work by disrupting insect and mite growth, often the molting process. IGRs are used at very low rates and, because of their mode of action, mortality is relatively slow. In turf insect management, IGRs have to be used when susceptible immatures are early in their development.

B. **Repellents** are compounds, both natural and synthetic, that cause a pest to stop feeding or move away. Most repellents are used as products applied to the skin or clothing to repel biting flies and ticks. Neem (see above) products appear to repel some insect feeding on plants. Various types of pepper, onion, and garlic extracts have been used to repel insects, but none of these products have demonstrated usefulness in turf.

C. **Attractants** and **pheromones** are compounds that attract a pest "thinking" that the compound is food or another of the species (aggregation and sex pheromones). Most of the compounds in this group have not been used effectively to reduce pests, but are used in traps to sample pest activity.

1. **Floral scents** such as geraniol + eugenol and pine oils have been used to trap various insects. The geraniol + eugenol mix is used in Japanese beetle traps (along with the sex pheromone). These traps do not reduce actual beetle damage to surrounding plants nor reduce grub populations. Though users of these traps get great satisfaction out of disposing of a bag of beetles, more beetles are attracted to the area than would normally be present, and only a fraction of the population is trapped. Assume that only three Japanese beetle grubs per square foot were present in the acre of turf surrounding the trap. What percent of the population was removed by the trap if 2000 beetles were captured? (Hint: There are 43,560 ft^2 in an acre.) Japanese beetle traps may be useful for monitoring adult activity.

2. **Moth sex pheromones** have been identified for a wide variety of field crop, ornamental plant, and turf pests. These pheromones have been used to disrupt mating or in traps for monitoring. Mating disruption occurs by releasing the pheromone, either as simple sprays or contained in plastic filaments or capsules, over a large area in order to confuse the males so that they cannot find the females also releasing the pheromone. This technique has been successful in several field crops. Sex pheromones are available for several turf-infesting caterpillars such as the armyworm complex, black cutworm, and cranberry girdler. Other moth pheromones are being discovered and synthesized for other turf moth pests.

3. **Scarab sex pheromones** have been identified for several of the adult of white grubs. The Japanese beetle sex pheromone has been added to the Japanese beetle trap to enhance the trap's efficiency. Pheromones are being developed for masked chafers, Oriental beetles, Asiatic garden beetles, and the green June beetle. These pheromones should be useful in traps for monitoring the flight of these pests.

D. **Desiccants** are materials that cause the insect pests to lose water faster than they can replace it. Since insects are very small, this water loss is rapidly lethal. Unfortunately, most desiccants must be kept dry, so turf use is not possible.

1. **Silica gel** is the same drying agent used in packing or flower drying. It can be ground into a powder or dust which is placed in dry areas where ants or cockroaches frequent.

2. **Diatomaceous earth** can act like a desiccant when dusted on the exterior of insects. The sharp edges of this product abrades away the thin coat or waterproofing wax on the exoskeleton of insects.

Using Pesticides to Manage Insects and Mites in Turf

Using insecticides and miticides appears to be a relative simple process, but many problems can be encountered, especially in the unique turf environment. Besides targeting the correct "window of opportunity," these pesticides have to deal with thatch and soil organic matter, volatilization, degradation from ultraviolet light or microbes, spray hydrolysis, tank mix incompatibilities, and pest resistance (Figure 3.34).

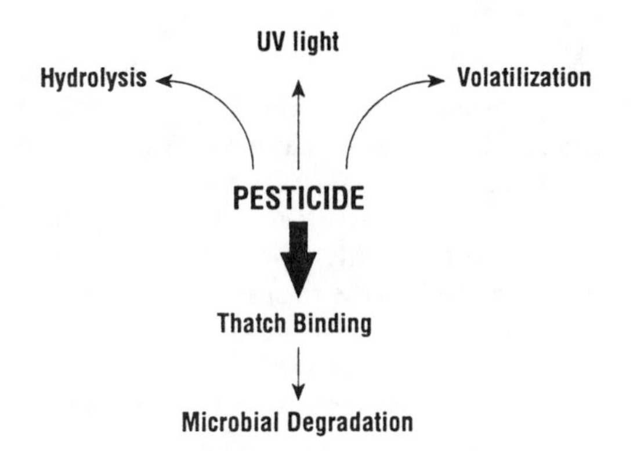

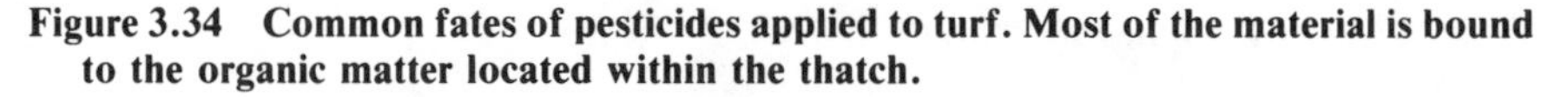
Figure 3.34 Common fates of pesticides applied to turf. Most of the material is bound to the organic matter located within the thatch.

Influence of Thatch

Most turf varieties tend to form a layer of dead leaves and stems held together by grass roots and stolons—thatch. Though a thin layer of thatch seems to be beneficial to the growth of the turf, thick layers can cause the turf to grow aboveground and can contribute to disease and insect problems. This layer of living and dead organic material can severely restrict the movement of pesticides. In recent studies on the movement of insecticides through this layer, over 95% of all insecticide types get bound to this layer. Therefore, if the target pest is white grubs or mole crickets, very little insecticide will reach the target if thatch is in the way.

Studies which have used pre- and postapplication irrigation, wetting agents, and core aeration as aids for moving the insecticide through the thatch layer have met with mixed or no improvement in efficacy. Turf managers should attempt to keep thatch to a minimum. Thatch can also be compacted with traffic, and partially decomposed thatch with a dry fungal layering can be totally impervious to water based insecticide applications.

Volatilization

Many insecticides and miticides are liquids at room temperatures. These compounds can volatilize (evaporate), and volatilization generally increases with increased temperature. Emulsible concentrates, when applied to turf foliage and not irrigated shortly after application, may have considerable product loss. Flowable and granular formulations generally reduce these problems. In any case, irrigation as soon as possible after a pesticide application will help move more product into the thatch/soil zone. On the other hand, pesticides applied for foliar feeders such as sod webworms or cutworms should be left on the turf foliage. Ideally, these products should be applied as late in the day as possible, after the turf canopy has cooled and air movement is reduced.

Ultraviolet Light Degradation

Exposure of many chemical compounds to ultraviolet radiation causes changes in the chemical bonds. This may break the compound into inactive molecules or cause it to combine with other chemicals which also inactivate the pesticide. Pyrethroids, insect growth regulators, microbial pesticides, and botanical insecticides are often highly susceptible to UV degradation. Manufacturers may add UV blockers to their formulations, but it is wise to water in applications of susceptible products before significant exposure to direct sunlight occurs. Again, if these products are being targeted for control of foliar pests, apply them in the late afternoon so as to avoid direct sun exposure.

Accelerated Microbial Degradation

Numerous microbes, usually bacteria and single celled fungi, are active in the thatch and soil, where they help recycle complex organic compounds. These microbes break down complex compounds into components which can be used as energy-supplying food or building blocks of growth. These microbes are usually divided into two broad categories—anaerobic and aerobic. Anaerobic microbes have the ability to be active in the absence of oxygen, while aerobic microbes must have access to oxygen. Fortunately, most anaerobic microbes should not be encountered in the turf habitat unless the turf or soil become water saturated for a considerable time. The "black layer" found in certain golf course soils is the result of this anaerobic condition. Of more importance to pesticide degradation are aerobic microbes. When these microbes are constantly challenged with the same organic compound (a pesticide), especially in a moist, highly organic environment (thatch/soil), the microbes often "learn" how to break down the organic compound and use the constituent parts.

At present, certain pesticides appear to be very susceptible to accelerated microbial degradation, while others seem to be fairly resistant. In any case, accelerated degradation is always a threat, and wise and judicious use of insecticides should be exercised.

Degradation management usually includes: using the pesticide no more than once in a season, alternating the pesticide class with other chemical classes, and applying only the recommended amount at the best time.

Chemical Hydrolysis

Hydrolysis is the general term used in chemistry to indicate that hydrogen atoms are being added to a compound. The addition of hydrogen usually breaks double bonds in a molecule, replaces other atoms, or generally alters the chemical structure. This alteration often causes the chemical pesticide to be much lower in toxicity or completely nontoxic. Most pesticides begin hydrolysis when subjected to extremes of alkalinity (i.e., sweet or high pH) or acidity (i.e., sour or low pH). Alkaline hydrolysis in a tank mix is one of the most common reasons for pesticide deactivation. Though most pesticides that are extremely susceptible to alkaline hydrolysis have buffers added to their formulations, periodic measurement of a tank mix is a wise effort. Most tank mixes should be kept near normal (pH = 7) or slightly acidic (pH < 7).

Generally, if you are using highly alkaline water, mix the pesticide with the water, agitate for a few minutes, and take a pH reading. Meters for measuring pH are relatively inexpensive, or pH indicator papers can be used. If the tank mix remains near pH 8 or above, the addition of a commercial buffer or acidifier may be in order. Most pesticide labels now contain information on when a pH adjustment is needed. If other chemicals have been added to the tank, especially fertilizers, shifts in the pH can occur as agitation and warming of the tank mix progresses during the day. If the tank mix is not to be used completely, shortly after mixing, take periodic pH readings to see if the mix needs to be adjusted.

Tank Mixing

In order to save time, insecticides are often mixed with fertilizers, fungicides, herbicides, or wetting agents. Occasionally, pesticides are not compatible with some of these other materials, and either deactivation of the pesticide occurs or some kind of undesirable settling or gelling can occur. Since most pesticide companies no longer provide mixing compatibility charts, it is wise to perform a formulation compatibility test before mixing a large quantity. In a quart container, mix a test batch of the materials to be combined. Stir and shake well and let set for 10–15 minutes. If a precipitate settles out (especially if wettable powders or flowables were not used) which is difficult to suspend, do not use the mix. Likewise, if a gel or greasy mass forms on the edge of the container, the mix is probably incompatible. Unfortunately, mixing compatibility tests do not indicate whether any pesticide deactivation has occurred. If you do not obtain expected results from a pesticide mixed with other materials, suspect chemical incompatibility.

Pest Resistance

Pest resistance to pesticides is a well-documented phenomenon. When enough of a population is challenged by a pesticide, those few animals that survive often have a genetic factor which allowed them to survive. These survivors pass on this factor to their offspring and the next generation has more individuals carrying the resistance factor. If this is allowed to continue, eventually most of a pest population carries one or more resistance factors to the pesticide, and they are no longer susceptible to the compound.

Resistance seems to occur where pesticides remain in the environment for a long time (thus the pest is constantly under pressure to change) or a large proportion of the population is continually exposed (all the crop, or turf, is sprayed with the same material several times in a season). Obviously, good methods of managing this problem are to use pesticides with relatively short residual periods, apply a pesticide only when needed, apply the pesticide only to the area needing pest reduction, and alternate products.

At present, since the loss of the long residual organochlorine pesticides, few turf insects have been found with resistance to other classes of pesticides. However, certain pests such as the southern chinch bug and green bug are notorious for having resistance to organophosphate and carbamate insecticides.

Insecticide/Miticide Effects on Nontarget Animals

Though herbicides, fungicides, insecticides, and miticides are all pesticides used to manage pests, insecticides and miticides have the unique position of acting on the nervous systems or metabolic pathways in animals. Because of their mode of action, insecticides and miticides may have dramatic effects on nontarget animals. When used according to label instructions, most insecticides and miticides will have only temporary and minimal adverse effects on nontarget animals. However, turf managers need to be constantly aware of the potential for problems.

Most beneficial insects found in the turf environment, especially ground beetles and rove beetles, are very sensitive to insecticides. Fortunately, if insecticides are not applied over the entire area (a complete golf course, including the roughs, or an entire neighborhood), these highly mobile predators rapidly recolonize the treated area. However, if periodic, repeat applications are made, these predators can eventually lose their effectiveness. This is another reason why general insecticide cover applications should be replaced with targeted applications to those areas needing management.

Several pesticides, especially carbamates, appear to have adverse effects on earthworm populations. Earthworms are significant actors in the environment's decomposition of thatch. Therefore, even though earthworms can be periodic nuisance pests which cause "lumpy" lawns from their castings left on golf course greens, every attempt should be made to reduce the continuous, regular usage of earthworm destroying pesticides. Current research indicates that earthworm populations can recover rather rapidly if applications of such pesticides are restricted to once per season.

Several insecticides have warnings concerning bird and fish toxicities. For birds, the greatest concern is for herbivorous waterfowl such as geese and some ducks. Turf managers need to survey areas to be treated in order to see if lakes, ponds, or streams are nearby or waterfowl are known to forage in the area. Of greatest danger are granular insecticides which may be picked up as the waterfowl work the edge of a waterway, and surface and liquid applied insecticides which are allowed to remain on the turf foliage.

Fish and other aquatic inhabitants are additional problems for turf managers trying to deal with pests in turf near waterways. Most contamination of ponds, lakes, and streams comes from contaminated surface water. The two most common causes are applications of insecticides to areas surrounding water followed by too much irrigation or sudden rain storms, or applications of pesticides to water saturated soils followed by additional irrigation or sudden rain. Insecticides which require posttreatment irrigation for safety or efficacy reasons should be applied to turf with underlying soils which can absorb the required irrigation. If the soils are saturated, wait a few days until conditions improve. Though light rain can assist in the movement of insecticides into the thatch/soil zone, sudden summer downpours can be very unpredictable. It is often wiser to wait until the potential threat is gone before making the pesticide application.

Equipment for Making Insecticide/Miticide Applications

In order for insecticides and miticides to have their maximum efficacy, they must be applied at the right time and should be applied evenly at the correct rate. Therefore, selection of the appropriate application equipment by the turf manager is very important for consistent results. Most insecticides are applied as liquid sprays or dry, granular materials.

Liquid sprays are applied through large area applicators (usually agricultural-type spray booms) or rain droplet sized hand applicators. Boom sprayers usually apply at 20–40 gallons of spray per acre [187–374 l/ha], and hand applicators apply at 45–200 [420–1870 l/ha] gallons per acre (= 1–4 gallons/1000 ft^2). Boom sprayers and low-gallonage hand applicators often leave most of their application on the turf foliage. When emulsible concentrates (EC) and soluble powders are used with these applicators, most of the insecticide will remain on the turf foliage if allowed to dry before irrigation occurs. Therefore, a liquid applied EC targeted for white grubs should be irrigated in immediately after application, and preferably before the application is allowed to dry. Wettable powders and liquid or dry flowable formulation are somewhat less adversely affected by drying before irrigation occurs. On the other hand, if surface or upper thatch dwelling pests are the targets, boom sprayers and low gallonage hand applicators are entirely appropriate.

Granular products are generally applied using drop or broadcast spreaders. Granulars need to be applied uniformly over the area, and both drop and broadcast spreaders can cause problems in getting a uniform application. The drop spreader requires that each granule be calibrated for its individual size and weight. Drop spreaders get very uniform coverage but making sure that the application swaths are exactly aligned can be difficult, especially in large areas. Broadcast applicators (rotary, flaying arm, etc.) require granules of rather uniform size and density. Otherwise, larger and heavier granules are liable to travel further than smaller and lighter granules. Each broadcast applicator produces a unique pattern with each granular product. Careful calibration is needed in order to ensure a uniform pattern. Turf managers often use half rates and cover the turf twice, at right angles, in order to get the most uniform application.

Subsurface applicators are being used to inject or slit-insert both liquid and granular insecticides into the soil/thatch zone. In southern states, this has appeared to improve activity against mole crickets. However, applications for white grub management have been less consistent or even dramatically less in control than conventional surface applications. Subsurface applicators should reduce the chances of leaving insecticide residues on the turf surface, and should place more insecticide in the zone where the pest may make contact. Considerably more research is needed to bring this technology to its full usefulness.

In summary, there are multiple alternative control methods that can be used in the urban landscape. The concept of integrated pest management provides a framework in which to use all of the alternatives in a systematic fashion. Of most importance is the idea that we must monitor for pest problems and then select the best targeted control available.

LEAF AND STEM INFESTING INSECT AND MITE PESTS

This category includes those arthropods which feed on the upper leaves and stems of turfgrass plants. Many of these pests often hide in the thatch, others remain exposed on the leaf surfaces, and the rest hide in the spaces beneath leaf sheaths and between nodes. Those insects with chewing mouthparts eat entire leaves and stems. These are usually the larvae of various moths and butterflies. The rest of the pests have rasping or sucking mouthparts and include the mites, thrips, aphids, and mealybugs.

Leaf chewing pests leave behind ragged edges on the leaves, sunken spots in the turf and, in the case of severe infestations, they eat all the green material down to the brown thatch. Pests with sucking mouthparts tend to discolor the turf, leaving it yellowed, rusted, or blanched white in color.

BERMUDAGRASS MITE

Species: *Eriophyes cynodoniensis* Sayed. [Phylum Arthropoda: Class Arachnida: Order Acarina: Family Eriophyidae] (Figure 3.35; Plate 3–1).

Distribution: In North America, this pest is found in all of the states where bermudagrass is grown. Is has also been spread worldwide.

Hosts: Bermudagrass is the only known host.

Damage Symptoms: Damage is first noticed when bermudagrass does not have vigorous growth in the spring and is often yellowed. The turf appears stunted,

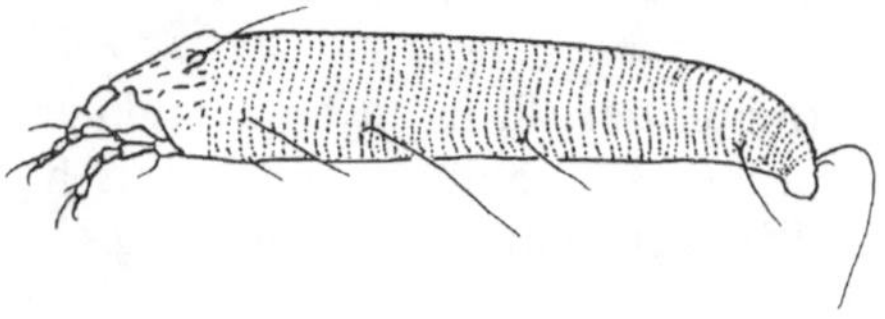

Figure 3.35 Bermudagrass mite, *Eriophyes cynodoniensis* Sayed. (Ariz. Coop. Ext.)

and close inspection reveals that the stem length between nodes is greatly reduced. Leaves and buds become bushy, forming a rosette or tuft which is called "witchesbrooming." Heavy infestations produce clumps which eventually turn brown and die.

Description of Stages: This mite has stages typical of the eriophyid mite group. These are extremely small, pear-shaped mites with wormlike, soft bodies.

Eggs: Round, translucent white eggs are about 0.002 inch [0.06 mm] in diameter.

Nymphs: The first nymph has the tapered shape of the adult but is almost clear. These mites have only two pairs of short legs, rather than the four pairs of legs found on most mites. The abdomen has minute rings which look like segments. The second nymph is about 0.005 inch [0.12 mm] long, and more white in color.

Adults: Only females are known. These look like the nymphs and are only 0.006 inch [0.2 mm] long when fully grown, but have a creamy white color.

Life Cycle and Habits: Because of their small size, these mites are very difficult to study, and little is understood about their life cycles and habits. Most eriophyids lay less than a dozen eggs during their adult span and these usually hatch in 2 to 3 days. At 75°F (23.9°C) it is estimated that adulthood is reached in 7 to 10 days, and eggs are laid for 2 to 5 days. Thus, a cycle can be completed in 1.5 to 2 weeks. This short time period allows for a rapid buildup of a population during summer temperatures. The bermudagrass mite seems to be quite tolerant of high temperatures, having moderate mortality at 120°F (48.9°C). Cold temperatures tend to stop development, though survival during the winter can take place where bermudagrass remains green at the soil surface. This mite can apparently spread by being blown or carried on the bodies of other insects. However, the most common method of spreading is by transportation of infested turf. The mites cannot survive on bermudagrass seed.

Control Options: Eriophyid mites usually are not killed by normal miticides, but are killed by many insecticides. Controls are probably warranted when 4 to 8 witchesbroomed tufts per ft^2 are encountered in an area.

Sampling

A 3 × 4 foot [0.9 × 1.2 m] plastic rectangle should be strung on one ft^2 grid (12 ft^2 total). The sampling hoop consists of four sections of 1-inch PVC pipe. Two sides are 3-foot segments and the others are 4-foot segments. The segments are joined with right angle connecters and glued. Holes are drilled through the sides at one, two, and three (on the 4-foot segments only) feet from each corner. A monofilament string is then threaded, in a grid pattern, through the holes. This hoop is tossed onto the turf and 10 of the ft^2 grids are rated for mite activity. Each tuft of turf with a witchesbroom is counted. Fairways and roughs should be sampled every 50 yards, and four hoop samples should be taken for each green and apron. Tees should have two hoop samples. When presence of the mite activity is noted, samples should be taken every month. If the activity is increasing and exceeds four to eight tufts per ft^2, chemical controls are probably warranted. Below four tufts per ft^2, cultural controls should be used.

Option 1: *Cultural Control—Use Resistant Varieties*—Common bermudagrass is often attacked, but improved varieties such as Tifgreen (238) and Tifway (419) have shown considerable resistance.

Option 2: *Cultural Control—Turf Maintenance*—The bermudagrass mite does not do well in short turf. However, mowing too short may scalp the turf. Good fertilization and water will help reduce stress and mask mite populations. Mite attacks are seldom damaging during wet periods.

Option 3: *Chemical Control—Soft Pesticides*—Though not specifically registered for this pest, some of the insecticidal soaps can be used on turf. Industry reports indicate that these soaps, when used with sufficient water to thoroughly wet the turfgrass blades and stems, are effective in controlling turf attacking mites.

Option 4: *Chemical Control—Traditional Pesticides*—Proper identification is needed because water stress can look like early mite damage. A microscope with at least 30× magnification will be needed to adequately see the mites. Short residual pesticides may have to be reapplied in 7 to 10 days to kill mites hatching from eggs. At present, diazinon (not on golf courses or sod farms), dicofol (=Kelthane®) and fluvalinate (=Mavrik®) are the only products registered for this pest. In the literature, chlorpyrifos (=Dursban®) has also been effective.

CLOVER MITE

Species: *Bryobia praetiosa* Koch, with several biotypes recognized. [Phylum Arthropoda: Class Arachnida: Order Acari: Family Tetranychidae] (Figure 3.36; Plate 3-2).

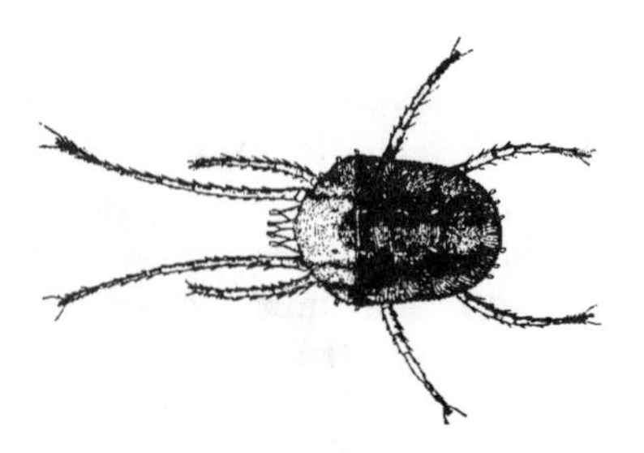

Figure 3.36 Clover mite, *Bryobia praetiosa* Koch, adult. (USDA)

Distribution: A cosmopolitan species found in North and South America, Europe, Asia, Africa, and Australia.

Hosts: This pest attacks a wide variety of plants, including several turfgrasses such as Kentucky bluegrass and ryegrasses.

Damage Symptoms: These mites rasp the surface of grass blades and leave a silvery appearance to the upper surface. The major problem with these mites is their nuisance activities. They tend to migrate into houses during population flushes in the spring and fall. Though they do not transmit any disease, nor do they bite, they leave a red stain when crushed which is difficult to remove.

Description of Stages: This mite has egg, larval, protonymphal, deutonymphal, and adult stages.

Eggs: The small round eggs are shiny red-orange colored and about 0.005 inch [0.12 mm] in diameter.

Larvae: The larvae have only three pairs of legs and are reddish in color.

Nymphs: The nymphs develop another pair of legs and have the typical adult form of a slightly depressed, oval body with elongate front legs. Under high magnification, the body is covered with tiny fingerprint-like ridges.

Adults: Only females are known, and these are reddish- to chestnut-brown in color. They have the front legs about twice the length of the other legs and are about 0.016 inch [0.4 mm] long.

Life Cycle and Habits: In the cool season turfgrass zones, this pest overwinters in the egg stage but adults can sometimes be found in protected areas. Clover mites also oversummer in the egg stage, being active during the cool of the spring and fall. In the southern zones, this pest oversummers in the egg stage but adults and nymphs are more common during the entire winter months. In the spring, when the temperatures get above freezing the eggs hatch, or in milder climates the adults become active and begin to lay eggs. This means that both eggs and adults may be present during the spring. The spring eggs hatch in a week at 28° to 48°F (-2.2–8.9°C), while overwintered eggs will hatch in 12 to 18 hours. Overwintering adults continue to lay eggs until mid-April, while new spring adults lay eggs which do not hatch until the following fall. The spring generation takes about a month to mature and is active from snow melt to mid-June. The oversummering eggs hatch in September and the mites mature in 25 to 35 days. These mites may lay additional eggs which hatch at this time, or the eggs may delay hatch until the following spring. The mites are strongly attracted to warm surfaces during cool weather and will climb up the sides of buildings and trees. On buildings, they may enter doors or windows and become a nuisance inside.

Control Options: This is a sporadic pest which has nuisance populations during favorable years when long cool springs or falls help build up populations. Generally little turf damage occurs, but home invasions may become unacceptable.

Option 1: *Cultural Control—Reduce Oviposition Sites*—Since this mite prefers to lay eggs away from the turf and on the sides of buildings and on the trunks of trees, creating wide mulch areas or making a border of small stone or gravel will reduce oviposition.

Option 2: *Chemical Control—Barrier Treatments*—Since the mites may enter houses, and populations often build up in the shady sides of buildings, treat the parameter turf of buildings with an appropriate miticide.

Option 3: *Chemical Control—General Turf Sprays*—Monitor mite populations during spring or fall when cool temperatures have lingered. If the mite populations are building, general cover sprays of a miticide will reduce the population and prevent house invasion. Use selective miticides which do not harm predatory insects and do not delay treatments in the spring, as the mites will lay summer eggs which are not susceptible to the miticides. Chlorpyrifos (=Dursban®), diazinon (not on golf courses or sod farms), dicofol (=Kelthane®), fluvalinate (=Mavrik®) and isazofos (=Triumph®) are registered for this purpose.

BANKS GRASS MITE

Species: *Oligonychus pratensis* (Banks). [Phylum Arthropoda: Class Arachnida: Order Acarina: Family Tetranychidae].

Distribution: Originally described from the Pacific Northwest, but now known to occur from Washington State to Florida and south. Also known from Hawaii, Puerto Rico, Central America, Mexico, and Africa.

Hosts: Commonly attacks Kentucky bluegrass in Washington State, Oregon, and Colorado, but switches to bermudagrass and St. Augustinegrass in the southern states.

Damage Symptoms: Lightly infested plants have small yellow speckles along the grass blades. As damage progresses the leaves become more straw colored and they eventually wither and die. This often happens in hot, dry spells and the damage may be mistaken for summer dormancy. This mite overwinters in all stages and in warm winters large numbers of mites may cause significant damage by the following spring. This mite produces considerable webbing at the bases of turf tillers and this may be easily seen in the morning dew. In St. Augustinegrass, the mite egg shells and webbing are sometimes mistaken for molds or dust. Concentrations of the mites on the tips of southern grasses cause general yellowing and dieback similar to heat or drought scorch.

Description of Stages: This is a true spider mite which has egg, larval, protonymphal, deutonymphal, and adult stages.

Eggs: The eggs are spherical and about 0.005 inch [0.125 mm] in diameter.

They are first pearly white but change to a light straw-yellow color before hatching.

Larvae: The larvae are oval in shape and have only three pairs of legs. When newly hatched, the larvae have red eye spots and salmon colored bodies. After feeding, the gut contents cause the body to become light green. The front pair of legs remain light orange in color.

Nymphs: Both the protonymphs and deutonymphs have four pairs of legs and continue to be bright green as they feed.

Adults: The adults are quite sexually dimorphic: the females are broadly oval and about 0.016 inch [0.40–0.45 mm] long, while the males have a strongly tapered abdomen and are only about 0.013 inch [0.33 mm] long. During the spring to fall feeding periods, the adults are bright green with light orange legs, but during the winter the green fades and the mites take on a bright orange-salmon color.

Life Cycle and Habits: This spider mite can be active any time during the year when temperatures are high enough. Usually, mature, mated females overwinter at the base of grass plants and in the soil, though a few males and nymphs may also be present. These overwintering females lose their green color and become orange-salmon colored. As spring arrives, the surviving mites begin to feed on emerging grass and gradually the mites turn greenish. After the normal color is established, the females begin to lay eggs in their copious webbing or on the plants. The females lay 50 to 70 eggs during their life span, though females laying over 100 eggs are known. The eggs take 4 to 25 days to hatch, depending upon the temperature. If turf temperatures are above 70°F [21.1°C], the larval stage takes only two days, the protonymph takes one day, and the deutonymph takes two days to mature. This means that during hot weather, eggs can hatch in four days and development may be completed in five days. During cool spring and fall temperatures, complete development of the immatures may take 25 to 37 days. This means that 6 to 9 overlapping generations may occur in a season. As soon as the adult females emerge from the deutonymph stage, males begin copulation. If no males are present, the females have the ability to lay unfertilized eggs which develop only into males. These males, in turn, can mate with the female, their mother, and she can then produce female offspring. Only mated females can produce female mites. During hot, dry summer weather, immature mites and sometimes adults migrate to the center of dormant grass clumps and rest until the grass returns to active growth following rains.

Control Options: Since this mite is a true spider mite, general miticides are needed for control; most insecticides are not effective.

Option 1: *Cultural Control—Watering the Turf*—This mite requires warm, dry conditions for optimum survival and reproduction. Irrigation at regular intervals seems to greatly reduce the mite populations.

Option 2: *Chemical Control—Spray Scheduling*—Since the eggs may take 2 to 3 weeks to hatch in cooler weather, reapplications of miticides based on their residual activity periods may be necessary. Set up a reapplication schedule which will catch any immatures that hatched since the last application. Also keep in mind that you want to control these immatures before they reach the adult stage and additional eggs are laid. Chlorpyrifos (=Dursban®), diazinon (not on golf courses or sod farms), dicofol (=Kelthane®) and fluvalinate (=Mavrik,Yardex®) are registered for mite control on turf. Other miticides may have special state registrations, so check with local authorities if this mite becomes a pest.

WINTER GRAIN MITE

Species: *Penthaleus major* (Duges). [Phylum Arthropoda: Class Arachnida: Order Acari: Family Eupodidae] (Figure 3.37; Plate 3–2).

Distribution: Common in small, grain-producing states west of the Mississippi River. Recorded attacking bluegrass and fescue in the Northeast. Also found in Australia, China, Europe, New Zealand, South Africa, South America, and Taiwan.

Hosts: Cool-season grasses such as bluegrass, ryegrass, and fescue; occasionally bentgrass.

Damage Symptoms: Damage is often misidentified as winter kill or snow mold. Turf leaves first appear silvered on the tips with more severe damage producing scorching and browning. Damage often occurs under snow, and brown patches appear when the snow melts.

Description of Stages: Typical mite life cycle except the eggs oversummer.

Eggs: Freshly laid eggs have a glistening reddish-orange color. These are glued to the bases of grass plants on the roots or on pieces of thatch. After drying for a day the eggs become wrinkled and more straw colored.

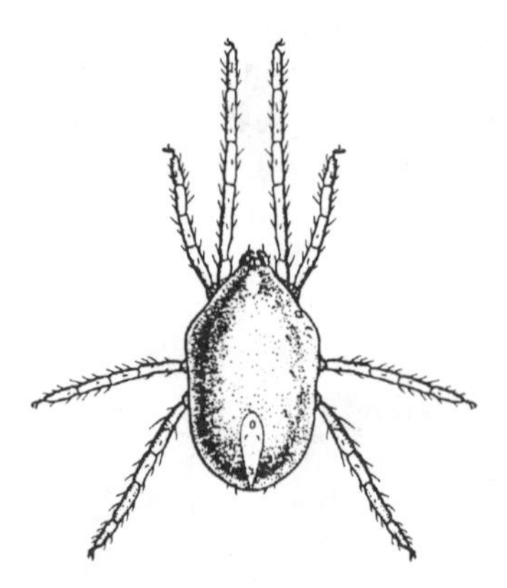

Figure 3.37 Winter grain mite, *Penthaleus major* (Duges), adult. (redrawn from Swan)

Larvae: As with all mites, the larvae have only three pairs of legs. Larvae are reddish-orange just after hatching but turn dark brown to black as they feed. The mouthparts and legs remain reddish-orange.

Nymphs: Two nymphal instars are found. When the larva molts into the nymphal stage, a fourth pair of legs is gained. The nymphs look like the adults, olive-black with reddish-orange legs, but are smaller. The nymphs also generally have a more tapered abdomen.

Adults: The adults are relatively large for mites, up to 3/64 inch [1 mm] long. They are the only turf-inhabiting mites with olive-black bodies, reddish-orange legs and mouthparts, a pair of white eye spots, and a dorsal anus. Only females are found, though males have been reported.

Life Cycle and Habits: The most distinctive feature about the winter grain mites' life cycle is the oversummering eggs and winter mite activity. In the northern United States, the mites appear to hatch in mid to late October when soil surface temperatures are approaching 50°F [10°C]. The larvae feed by rasping the surface of grass blades and sucking up the cell contents. Within a few days, the larva molts into the nymphal stage which feeds in the same manner for a week or two. The mites tend to hide during daylight and can be found clustered on the crown of grass plants, in the hatch, and at the soil surface during warm bright winter days. Apparently snow cover does not inhibit feeding and may actually afford protection. Females can live up to five weeks, during which they may lay 30 to 65 eggs. Eggs laid from November through March usually hatch that winter, but eggs laid from March on usually oversummer to hatch the following fall. It appears that two overlapping generations may occur during the winter, with peak populations being found in late December and late February. Winter grain mites often produce a droplet of liquid from the anus if disturbed. This may be a defensive action though undisturbed, feeding individuals also produce droplets. This mite is also very susceptible to desiccation, often becoming inactive and rapidly shriveling if moved to a dry spot. This mite also seems to be reactive to carbamate insecticides. Carbamates seem to kill natural mite predators as well as stimulate reproduction of some mites.

Control Options: Damage by this pest is almost impossible to predict. When turf has been treated in summer with a carbamate, such as in cutworm or webworm control, winter grain mites often attack during the winter.

Option 1: *Cultural Control—Mask Damage*—Since this pest rarely kills the turf, the normal damage is spring silvering of the turf. This can be rapidly masked by applying a dormant fall fertilization or applying a light spring fertilizer application in order to encourage rapid recovery of the turf.

Option 2 : *Chemical Control—Do Not Regularly Use Carbamates*—Alternate or use other types of pesticides to control summer turfgrass pests. This may reduce the amount of destruction of natural mite predators.

Option 3: *Chemical Control—Early Applications of Miticides*—Since it is hard to apply pesticides to turf under snow or in cold temperatures, treatment of the turf in the fall has been suggested. This has not worked well in the past, though new mite ovicides (egg poison) are being developed which may prove useful in the future.

Option 4: *Chemical Control—Spring Applications of Pesticides—* Most organophosphate insecticides will satisfactorily kill this pest in the spring when temperatures allow spraying. Usual miticides do not always kill this pest. Chlorpyrifos (=Dursban®) is currently registered for control of this pest.

GREENBUG

Species: *Schizaphis graminum* (Rondani) with several biotypes. [Phylum Arthropoda: Class Insecta: Order Homoptera: Family Aphidae] (Figure 3.38; Plate 3-6).

Distribution: Reported to have damaged turfgrass from Kansas to New York and south into Kentucky and Maryland; worldwide pest of cereal grains in Europe, Africa, and North America.

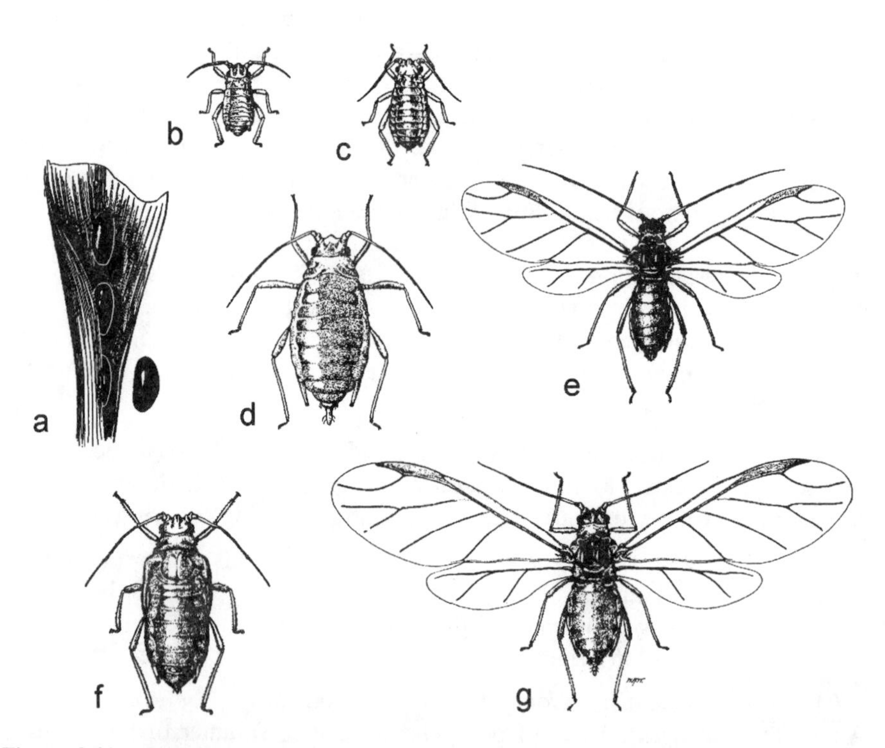

Figure 3.38 Greenbug, *Schizaphis graminum* (Rondani) life stages, a) eggs on leaf blade; b) first instar nymph; c) second instar nymph; d) "pupa" of viviparous female; e) wingless viviparous female; f) winged viviparous female; and g) winged male. (USDA)

Hosts: Different biotypes prefer wheat, sorghum, oats, and over 60 members of the grass family; prefers Kentucky bluegrass but will survive and reproduce on Chewings fescue and tall fescue.

Damage Symptoms: Individuals suck plant juices which rob the plant of nutrients and water. However, this aphid has a toxic agent in its saliva which causes the leaf tissue around the feeding site to turn yellow and then burnt-orange. Injury usually begins in shaded areas under trees or next to buildings, often along the north and east sides. This damage may then spread toward sunny areas with the general turf color changing from green to yellow to brown. Damage begins in June and may continue until a killing frost occurs. This pest seems to appreciate hot dry conditions, as rainy periods reduce population outbreaks. Severe damage is most evident in November but may also show up from late July through September.

Description of Stages: Though aphids technically undergo gradual metamorphosis, they have complicated life cycles in which the females give live birth to nymphs asexually (ovoviviparous, parthenogenesis) until late fall. At this time, sexual forms are produced and overwintering eggs are laid. Several other species of aphids may be found on turfgrasses, but most of these are dark or not green in color.

Eggs: The eggs are elongate oval, about 1/32 inch [0.8 mm] long, and are green when first attached to grass blades. They turn a shiny black in a couple of days.

Nymphs: These look like adults except they are smaller. The pear-shaped body is light green and usually has a darker green stripe down the back. The tips of the legs, antennae, and cornicles (the tail-pipe like structures on the abdomen) are black. Nymphs destined to become winged forms have obvious wing pads in the last instar and are called "pupae."

Adults: Adults are about 5/64 inch [2 mm] long and have the same green color of the nymphs, as well as black markings. Winged forms usually appear whenever crowding occurs, often after considerable damage has taken place. Winged adults are usually darker green and have the wing veins marked with black.

Life Cycle and Habits: Greenbugs were suspected to migrate into turfgrass from small grain, agricultural fields in the south and west. However, eggs of some of the biotypes have been found which can overwinter in northern climates, and nymphs have been found in northern turfgrasses too early in the spring to have been blown in. Also, certain lawns will be infested year after year, while adjacent lawns are not attacked. All aphids that hatch in the spring are females which reproduce asexually (parthenogenesis) by giving birth to first instar nymphs (ovoviviparity). In cool weather, these nymphs may take two weeks or more to mature but as summer temperatures rise, only 7 to 10 days are needed. Thus, populations can literally explode in a few weeks as mature females may produce 2 to 3 nymphs per day. Greenbugs seem to prefer the shade and build up populations in the shade of trees, buildings, and fences. Though greenbugs do best in the shade, they reproduce most rapidly in warm, dry weather. As daylight periods decrease in the fall, nymphs are produced which grow into winged sexual forms that can fly to new areas. Apparently these forms seek out suitable places to lay overwintering eggs. Greenbugs produce less honeydew than most of their relatives, but enough is often present to attract ants, bees, and other sugar-seeking insects.

Control Options: Many natural controls normally keep this pest in check, but these may be missing in turfgrass habitats. Some populations of this pest may also be resistant to organophosphate insecticides such as chlorpyrifos and diazinon.

Option 1: *Natural Control—Allow Natural Control Agents to Attack Greenbugs*—Lady beetles, lacewings, and parasitic wasps often seek out the greenbugs and effectively reduce populations if preventive insecticide sprays used to control chinch bugs, sod webworms, and cutworms have not been overused.

Option 2: *Cultural Control—Use Resistant Turfgrasses*—Perennial ryegrass, zoysiagrass, and bermudagrass are not attacked. New research indicates that endophyte-infected turfgrasses are generally resistant to greenbugs. Kentucky bluegrass clones which have been identified as being resistant to two greenbug biotypes are: Kenblue 46, A-34 (GB-3), Wabash 21, Delta 20, and Cougar 6. Others are being identified, so contact seed source companies for new releases. If a lawn has been so severely damaged by greenbugs as to require renovation, consider using a resistant turf variety.

Option 3: *Chemical Control*—Greenbugs are relatively easy to control with contact and systemic insecticides applied to active populations. However, some biotypes appear to have some resistance to organophosphate insecticides. If this resistance has been identified locally, application of carbamates, synthetic pyrethroids, or insect growth regulators may be useful.

Sampling

Greenbugs are most commonly detected by looking for the yellow-orange discoloration of turf around trees and next to buildings. By looking closely at this turf, the aphids can be seen lined up along the upper surface of leaf blades. Color-blind individuals may do better by taking an insect sweep net and using it in the area. The darker aphids readily show up next to the white cloth of the net. Sweep netting will also help detect the presence of lady beetles and lacewings.

Insecticides and Application

Currently, the systemic insecticide, acephate (=Orthene®), seems to produce the best results when applied to an active infestation.

SOD WEBWORMS (=LAWN MOTH)–INTRODUCTION

Species: Many species of sod webworms attack turfgrasses in North America. The most common type, several species which used to belong to the large genus *Crambus* (subfamily Crambinae), rest on grass blades during the day and characteristically roll the forewings tube-like around the body. The imported tropical sod webworms are more devastating to southern turfgrasses, hold the forewings roof-like over the body, and belong to the subfamily Pyralinae. Usually two or three species may be causing damage in any given area and species complexes vary across the continent. [Phylum Arthropoda: Class Insecta: Order Lepidoptera: Family Pyralidae].

Distribution: Different species are common across North America. Species which seem to prefer cool-season grasses are: bluegrass sod webworm, *Parapediasia teterrella* (Zincken); larger sod webworm, *Pediasia trisecta* (Walker); the western sod webworm, *P. bonifatellus* (Hulst); striped sod webworm, *Fissicrambus mutabilis* (Clemens); the elegant sod webworm, *Microcrambus elegans* (Clemens); and the cranberry girdler, *Chrysoteuchia topiaria* (Zeller). Some of these also occur in the warm season zones. Additional species of crambids are more common in warm season turfgrasses but the imported tropical sod webworm, *Herpetogramma phaeopteralis* Guerne, is the principal pest.

Hosts: All species of turfgrasses are attacked.

Damage Symptoms: The crambid types generally construct tunnels in the soil and thatch, lining them with silk. From these hiding places, they cut down individual blades of grass. This eventually gives a sparse and ragged appearance to the turf. Extensive infestations may lead to irregular brown patches of turf, especially in dry periods. The tropical sod webworm will web foliage and often feeds along the tips and edges of grass blades. High populations can literally mow down turf. Birds are commonly seen feeding where sod webworm populations are high.

Life Cycles: These pests have complete life cycles with eggs, larval, pupal, and adult stages. Species in northern areas have one to three generations per year, while southern species are inactive only during cold weather.

Identification of Species: Over 30 species of sod webworms have been identified in North America. However, only about half of these species commonly occur in turfgrass areas and even fewer species reach pest status. The adults are fairly easy to identify to species by using wing color patterns and male genitalia. The larvae are quite difficult to identify to species, and an expert should be consulted if larval identification is needed. The crambid types lay ribbed eggs by dropping them into the turf, and the tropical sod webworms attach flat scale-like eggs to blades of grass.

Control Options: Most sod webworms are easy to control, though they may be difficult to reach within their silken tunnels. Sampling can be done by using a disclosing solution over a square yard of turf and counting the number of emerging larvae. Adults may be captured using an insect net or light trap.

Option 1: *Cultural Control—Use Fertilizer and Water*—Damage can often be outgrown if water is continually available. Considerable damage may occur if irrigation is not possible during periods of drought, or close mowing is used.

Option 2: *Biological Controls*—Natural parasites are known, but ground beetles and rove beetles are major predators of eggs and smaller larvae. Fungal and viral diseases have also been identified, but these usually do not provide consistent control. The insect parasitic nematodes, *Steinernema* spp., seem to provide adequate control of this group when used at 1×10^9 juveniles per acre. Nematode efficacy can be improved by applying them in the early morning or late afternoon when sunlight is at a minimum, the thatch has been thoroughly moistened, and irrigation occurs immediately after application (before the spray droplets dry).

Option 3: *Cultural Control—Use Resistant Turfgrass Varieties*—Resistance against the tropical sod webworm has been observed in bermudagrass selections, and resistance to crambids has been demonstrated in bluegrass cultivars. Perennial ryegrasses, tall fescues, and fine fescues with fungal endophytes are also highly resistant to sod webworm attacks.

Option 4: *Chemical Control—Microbial Toxins, BT*—Several strains of the bacterium, *Bacillus thuringiensis*, have been shown to control sod webworm larvae. Bt products are most effective against young larvae.

Option 5: *Chemical Control—Use Contact and/or Stomach Pesticides*—Most webworms are easily controlled if the pesticides are ingested or penetration of the webbing tunnels is achieved. Since the larvae feed shortly after dark, best control is achieved by spraying in the late afternoon. Late fall or early spring applications are often not effective because many larvae are hiding in deeper soil chambers. Some species may require additional treatments to control second generation larvae produced by migrating adults.

Sampling

Sod webworm larvae (Plate 3-15) are sometimes difficult to confirm in turf. They may build webbed tunnels through the thatch or into holes in the ground. Visual inspection often reveals larger, sawdust-like fecal pellets (=frass) with silk webbing. Green frass indicates recent or current activity. A soap disclosing solution should use two gallons sprinkled over a one yd^2 area in order to force any caterpillars to the surface. Generally, 5 to 10 larvae per ft^2 may warrant control. Bird feeding may indicate sod webworms, but is not a confirmation of their presence.

Insecticides and Application

Acephate (=Orthene®), azadirachtin (=Turplex®), bendiocarb (=Turcam®, Ficam®), carbaryl (=Sevin®), chlorpyrifos (=Dursban®), cyfluthrin

(=Tempo®), diazinon (not on golf courses or sod farms), ethoprop (=Mocap®), fluvalinate (=Mavrik®, Yardex®), isazofos (=Triumph®), isofenphos (=Oftanol®), and trichlorfon (=Dylox®, Proxol®) are registered for sod webworm control. Liquid applications which are not irrigated have performed better than granular applications.

BLUEGRASS SOD WEBWORM

Species: *Parapediasia teterrella* (Zincken). [Phylum Arthropoda: Class Insecta: Order Lepidoptera: Family Pyralidae] (Figure 3.39).

Distribution: Found in the eastern half of North America.

Hosts: Prefers Kentucky bluegrass but also feeds on ryegrass, fine and tall fescues, as well as weed grasses such as crabgrass and orchardgrass.

Damage Symptoms: Closely mown turf shows symptoms more rapidly than poorly maintained turf. Individual larvae can cause small depressed pot marks of brown grass. As the larvae grow, these spots may enlarge into individual areas of several inches in diameter. Heavier infestations have the enlarging dead patches touching to form irregular patterns of sparse turf. Poorly maintained turf may have a general ragged appearance with many scattered dead stems. Birds often make probing holes in the areas of dead turf. Many infestations in golf course fairways and roughs as well as in home lawns are passed off as being summer dormancy.

Description of Stages: The stages are typical of moth complete life cycles.

Eggs: Cylindrical, with ends bluntly rounded and with 16 to 18 longitudinal ridges. These 0.02 × 0.012 inch [0.5 × 0.3 mm] in diameter eggs also have smaller cross ridges between the longitudinal ones. The eggs are pure white when laid and gradually change to a deep straw-yellow until they hatch.

Larvae: Freshly hatched larvae are 3/64–5/64 inch [1–2 mm] long, have blackish brown head capsules and a translucent yellowish body. After feeding, the body becomes greenish from food in the gut. The body spots are barely

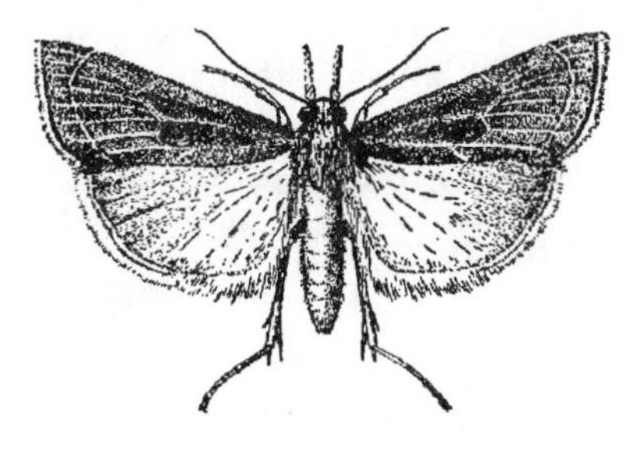

Figure 3.39 Bluegrass sod webworm, *Parapediasia teterrella* (Zincken), adult. (USDA)

visible, though minute hairs are conspicuous. As the larva molts and grows through a minimum of seven instars, the head capsule turns brownish yellow tinged with green, the body ground color becomes straw yellow, and the segmental spots are reddish brown. Mature larvae are 3/8–1/2 inch [9–13 mm] long.

Pupae: The amber-yellow pupa is nondistinctive and is 3/8 × 3/32 inch [8–10 × 2.5 mm].

Adults: The adults have a wing span of 9/16–13/16 inch [15–21 mm]. The palps and head are white above, and grade to a light brown below. The most distinctive feature of the forewings are the seven spots along the tip. A curved brownish-orange line runs across the wing just inside the tip, and the veins are usually distinctly lighter in color. The hind wings are lighter than the forewing, and have a narrow brown line around the margin.

Life Cycle and Habits: Adults can be found during most of the summer months, but peaks in adult flights in June and August suggest two generations are normal in the middle states. Larvae overwinter in silk-lined chambers placed in the soil or thick thatch. Mature larvae may feed briefly before pupating in mid-May. Partially grown larvae feed rapidly to mature, and pupate in late May and early June. The pupa may take 5 to 15 days to mature, depending on temperatures. The adults emerge from the pupa after dark, expand their wings, and usually mate in the middle of the night. Some couples may remain together after sunrise but most finish mating by daylight. Those who have not mated the night of emergence usually accomplish this task the following night. Males often die within a day and adult females live only 5 to 7 days. Dry conditions may contribute to shorter life spans. Females begin laying eggs the night after mating by hovering over the turf. They rarely fly higher than two feet. Maximum egg laying occurs about an hour after sunset and may continue for a couple of hours. A female may lay 200 eggs before expiring. The eggs have no adhesive and tend to work into the thatch. The eggs can hatch in 5 to 6 days at 70°F [21.1°C] or above, but take longer at lower temperatures. The larvae hatch by breaking open the end of the egg and soon spin some webbing along a leaf blade. Here the larvae eat the surface tissues, and after a molt or two drop to the ground to form a larger tubelike silken tunnel. This tunnel has pieces of thatch attached and often has piles of green fecal pellets near its opening. Older larvae usually feed at night and come to the opening of their tunnel to clip off blades of grass or entire stems. The larvae usually take about 40 to 45 days to mature during the summer. Second generation larvae maturing in the fall dig deeper into the thatch or soil to overwinter in a thicker silk-lined chamber. Extra molts may occur in the spring if larvae do not obtain enough food during the fall feeding period.

Control Options: See: Sod Webworms—Introduction.

LARGER SOD WEBWORM

Species: *Pediasia trisecta* (Walker) [Phylum Arthropoda: Class Insecta: Order Lepidoptera: Family Pyralidae] (Figures 3.40 and 3.41; Plate 3-14).

Distribution: Common in the northern half of North America. Most common in the bluegrass and tall fescue growing regions.

Hosts: Seems to prefer Kentucky bluegrass but may attack ryegrass, fine and tall fescues.

Damage Symptoms: See: Bluegrass Sod Webworm.

Description of Stages: The stages are typical of the moth complete life cycles.

Eggs: Elongate oval, with ends rounded and with 14 to 19 longitudinal ridges. The 0.02 × 0.012 inch [0.5 × 0.3 mm] eggs also have faint cross striations between the long ridges. The eggs are light yellow when laid, and turn brownish-yellow before hatching.

Larvae: Newly hatched larvae are 1/16 inch [1-2 mm] long and have a reddish-brown head capsule marked with black. The body is pale yellow but turns reddish, especially toward the posterior. As the larva grows through a minimum of seven instars, the head capsule becomes brownish-yellow with darker markings, and the general body color becomes yellowish with green food contents. The body spots are distinctly chocolate-brown. Mature larvae are 1.0 inch [24–28 mm] long.

Pupae: The light brown pupa is nondistinctive and about 7/16 × 1/8 [11 × 3 mm].

Adult: The adults have a wing span of 7/8–1 3/8 inch [21–35 mm]. The head,

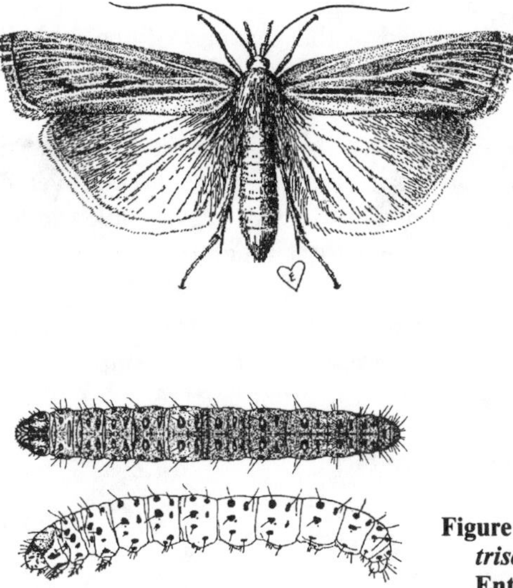

Figure 3.40 Larger sod webworm, *Pediasia trisecta* (Walker), adult. (USDA)

Figure 3.41 Larger sod webworm, *Pediasia trisecta* (Walker), larva. (Rept. Ill. Sta. Ent.)

palps, and thorax are straw colored and speckled with brown tipped scales. The forewing color is rather nondistinctive and varies considerably from almost solid cream to grayish with light colored veins. The tip has silvery grey scales interrupted with white scales where the veins end. Usually three small black spots are present at the tip.

Life Cycle and Habits: This webworm appears to have 2 to 3 generations per year, with peak adult flights in mid-June, late July, and mid-September. However, some adults may be found from mid-May to early October. In the Pacific Northwest, this species is listed as being univoltine. The larvae overwinter in silk-lined chambers dug into the ground beside turf roots. Feeding resumes for a short time in the spring, and pupation occurs in May. Within 10 to 20 days the adults emerge. The adults emerge at night, with males appearing early after dark and females emerging a couple of hours later. Most emergence occurs from 9:00 PM to 1:00 AM and by 1:00 AM males are swarming about the turf in search of receptive females. Swarming and mating continues until dawn. Mated females begin laying eggs the following night, beginning a half hour after sunset and continuing for a couple of hours. Females lay an average of 180 eggs, ranging from 80 to 220, over 4 to 6 nights. Eggs dropped by the females hatch in 5 to 8 days, depending on temperature. The young larvae crawl about until finding a suitable grass blade upon which to build a first home and begin feeding. A loose net-like web is spun in the fold of the blade and feeding is done by eating the tissue in a trough between two veins. After feeding, the larva builds a tighter web over the feeding area and attaches its fecal pellets. After molting three times, the larva no longer can nest in the blade's groove and soon drops to the base of the plant. Here it forms a new webbing tunnel running along the ground and incorporating soil and thatch. The larva feeds at night by cutting off blades of grass and dragging them into the tunnel. While feeding on a fresh grass blade at one end of the tunnel, green sawdust-like fecal pellets (frass) are deposited at the free end of the tunnel. When this tunnel becomes too fouled with frass, a new tunnel is often constructed in another direction. Several of these abandoned tunnels may be constructed before the larva matures. At maturity, the larva forms a cocoon of silk spun near its feeding tubes. The cocoon is elongate oval and is covered with soil and thatch pieces. The cocoon is lined inside with grayish silk and an opening exists at one end. Larvae not maturing early enough to emerge in the fall overwinter in cells constructed in the soil or thatch. These cells are roughly spheres of silk in which the tightly coiled larvae rest. In summer, the larvae takes 30 to 50 days to mature, depending on temperatures. The pupal stage then develops for another 10 to 20 days before the adult emerges.

Control Options: See: Sod Webworms—Introduction.

WESTERN LAWN MOTH

Species: *Tehama bonifatella* (Hulst). [Phylum Arthropoda: Class Insecta: Order Lepidoptera: Family Pyralidae].

Distribution: Commonly found in the Rocky Mountain plateau west to the Pacific Coast. It appears to be most common from Washington through California.

Hosts: Prefers Kentucky bluegrass, perennial ryegrass, fine fescue, and bentgrass. It may move over to bermudagrass in southern climates.

Damage Symptoms: See: Bluegrass Sod Webworm.

Description of Stages: The stages are typical of moth complete life cycles.

Eggs: Slightly barrel-shaped, with about 35 longitudinal rows of depressed plates, approximately 1/16 × 7/128 inch [1.58 × 1.35 mm]. The eggs are yellow when laid and gradually change to a light orange-purple at hatch.

Larvae: Freshly hatched larvae are 3/64 inch [1.2 mm] long, have brown head capsules and a translucent yellowish body. After feeding, the body becomes greenish from food in the gut. The body spots become distinctly dark by the fourth or fifth instar. As the larva molts and grows through a minimum of seven instars, the head capsule turns dark brown with darker mottling. Mature larvae are about 5/8 inch [15 mm] long.

Pupae: The yellow, turning to brown, pupa is nondistinctive and is 5/16 × 3/32 inch [8 × 2.5 mm].

Adults: The adults have a wing span of 11/16–15/16 inch [17–23 mm]. The palps and head are white above and grade to buff below. The most distinctive feature of the buff colored forewings are the three elongate spots running along the middle. The forewing ends with a chevron mark, followed with a series of black dots. The hind wings are darker than the forewing and have a silvery cast.

Life Cycle and Habits: Adults can be found during most of the summer months, but peaks in adult flights indicate that four to five broods occur each summer. The second and third broods, in July and August, appear to cause the most significant damage. Late fall larvae overwinter in silk-lined chambers placed in the soil or thick thatch. Mature larvae may feed briefly before pupating in mid-May. Partially grown larvae feed rapidly to mature, and pupate in mid- to late April. The pupa may take 5 to 15 days to mature, depending on temperatures. The adults emerge from the pupae after dark, expand their wings, and mate during the first night. Males often die within a day or two, while adult females live only 5 to 7 days. Dry, hot conditions may contribute to shorter life spans. Females begin laying eggs the night after emerging. They drop eggs into the turf canopy while hovering between 12 and 24 inches. Maximum egg laying occurs about an hour after sunset and may continue for a couple of hours. A female may lay 100 to 300 eggs before expiring. The eggs have no adhesive and tend to work into the thatch. The eggs can hatch in 5 to 10 days, depending on the temperature. Newly hatched larvae eat the surface tissues of leaves, and after a molt or two, drop to the ground to form a larger tubelike silken tunnel. This silken tunnel may have plant debris attached and often has piles of green fecal pellets near its opening. Older larvae prefer to feed at dusk or dawn. The larvae usually take about 30 to 40 days to mature during the summer. A complete generation takes about 45 days, and three to four generations can occur over the summer. The second and third generations appear to be the largest in numbers. These larvae are feeding in June and late July to early August.

Control Options: See: Sod Webworms—Introduction.

TROPICAL SOD WEBWORM

Species: *Herpetogramma phaeopteralis* (Guenee). [Phylum Arthropoda: Class Insecta: Order Lepidoptera: Family Pyralidae] (Plate 3–16).

Distribution: Found worldwide in the tropical zones. Attacks turfgrasses in Florida and the Gulf States, where it may not freeze during the winter.

Hosts: Known to attack St. Augustinegrass, bermudagrass, and centipedegrass. Also may attack zoysiagrass and bahiagrass.

Damage Symptoms: Outbreaks seem to appear overnight when the larger larvae begin to forage on the aboveground turf leaves. Early symptoms are ragged edges on the leaves, followed by stripping of the foliage much like armyworm damage.

Description of Stages: The stages are typical of the moth complete life cycles.

Eggs: Unlike the barrel-shaped eggs of crambids, these eggs are flat and scale-like. The eggs are oval, 0.02 × 0.028 inch [0.55 × 0.7 mm], and clustered in small masses. They appear translucent white, and become brownish red before hatching.

Larvae: Look very much like crambid larvae, but can be distinguished by the arrangement of spots and head markings. The head capsule is dark, yellowish brown and the body is a cream color which is tinted with green by the food in the gut. Mature larvae are $^{5}/_{8}$–$^{3}/_{4}$ inch [15–20 mm] long.

Pupae: The reddish-brown pupa is nondistinctive and about $^{3}/_{8} \times {}^{3}/_{32}$ inch [9.0 × 2.6 mm].

Adults: The adults are very different from crambids because they do not roll the forewings around the body. Instead the wings are held roof-like over the body, and the general body shape looks much like a swept-wing jet. Adults may be dingy brown to straw colored and have irregular darker marks. The wing span is about $^{13}/_{16}$ inch [20 mm].

Life Cycle and Habits: This webworm has continuous generations during the year and only slows down its cycle during cooler temperatures. However, outbreaks seem to occur most commonly during the hotter summer months following rainy spells. Females lay eggs on grass blades or overhanging vegetation of ornamentals during the early night and these eggs generally hatch in 6 to 10 days when temperatures are in the 70's°F [21.1°C]. Young larvae usually spin webbing in the V-shaped depression of a blade of grass and feed by removing the tissues between veins. The larvae continue feeding in this manner for the first three or four instars and move to new blades of grass if needed. These early instars take from 13 to 18 days, depending on the temperature. At this stage of development the larvae begin to feed on the edges of the grass blades and leave a ragged edge. However, this is still rather hard to detect in healthy turf. The last instars, 7 and 8, feed for 10 to 12 days by eating entire leaves from the grass plants. This is the time when they begin to be noticed, as the larvae seem to work like a living mowing machine radiating from the area where the eggs were deposited. In fact, the last instar larvae can eat over ten times the amount of grass per day compared to the amount eaten by the fourth instar larvae. The larger larvae are active only at night and hide below the grass surface during the day. By parting the turf at the edge of the damage margin, the resting larvae can be found curled up at the soil surface. The larvae also tend to leave trails of silk when they move from one grass blade to another and this is easily seen in the morning if a dew is present. At maturity, the larvae spin a loose bag of silk in which they pupate. This bag usually incorporates bits of grass and debris on the surface, rendering it hard to detect. Within this cocoon, the pupa rests for 7 to 14 days, depending on the temperature. The adult moths emerge at night and do not fly during the day unless disturbed. The adults mate in the evening and seem to require a nectar meal in order to survive for any time, usually no more than two weeks. The adults are strongly attracted to lights but do not become active at low temperatures. In fact, this insect is very sensitive to low temperatures. Larvae which mature in 25 days at 78°F [25.6°C] take up to 50 days to mature at 72°F [22.2°C]. Likewise, pupae take twice as long to emerge at the lower temperature. This pest seems to overwinter with difficulty where temperatures get to freezing during the winter. Populations seem to decline when soil temperatures drop below 60°F [15.6°C], and this may account for the absence of any major damage until late summer populations build up.

Control Options: Also see: Sod Webworms—Introduction.

Option 1 : *Cultural Control—Use Resistant Turfgrass Varieties*—'Common', PI-289922, Fb-119 and 'Ormond' varieties of bermudagrass (*C. dactylon*) had much less foliar damage than 'Tifway' and 'Tifgreen' (*C.* × *magenissi*). Other varieties should be checked for resistance. No information is available on resistance in other southern turfgrass species.

Option 2: *Biological Control—Use BT Microbial Toxins*—the use of *Bacillus thuringiensis* has had mixed results against this pest, and repeat applications may be needed to prohibit damage from reinvading larvae. Some of the *B.t.* var. *spodoptera* strains, such as Javelin® and Steward®, have had better activity against this pest.

GRASS WEBWORM

Species: *Herpetogramma licarsisalis* (Walker). [Phylum Arthropoda: Class Insecta: Order Lepidoptera: Family Pyralidae].

Distribution: Suspected to have originated from southeast Asia. It is currently found in those countries and adjoining islands and Australia. The pest was first found in Hawaii in 1967 on Oahu. It has since spread to the other islands of Hawaii—Hawaii, Kauai, Maui, and Molokai.

Hosts: Kikuyugrass, *Pennisitum claudestimum* Hochst ex Chior, seems to be the major preferred host, though 13 other grass hosts are known in Hawaii. Bermudagrass, sunturf bermuda, centipedegrass, and St. Augustinegrass are common turfgrasses attacked.

Damage Symptoms: The larvae feed on leaves, stems, and crowns of turfgrasses. Initial feeding produces grass blades with a ragged edge, but continued feeding produces irregular brown patches. Considerable webbing and frass (fecal pellets) are usually very evident.

Description of Stages: This insect's stages are typical of moths with complete life cycles.

Eggs: The eggs are flat, oval, and scalelike. They measure 0.02 × 0.35 inch [0.6 × 0.9 mm] and can be laid singly or in groups. They are usually attached to the upper surface of leaves, along the midrib. They start out being creamy white and change to dark orange before hatching.

Larvae: The larvae are light green to brown in color and have the conspicuous rings of dark brown spots, typical of pyralid larvae. There are five instars, with the first being about $^{3}/_{32}$ inch [2.3 mm] long and the mature larva about $^{13}/_{16}$ inch [20 mm] long.

Pupae: The light brown pupae turn dark brown just before the adults emerge. They are placed in a tightly woven silken case and are about $^{13}/_{32}$ inch [10.5 mm] long.

Adults: These moths are fawn to light brown in color, and the wings have faint darker spots and a zig-zag line near the outer margin. The body is about $^{3}/_{8}$ inch [10 mm] long and the wing span is $^{15}/_{16}$ inch [23–24 mm].

Life Cycle and Habits: Adults apparently emerge from their pupal cases at night, and mating occurs during the same night. The females need 3 to 6 days before they are ready to lay eggs. They usually feed on flower nectar and moisture on leaves. Once the preoviposition period is over, an average of 250 eggs are laid over 5 to 7 nights. The eggs take 4 to 6 days to hatch. The newly hatched larvae feed on leaf upper surfaces. They strip away strips of the upper tissues, leaving the lower epidermis intact. The larvae molt every two days, and the third instars begin consuming entire leaf margins. The last instar, the fifth, takes over four days to mature. The larvae hide in silken tunnels in the turf thatch during the day and emerge to feed at night. The larval development takes 10 to 14 days. Mature larvae construct a silken hibernaculum-like cocoon. This structure has frass and plant debris attached. The pupae take 6 to 7 days to mature before the adults emerge. A generation takes about 32 days to complete. The adults can be found in large numbers resting in tall grasses or in low shrubs and plants alongside turf.

Control Strategies: See: Sod Webworms – Introduction. Grass webworm outbreaks can occur at any time during the season but are usually associated with rainy weather. Periodic sampling for the larvae should be performed if adults are noticed flying about. Remedial controls may be needed when 10 to 15 larvae per yd^2 are found.

Option 1: *Cultural Control – Resistant Turfgrasses* – Egg laying preference and larval feeding studies indicate that Sunturf bermudagrass, Tifdwarf bermuda, St. Augustinegrass, and centipedegrass are the most attractive to this pest. Common bermudagrass and Tifway bermuda are more resistant to attack.

Option 2: *Biological Control – Conserve Parasites* – The tiny wasp, *Trichogramma semifumatum* (Perkins), attacks the grass webworm's eggs. Up to 96% of the eggs in an area may be attacked. Various other larval parasites have lesser effects. Delay or reduce applications of traditional insecticides in order to conserve these parasites.

CUTWORMS AND ARMYWORMS – INTRODUCTION

Species: Several species of thick bodied, nonhairy caterpillars may attack turfgrasses. In the northern zones, the black cutworm, *Agrotis ipsilon* (Hufnagel), bronzed cutworm, *Nephelodes minians* Guenee, and variegated cutworm, *Peridroma saucia* (Hubner), are the most common. In southern and transition zones, the above species as well as the granulate cutworm, *Felta subterranea* (Fabricius), the armyworm, *Pseudaletia unipuncta* (Haworth), fall armyworm, *Spodoptera frugiperda* (Smith), and yellowstriped armyworm, *Spodoptera ornithogalli* (Guenee) are found attacking turfgrasses. Other species are occasionally found in turfgrasses, though often in lesser numbers. [Phylum Arthropoda: Class Insecta: Order Lepidoptera: Family Noctuidae] (Figures 3.42 to 3.47).

Distribution: The above-mentioned cutworms are found all across North America, and some are worldwide in distribution. The armyworm is also

Figure 3.42 Variegated cutworm, *Peridroma saucia* (Hubner), adult. (after Rept. Ill State Entomol.)

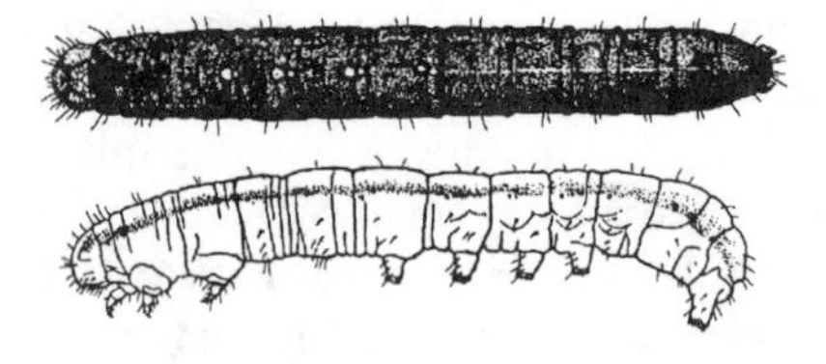

Figure 3.43 Variegated cutworm, *Peridroma saucia* (Hubner), larva. (after Rept. Ill State Entomol.)

Figure 3.44 Bronzed cutworm, *Nephelodes minians,* Guenee, adult. (after Rept. Ill State Entomol.)

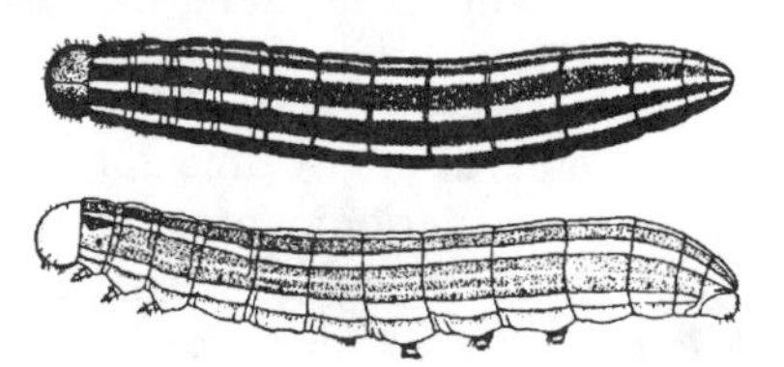

Figure 3.45 Bronzed cutworm, *Nephelodes minians* Guenee, larva. Dorsal view (upper) and lateral view (lower) (after Rept. Ill State Entomol.)

found worldwide, but is most common in the United States east of the Rocky Mountains. The fall armyworm is a tropical and semitropical species that migrates from Central America and the Gulf States northward every summer; it can be found as far north as Montana, Michigan, and the New England states by summer's end.

Hosts: All species of turfgrasses may be attacked.

Damage Symptoms: The true cutworms are semi-subterranean pests. They usually dig a burrow into the ground or thatch and emerge at night to chew off grass blades and shoots. This feeding damage often shows up as circular spots of dead grass or depressed spots which resemble ball marks on golf greens. The armyworm and fall armyworm feed on the turf surface, leaving a ragged, torn

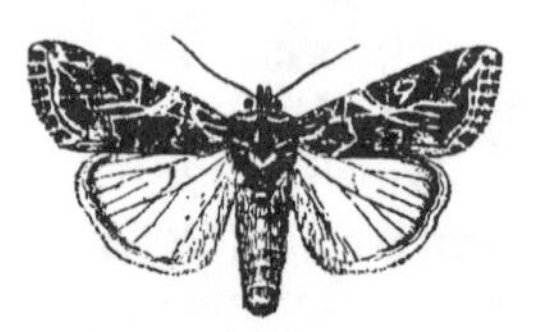

Figure 3.46 Yellowstriped armyworm, *Spodoptera ornithogalli* (Guenee), adult. (after Rept. Ill State Entomol.)

Figure 3.47 Yellowstriped armyworm, *Spodoptera ornithogalli* (Guenee), larva. Notice typical dark spot on side of first abdominal segment. (USDA)

appearance to the turf. Heavy infestations can result in the turf being eaten down to the ground by a moving "wave" of caterpillars.

Description of Stages: Cutworms and armyworms are the larvae of moths and thus have complete life cycles with eggs, larvae (caterpillars), pupae, and adults.

Eggs: The eggs are usually round, 0.02–0.03 inch [0.5–0.75 mm] in diameter, flat on the lower surface, bluntly pointed at the top, and often with sculpturing, lines and ridges, on the surface.

Larvae: The larvae have generally hairless bodies except for a few bristles scattered over the body. Besides the three pairs of true legs, these caterpillars have five pairs of fleshy prolegs on the abdomen. Most cutworms have characteristic markings on the head and body which aid in species identification. Full grown cutworm larvae are $1/4$ inch [6.0 mm] wide and $1\,3/8$–2.0 inches [35–50 mm] long. Most cutworms and the armyworm coil into a spiral when disturbed.

Pupae: The pupae are brown, reddish-brown, or black and $1/2$–$7/8$ inch [13 to 22 mm] long. The antennae, wingpads, and legs are firmly joined together, but the abdomen is free to twist around if the pupa is disturbed.

Adults: The adults are generally dull colored moths with wing spans of $1\,3/8$–$1\,7/8$ inches [35 to 45 mm]. At rest, the wings are folded flat over the abdomen.

Life Cycle and Habits: The armyworm, black, bronzed, variegated, and granulate cutworms appear to overwinter as larvae or pupae in the northern states. The fall armyworm overwinters only in southern states in all stages of development. In the northern states, the nonmigrating caterpillars may have 2 to 4 generations per year, while in the southern states 3 to 7 generations may be found, depending upon the length of the season. The bronzed cutworm has a single generation per year. Females mate and feed at night on flowers of trees, shrubs, and weeds. Mated females seek out crops or grasses and lay clusters of eggs on the leaf blades. A single female may lay 300 to 2000 eggs over several days. Under optimum conditions, these eggs hatch in 3 to 10 days and the young larvae begin to feed on the leaves. The cutworms excavate a hole into the ground or use existing holes, such as aeration holes, to hide during the day. From this retreat, the larvae venture forth at night to feed on plant material. Often they will drag a leaf or stem back to their burrow to feed on during the daytime. The armyworms do not remain in a hiding place but may hide during the day under a leaf or in the thatch. Older armyworms do not even hide during the day but feed continuously. Most of these species take 20 to 40 days to complete their larval development. The pupae may be located in the cutworm retreat or in the case of the armyworms, in the thatch. The pupa takes about two weeks to mature. These developmental times may be greatly lengthened during the cooler parts of the season.

Control Options: These pests can generally be controlled by using one of the contact or stomach pesticides. However, if populations are high in surrounding areas, such as in field crops, continual reinfestations may occur. Because of this, contact with local cooperative extension services or pest management consultants may be beneficial in determining local pest activity.

Sampling

Young cutworms and armyworms are difficult to locate, so use of a disclosing solution is beneficial in determining population pressure. Use two gallons of soapy water applied to one square yard (1.0 yd^2) [4.5 L/m^2] area. Within 3 to 5 minutes the caterpillars will come to the surface and can be easily counted. Remedial controls are needed if 5 to 10 larvae per yd^2 of fairway or lawn turf are found. Thresholds on golf putting greens and tees may only allow tolerance of less than one cutworm or armyworm per yd^2.

Option 1: *Cultural Control—Use Resistant Turfgrasses*—In transition and northern turf, using endophyte-enhanced tall and fine fescues or perennial ryegrasses can eliminate cutworm and armyworm problems. Several cultivars of bermudagrass are also resistant to yellowstriped and fall armyworm attack. If turf is to be renovated because of past cutworm or armyworm attack, consider using resistant turfs.

Option 2: *Biological Control—Use BT*—The bacterium, *Bacillus thuringiensis*, is commercially available and adequately controls caterpillars if enough material is ingested by young larvae. Use of this product has had mixed results in turfgrass management because of the problems involved with mixed sizes of larvae. Use of this microbial toxin is best when the caterpillars are in the first to third instars. Do not wash the material off of the turf foliage after the application.

Option 3: *Chemical Control—Use Poison Baits*—These caterpillars often will feed on grain baits, and these can be used when populations are causing reinfestations. Be sure to have the cutworms identified, because some species do not accept grain baits. This technique is more commonly used in agricultural fields or pastures.

Option 4: *Chemical Control—Sample Populations and Spraying*—If a disclosing solution test has indicated considerable activity (5 to 10 or more larvae per yd^2), a pesticide application may be needed. On golf course putting greens and tees, only a few cutworms or their damage spots may be tolerated. The armyworms are easily controlled by contact/stomach poisons as long as the larvae are actively feeding. Some problems with cutworm control has been noted. Apparently mixed age populations may be present, and short residual insecticides will not kill any larvae that hatch several days after the initial application. Observe the sizes of larvae flushed out and if small to mature larvae are present, reapplication may be necessary in a week. For improved efficacy, apply pesticides in the afternoon or evening and do not irrigate after the application. Since most of the control is obtained through the caterpillars

Figure 3.48 Black cutworm, *Agrotis ipsilon* (Hufnagel), adult. (after Rept. Ill State Entomol.)

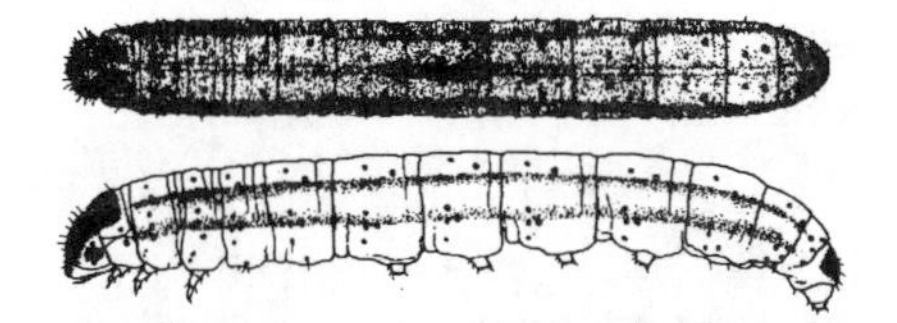

Figure 3.49 Black cutworm, *Agrotis ipsilon* (Hufnagel), larva. Dorsal view (upper) and lateral view (lower) (after Rept. Ill State Entomol.)

ingesting the pesticide, liquid sprays are usually more effective than granular applications. Acephate (=Orthene®), azadirachtin (=Turplex®) carbaryl (=Sevin®), chlorpyrifos (=Dursban®), cyfluthrin (=Tempo®), diazinon (not on golf courses or sod farms), fluvalinate (=Mavrik®, Yardex®), isazofos (=Triumph®), isofenphos (=Oftanol®), and trichlorfon (=Dylox®, Proxol®) are registered for this group.

BLACK CUTWORM

Species: *Agrotis ipsilon* (Hufnagel), is also called the greasy cutworm and the dark sword grass moth. [Phylum Arthropoda: Class Insecta: Order Lepidoptera: Family Noctuidae] (Figures 3.48 and 3.49; Plate 3-17).

Distribution: The black cutworm is found all across North America but is also found in Europe, Asia, and Africa. Though this pest has trouble spending the winter where soil temperatures may reach below 15°F, adults are strongly migratory and northern regions can become reinfested each summer.

Hosts: All species of turfgrasses may be attacked.

Damage Symptoms: This is a true cutworm and has semisubterranean habits. They usually dig a burrow into the thatch/soil or use existing cracks and crevices or aeration holes, from which they emerge at night to clip off grass blades and shoots. This feeding damage often shows up as circular spots of dead grass or depressed spots which resemble ball marks on golf greens.

Description of Stages: Cutworms are the larvae of moths and thus have complete life cycles with eggs, larvae (caterpillars), pupae, and adults.

Eggs: The eggs are usually round, 0.02 inch [0.5 to 0.6 mm] in diameter, flat on the lower surface, and bluntly pointed at the top. When freshly laid, they are greenish but become tan to dark brown before hatching.

Larvae: The larvae have generally hairless bodies except for a few bristles scattered over the body. The general body color is gray to nearly black on the upper half, and slightly lighter gray below. Often, a pale and indistinct mid-dorsal line is visible. Under a microscope, the body integument appears to have a pebble-like surface. Mature larvae can be 1$^{3}/_{16}$–1$^{3}/_{4}$ inch [30–45 mm] long and $^{1}/_{4}$ inch [7.0 mm] wide.

Pupae: The pupae are brown, reddish-brown, or black and $^{1}/_{2}$–$^{7}/_{8}$ inch [13–22 mm] long. The antennae, wingpads, and legs are firmly joined together, but the abdomen is free to twist around if the pupa is disturbed.

Adults: The adults are generally dark gray to black colored moths with some brown markings. The forewings span 1$^{3}/_{8}$–1$^{3}/_{4}$ inch [35–45 mm] and a black, dagger-shaped marking appears at the outer edge.

Life Cycle and Habits: The black cutworm appears to overwinter as larvae or pupae in the northern states. At the level of Ohio, three generations occur but 5 to 6 generations can occur in Louisiana. Subtropical regions probably have continuous generations. Adults emerge at night from pupae formed in the soil or thatch. The adults usually mate within the first 2 to 4 nights after emergence. They actively feed on nectar from flowering plants at night. Each female has the capacity to lay 1200 to 1600 eggs over 5 to 10 days. Though the adults prefer to lay eggs on curly dock and mustard, they may lay eggs on any suitable host. They often get into turf where broadleaf weeds are present. Eggs are usually laid singly or in small clusters. The eggs hatch in 3 to 6 days, and the first instar larvae usually feed on the leaf surface. After several days and molting, the larvae work their way to the base of plants. The cutworms excavate a hole into the ground and line it with silk. However, on golf courses, the larvae readily set up in the aeration holes or in the holes left behind from the spikes on golf shoes. From this retreat, the larvae venture forth at night to feed on plant material. Often they will drag a leaf or stem back to their burrow to feed on during the daytime. Most of the larvae go through 6 to 7 molts in 20 to 40 days to complete their larval development. The pupa is usually formed in the cutworm retreat with some additional silk lining the chamber. The pupa takes about two weeks to mature. These developmental times may be greatly lengthened during the cooler parts of the season.

Control Options: See: Cutworms and Armyworms—Introduction. Black cutworms are generally controlled by using one of the contact or stomach pesticides. However, if populations are high in surrounding areas, such as in field crops, continual reinfestations may occur. Because of this, contact with local cooperative extension services or pest management consultants may be beneficial in determining local pest activity. Young cutworms are difficult to locate, so use of a disclosing solution is beneficial in determining population pressure. If a soap flush reveals 5 to 10 larvae per yd^2 on golf course fairways or lawn turf, remedial controls will be necessary. Only several cutworm spots on greens may require treatments. Adults can be monitored using the black cutworm pheromone trap or black light traps. Larvae can be expected about 7 to 10 days after peak trap catch.

Option 1: *Biological Control—Microbial Toxins*—Some of the *Bacillus thuringiensis* var. *spodoptera* toxins, such as Javelin® and Steward®, have had good activity against the black cutworm's relatives, the armyworms. For maximum efficacy against the black cutworm, these pesticides need to be applied when the young larvae (first and second instars) are active. Make applications about seven days after peak adult trap catches are realized.

Option 2: *Biological Control—Entomopathogenic Nematodes*—The insect parasitic nematodes are generally effective against the larvae of this pest. Products containing *Steinernema carpocapsae* (=Biosafe®, Biovector®, Exhibit®, Scanmask®) can be applied when the larval populations have been surveyed. Best efficacy is obtained by applying the nematodes in the late afternoon, just before sunset.

Option 3: *Cultural Control—Weed and Aeration Management*—Since this pest is attracted to various broadleaf weeds, reduction of these populations will reduce the attractiveness of the turf environment. Since the larvae have better survival in existing burrows, hold back aeration when adult trapping indicates significant activity.

ARMYWORM

Species: *Pseudaletia unipuncta* (Haworth) is often called the "common" armyworm. It has been listed in the past under the generic names, *Leucania* and *Cirphis*. [Phylum Arthropoda: Class Insecta: Order Lepidoptera: Family Noctuidae] (Figures 3.50 and 3.51; Plate 3–18).

Distribution: The armyworm is found throughout the United States, generally east of the Rocky Mountains. It is occasionally a turf pest in Utah and California.

Hosts: All grasses and grass crops may be attacked.

Damage Symptoms: As their name implies, armyworms can reach large numbers which literally "march" across turf, eating every bit of green leaf and stem. They often occur in large numbers where turf is being grown near small-grain crop fields. They will also build up large populations in grassy meadows

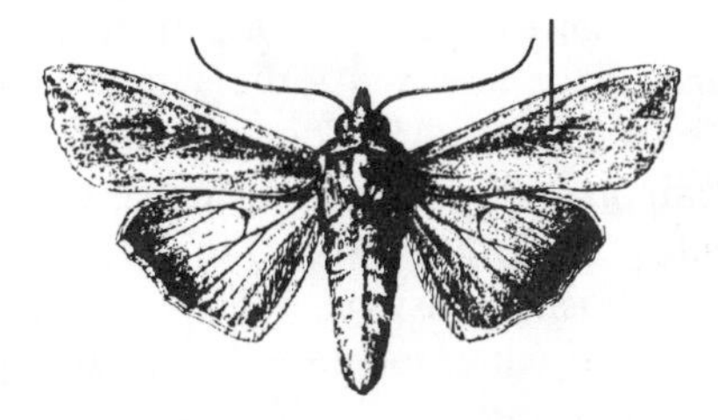

Figure 3.50 Armyworm, *Pseudaleta unipuncta* (Haworth), adult with characteristic white spot marked on right. (USDA)

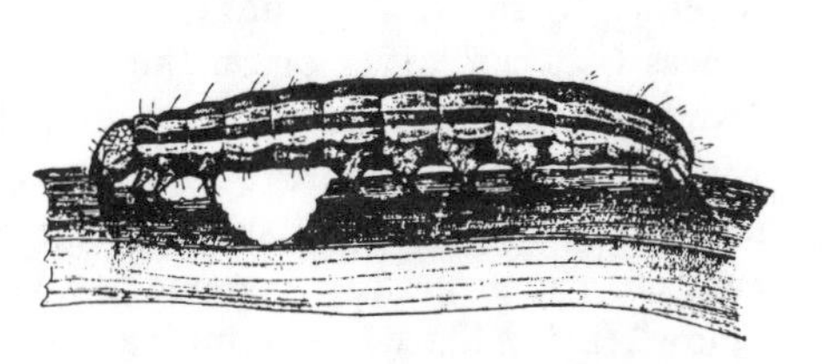

Figure 3.51 Armyworm, *Pseudaleta unipuncta* (Haworth), larva feeding on grass leaf. (USDA)

and along rights-of-way for roads or utilities. When small, the larvae leave a ragged leaf margin, but as the larvae grow, they devour everything to the thatch/soil level. This moving front of feeding caterpillars can consume many square feet of turf in a night. Most people describe the event as: "their turf disappeared overnight." The armyworm has been a pest of cereal and forage crops since the colonial times.

Description of Stages: The armyworm has a typical moth life cycle with egg, larval, pupal, and adult stages.

Eggs: The eggs are round, shiny white with a slight greenish case, and have no surface ridges. They are about 0.02 inch [0.5 mm] in diameter, and are laid in several rows with up to 500 in a mass.

Larvae: Newly hatched larvae are light green and about 1/16 inch [2 mm] long. As they grow, the larvae remain olive green to brown, and have a darker stripe along each side and a broad, darker stripe down the back. Mature larvae may be 1 3/16 inch [30 mm] long.

Pupae: The pupae are typical brown forms with the legs and wing pads showing through the pupal shell. They are about 9/16 inch [15 mm] long and can rotate the abdomen when disturbed.

Adults: The adults are light tan with darker shading. They have a distinct, small, light colored spot on the mid forewing, often with a tiny black dot in the middle. Adults are about 3/4 inch [20 mm] long and have a 1 3/16 inch [30 mm] wing span.

Life Cycle and Habits: The armyworm is able to withstand most winter temperatures in the larval and pupal stages located in the soil. It emerges as new adults in March to April and the first generation can cause crop damage in May to mid-June. A second generation runs through June and July, with the third generation active in August to the first frost. The females lay clusters of up to 500 eggs in rows lined up on grass surfaces. The eggs are often covered by a fold in the leaf or by another leaf stuck to their surface. Each female is capable of laying several thousand eggs. When the larvae first emerge, they use a looping motion to move across leaf surfaces. They tend to stay together and feed on the same plants until everything is devoured. The larvae then move together to the next plant. On sunny days, the caterpillars tend to hide in spaces between leaves or on the undersurfaces of leaves. As the caterpillars near maturity, they do not hide and can move, in mass, across grassy areas, with each caterpillar devouring individual plants. Armyworms most commonly invade turf from surrounding small grain crops. When the crops are eaten to the ground or become too large to easily consume, the caterpillars "mow" turf. Mild winter temperatures combined with moderate, moist spring and early summer conditions favor armyworm outbreaks. The second or third generation is the one most likely to cause damage on turf.

Control Options: See: Cutworms and Armyworms—Introduction. The true armyworm rarely kills turf, because the "hoard" tends to only eat the leaves and upper stems of turf. Since the crowns are not destroyed, the turf should soon recover with irrigation.

FALL ARMYWORM

Species: *Spodoptera frugiperda* (Smith) is probably the most important of the "spodopteran" complex. However, a close relative, the yellowstriped armyworm, *S. ornithogalli* (Guenee), is commonly found intermixed with the fall armyworm. [Phylum Arthropoda: Class Insecta: Order Lepidoptera: Family Noctuidae] (Figures 3.52 to 3.54).

Distribution: This pest is thought to be semitropical in origin, probably from Mexico or Central America. It is a permanent resident of the Gulf states and is able to migrate northward during the spring and summer. Since it is not able to withstand freezing temperatures, the northern populations die each fall.

Hosts: This pest attacks several hundred grassy and broadleaf plants.

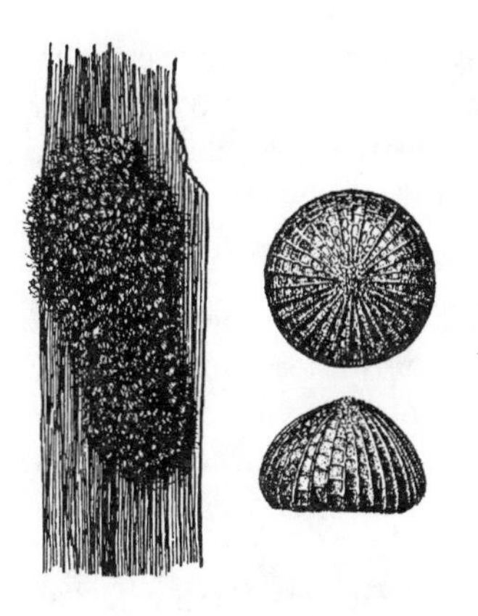

Figure 3.52 Fall armyworm, *Spodoptera frugiperda* (Smith), egg mass on leaf (left) and close up of egg top and side (right). (USDA)

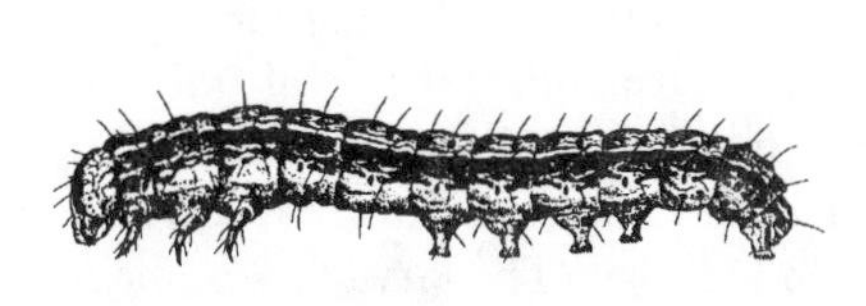

Figure 3.53 Fall armyworm, *Spodoptera frugiperda* (Smith), mature larva. (USDA)

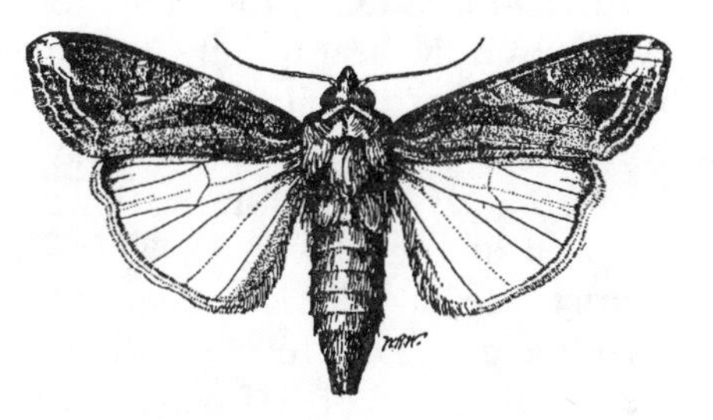

Figure 3.54 Fall armyworm, *Spodoptera frugiperda* (Smith), adult male. (USDA)

Damage Symptoms: The younger larvae feed on the margins of leaves, which produces a ragged look. As the larvae mature, they eat all the above ground leaves and stems. The larvae do not seem to be gregarious, and feeding damage in turf can appear as progressive thinning of the turf.

Description of Stages: The fall armyworm has typical egg, larval, pupal, and adult stages found in all moths.

Eggs: The eggs are laid in masses of 100 to 250 in two to three layers. They are about 0.016 inch [0.4 mm] in diameter, begin as greenish gray, turn blackish brown at hatch, and have fine square reticulations on the surface. Each mass is covered with scales from the abdomen of the female.

Larvae: First instar larvae are about 1/16 inch [2 mm] long and are light green in color. As they grow, they obtain distinctive lines down the body, range from light tan to olive green in color, have distinctive black spots at the base of larger hairs, and reach 1 3/16 inch [30 mm] in length. The head has a characteristic, white, inverted Y-mark.

Pupae: The pupae are droplet-shaped, with the narrow end at the abdomen. They are a dark chestnut brown and are about 5/8 × 3/16 inch [15 × 4.5 mm].

Adults: The male and female markings are fairly different. Both are generally shades of gray with white markings. The females can be almost entirely gray without much marking. Both males and females have a characteristic drop-shaped lighter mark running from the middle of the forewing and trailing to the hind tip margin. They range from 11/16 to 1.0 inch [17–25 mm] long with wingspans of 3/4 to 1 3/16 inch [20–30 mm].

Life Cycle and Habits: This pest is a constant threat to southern turf. It rarely damages transition or northern turf because it can not overwinter where the turf may freeze. All stages are present around the year in the areas of the Gulf states within 100 miles of the coast. With the arrival of warmer spring temperatures, adults lay eggs on preferred grasses and small grain crops. However, the adults may lay egg masses on trees, shrubs, and flowers. If the larvae cannot feed on this host, they drop to the ground and feed on grasses and weeds. The eggs take 7 to 10 days to hatch in cool weather and can hatch in 2 to 3 days in the heat of July and August. The young larvae eat the egg shell before turning to plant foliage. They feed together until they reach the fourth or fifth instar. At this time they may become cannibalistic. Larger larvae feed during the day and if disturbed by a shadow or touching, they will fall from the plant and remain tightly coiled where they fall. They soon recover and return to feeding. The larvae can take as little as 12 days to mature in July and August and up to 28 to 30 days to mature in October. The larvae consume an average of 13 cm^2 of leaf tissue, of which 12 cm^2 is consumed in the last two instars. This is why they appear to suddenly strip the turf down within a couple of days. Upon maturation, the sixth instar larvae move into the lower thatch area and form pupae within some loosely spun silk webbing. The pupa takes 9 days to mature in August and up to 20 days in October. The adults actively feed on nectar and overripe fruits. They can live for about two weeks, and the females lay 3 to 5 egg masses. The adults have strong migratory urges and commonly travel hundreds of miles on rapidly moving storm fronts. Though they commonly arrive in northern states by June and July, they rarely attack turf. Their main northern hosts are corn and soybeans.

Control Options: See: Cutworms and Armyworms—Introduction. The fall armyworm and the yellowstriped armyworm commonly reach plague proportions from July to October in the Gulf states. State agricultural agencies commonly release armyworm alerts when these great outbreaks are bound to occur. Turf managers should be looking for these notices because these are the times that turfgrasses will be under the greatest pressure.

LAWN ARMYWORM

Species: *Spodoptera mauritia* (Boisduval). [Phylum Arthropoda: Class Insecta: Order Lepidoptera: Family Noctuidae].

Distribution: This pest seems to be a native of the Oriental and IndoAustralian regions, but has been transported to many of the Pacific islands, including Hawaii. It is not known in North America.

Hosts: Like many of the true armyworms, this species feeds on a wide variety of grasses and sedges. It is especially fond of bermudagrass and zoysiagrass.

Damage Symptoms: As their name implies, armyworms can literally march across the turf, eating most of the leaves and stems. Moderate populations produce a ragged appearance to the turf. High populations produce a fairly well-defined line between green, undamaged turf and a completely eaten and brown area. This front can move forward by a foot per night.

Description of Stages: Armyworms are moths with complete life cycles.

Eggs: The eggs are laid in masses attached to the leaves of trees, on buildings or other objects overhanging or near turf. The mass is usually elongate-oval in outline and has 5 to 7 layers of eggs. The females usually cover the mass with their abdominal hairs. The masses contain 600 to 700 eggs and each egg is about 0.02 inch [0.5 mm] in diameter.

Larvae: The first instars are about 3/64 inch [1.2 mm] long and are greenish after feeding. Later instars develop patterns of brown to black stripes and spots. Each body segment has a pair of jet black dashes next to the longitudinal yellow stripes. Mature larvae are 1 3/8–1 3/4 inch [35–40 mm] long.

Pupae: Pupae are first light brown but soon turn a dark chestnut brown. They are about 5/8 to 3/16 inch [16 × 4.5 mm].

Adults: Adults resemble the fall armyworm. The male is more distinctly marked than the female. They have a drop-shaped light spot about midway down the wing, which is followed with a dark black spot. The wing is generally mottled, dark gray. The females have a wing span of 1 5/32 to 1 3/4 inch [34–40 mm], and the males are slightly smaller.

Life Cycle and Habits: This tropical species has continuous generations throughout the season. The adult moths emerge at night and the females are mated within a day. The adults may feed on nectar and water from rain or dew. After a four day preoviposition period, the females begin to lay masses of eggs. Egg laying begins at dusk and is generally completed before midnight. Egg masses are normally attached to the foliage of trees and shrubs near turf. They are almost never laid on grass blades. Since the adults are attracted to lights, many egg masses are often laid on nearby buildings or trees. The eggs take three days to hatch and the young larvae drop to the turf. During the first five instars the larvae feed during the day and night. They seem to prefer to stay together, often with several feeding on the same plant. As they strip off the foliage they move forward to new turf. The older larvae are easier to see and usually hide in the thatch and plant debris during the day. The 7 to 8 instars take nearly 28 days to complete development. The mature larvae burrow into the thatch and pupate. Pupation is completed in 11 days.

Control Strategies: Also see: Cutworms and Armyworms—Introduction. Considerable effort has been made to import biological controls for this pest in Hawaii. Under normal conditions, these parasites and predators do an adequate job but occasionally they do not build up fast enough to prohibit turf damage. If turf is suspected to have a damaging population of lawn armyworms, sampling should be performed. Controls are needed if 10 to 15 larvae per yd^2 of lawn turf are found with a soap disclosing flush.

Option 1: *Biological Control—Conserve Predators and Parasites*—Two tiny wasp egg parasites attack this pest (*Telenomus nawai* Ashmead and *Trichogramma minutum* Riley). The larval attacking wasp, *Apanteles marginiventris* (Cress.) seems to be the most important natural enemy of the lawn armyworm. By providing nectar-producing flowers and targeting insecticide sprays only to the areas needing them, these parasites can be conserved.

Option 2: *Cultural Control—Reduce Night Lighting*—Since the adults are attracted to lights at night and they oviposit nearby, place lights away from sensitive turf areas. Lights within the yellow range are also much less attractive to night flying insects.

OTHER TURF INFESTING CATERPILLARS

There are several other turf infesting caterpillars which are occasionally encountered damaging turfgrasses. Most of these can be managed as if they were cutworms or armyworms.

STRIPED GRASSWORMS (=GRASS LOOPERS)

Species: Several species of the genus *Mocus. Mocus latipes* (Guenee) is the most common species damaging turf, but *M. disseverans* (Walker) and *M. marcida* (Guenee) may be present. [Phylum Arthropoda: Class Insecta: Order Lepidoptera: Family Geometridae].

Distribution: These are natives of the New World ranging from northern Canada to Argentina east of the Rocky and Andes mountain ranges. Generally *M. latipes* is a periodic pest from Texas across to Florida.

Hosts: Bahiagrass, bermudagrass, and St. Augustinegrass are preferred.

Damage Symptoms: The larvae feed on grass blades and heavy populations can completely strip all the leaves so that only the stolons remain. Light infestations leave a ragged appearance to the turf.

Description of Stages: These lepidopterous pests have complete life cycles with egg, larval, pupal, and adult stages. The following descriptions are based on *M. latipes*, but the others are very similar.

Eggs: The round eggs are about 0.024 inch [0.6–0.7 mm] in diameter and have longitudinal ridges with faint cross ridges. The eggs are first light green and become mottled with dark brown with age.

Larvae: The larvae are typical loopers, that is inchworm-like. They have only three sets of prolegs on the abdomen. The first instar nymphs are about $^{3}/_{16}$ inch [5 mm] long, and slightly striped with brown and cream colors. Subsequent instars are more strikingly striped from the head capsule to the tip of the abdomen. Mature larvae are 1.0–1$^{3}/_{8}$ inch [25–35 mm] long.

Pupae: The pupa is chestnut brown in color and is often covered with a fine white powder. The cocoon is spindle-shaped and often has several grass leaves, often dead ones, attached.

Adults: The adults are about the size of a cutworm, but the wings are broader. The wings are mottled grey and brown and usually have a darker border along the tip. The wing span is usually 1$^{3}/_{4}$ to 2$^{3}/_{16}$ inch [45–55 mm].

Life Cycle and Habits: Several generations occur per year, but the major pressure from this group of pests occurs in late summer and early fall. Females emerging in April mate and begin to lay eggs within three days. Each female may lay 300 to 400 eggs individually attached to grass blades. Most of the eggs are laid within a week. The eggs hatch in 3 to 4 days, and the first instar larvae actively "inchworm" around until a suitable grass blade is found. Here the larvae spin some webbing over the V-shaped groove in the grass blade and begin to feed by stripping the epidermis from the top of the leaf. The larvae grow rapidly, molting every 2 to 3 days when temperatures are high. When the fourth instar is reached, the larvae begin to feed on the leaf margins, leaving ragged edges. The next several instars (up to 6 or 7) feed over 2 to 3 weeks and the larger larvae can completely strip the leaves from plants and often chew on young shoots. The mature larvae are about 5 mm and blend in well with the surrounding environment. The mature larvae, ready to pupate, seek out protected areas under the turf canopy and construct a spindle-shaped cocoon. This cocoon usually has pieces of grass clippings attached and is very difficult to find. The pupa matures in 8 to 10 days, and the new moths emerge in the late afternoon. At 80°F [26.7°C] the life cycle takes 30 to 33 days, and somewhat longer at temperatures down to 70°F [21.1°C]. Development is greatly slowed at temperatures below 60°F [15.6°C].

Control Options: See: Cutworms and Armyworms—Introduction. These insects are easily controlled by most of the methods used for surface feeding caterpillars.

FIERY SKIPPER

Species: *Hylephila phyleus* (Drury). [Phylum Arthropoda: Class Insecta: Order Lepidoptera: Family Hesperiidae] (Figure 3.55). Several species of skippers can be found feeding on turfgrasses. Proper identification probably requires the help of an expert.

Distribution: This butterfly is found over most of North and South America. It is most abundant in the Gulf states, and was detected in Hawaii in 1970.

Hosts: The larvae seem to prefer bermudagrass, but St. Augustinegrass is commonly fed upon. Turf with crabgrass appears especially attractive.

Damage Symptoms: The individual larvae feed in localized spots and cause isolated round spots, 1 to 2 inches [2.5–5 cm] in diameter. If an extensive infestation occurs, the spots join together and larger irregular brown patches result.

Description of Stages: This is a typical butterfly with a complete life cycle.

Eggs: The round eggs look like a small ball cut in half. They are about 0.03 × 0.02 inches [0.7 × 0.5 mm]. When freshly laid they are white, but change to a light blue-green in a day or two.

Larvae: Skipper larvae are unique in form. They have dark heads, an

Figure 3.55 Fiery skipper, *Hylephila phyleus* (Drury), larva (right) and adult (left).

obvious constriction of the neck, and plump bodies covered with tiny bristles. The first instar larva is a pale greenish yellow and about 3/32 inch [2.5 mm] long. Mature larvae become yellow brown to gray brown, with an indistinct median longitudinal stripe. Mature fifth instar larvae are 1.0 inch [24 to 25 mm] long.

Pupae: The pupa is formed in the thatch or in loosely webbed-together turf debris. The pupa is 5/8 inch [15–18 mm] long, and changes from a light greenish-tan to an overall brown just before adult emergence.

Adults: The adults are robust yellow butterflies with orange and brown markings. The wings span about 1.0 inch [25 mm], with a body about 5/8 inch [16 mm] long.

Life Cycle and Habits: In tropical and subtropical regions, this pest is active all year. In these areas, 3 to 5 generations are normal. In the northern part of its range, only 2 to 3 generations are completed, with the adults being most numerous in mid- to late summer. Adult fiery skippers are strongly attracted to nectar-producing flowers such as lantana and honeysuckle. They will also visit other annual flowers. The males appear to be slightly territorial and often perch near flowers being visited by the females. They strongly pursue females or other males flying into the area. Newly emerged females appear to take 3 to 4 days before oviposition begins. During the middle of the day, females alight on turf and place eggs at random. The females rarely lay more than one egg per stop, and the eggs are usually cemented to the underside of a leaf blade. Egg laying continues for 4 to 6 days, with each female laying between 50 and 150 eggs. At 80°F [26.7°C] and above, eggs can hatch in 2 to 3 days. Up to six days may be needed at lower temperatures. The newly hatched larvae feed on the margins of turfgrass blades, but soon drop into the turf canopy and spin a loose silk shelter. From within these silken shelters, the larvae emerge at night to remove and eat entire leaves. This produces small round damage spots in the turf. Once larval development is complete, they form their cocoon in the silken shelter. At 75°F [23.9°C], this pest took 48 days to complete development from egg to adult. Above 81°F [27.2°C], the cycle took only 23 days.

Control Options: See: Cutworms and Armyworms—Introduction. This insect rarely produces enough numbers to damage large areas of turf. However, the scattered damage spots on golf course putting greens or tees can contribute to poor playing surfaces. There are only a couple of parasites known from this pest, and their influence on populations is not known. If suspicious spots are noted in the turf, sampling with a soap disclosing solution should be performed. If 5 to 8 larvae per yd^2 of lawn turf or golf course fairway are found, treatments are warranted. Fewer numbers on putting greens may need attention.

Option 1: *Biological Control—Microbial Toxins*—Some of the strains of *Bacillus thuringiensis* (Bt) toxins are registered for sod webworms and armyworms in turf. These should also have activity against the fiery skipper larvae. For maximum efficacy against the fiery skipper larvae, these pesticides need to be applied when the young larvae (first and second instars) are active. Since the larvae are enclosed in silken tunnels and feed only at night, apply the material in the late afternoon or evening and do not irrigate until the next morning.

STEM AND THATCH INFESTING INSECT AND MITE PESTS

This complex and poorly defined area is composed of mainly stems, stolons, and organic material in various states of decay. Turf which is overwatered and fertilized may grow roots in this zone without having much contact with the soil. In this situation, true thatch/soil dwelling pests may invade this stem and thatch area. This zone provides high humidity, temperature mediation, and an abundant supply of living and dead organic matter for food.

Pests with sucking mouthparts seem to prefer this zone. They usually feed by removing plant fluids from the vascular system located in the stems, stolons, and lower leaf sheaths. Many of these sucking pests form salivary tubes or inject salivary materials which clog the vascular bundles. This results in discoloration, stunting, or death of the leaves or tip ends of stems. Of major concern is when these sucking pests attack the vegetative crowns of various grasses. When this happens, the entire plant may not survive. Sucking insects are often difficult to manage with traditional stomach insecticides. Since these pests do not chew and ingest plant matter, contact insecticides are more effective in management. A few pesticides have systemic action, and these are often very good for management of this group.

Pests with chewing mouthparts may enter the stems, stolons, and crowns, and are called borers during this phase. Others simply feed externally on the living and dead organic matter.

Pests located in the stem and thatch zone may escape detection until considerable damage has been done. Turf managers must train themselves to thoroughly search this thick matter area, either visually or with other monitoring tools. Soap flushes and water flotation methods are often successful in disclosing pests located in this area.

CHINCH BUG AND HAIRY CHINCH BUG

Species: Chinch bug, *Blissus leucopterus leucopterus* (Say); hairy chinch bug, *Blissus leucopterus hirtus* Montandon. [Phylum Arthropoda: Class Insecta: Order Hemiptera: Family Lygaeidae] (Figure 3.15).

Distribution: The common chinch bug is normally found from South Dakota across to Virginia and south to a line running from mid-Texas across to mid-Georgia. The hairy chinch bug cohabits some of the northern range of the common chinch bug, but also extends throughout the northeastern states and into southern Canada.

Hosts: The hairy chinch bug prefers turfgrass species such as fine fescues, perennial ryegrasses, Kentucky bluegrass, bentgrass, and zoysiagrass. The common chinch bug prefers grain crops but will attack turfgrasses such as bermudagrass, fescues, Kentucky bluegrass, perennial ryegrass and zoysiagrass.

Damage Symptoms: Irregular patches of turf begin to yellow, turn brown, and die. These patches continue to become larger in spite of watering. Apparently feeding by chinch bugs blocks the water and food conducting vessels of grass stems. By blocking the water, the leaves wither as in drought and the manufactured food doesn't get to the roots. The result is plant death. Damage generally occurs during hot, dry weather from June into September.

Description of Stages: These pests are true bugs and have a gradual life cycle with egg, nymphal, and adult stages. All the species of *Blissus* are very similar in form, and an expert is needed to separate species and subspecies.

Eggs: The eggs are elongate bean-shaped, approximately 0.03 × 0.01 inches [0.84 × 0.25 mm], and are roundly pointed at one end and blunt at the other. The blunt end has several small tubercles visible through a dissecting microscope. The eggs are first white and change to bright orange just before hatching.

Nymphs: There are five nymphal instars which change considerably in color and markings. The first instar has a bright orange abdomen with a cream colored stripe across it, a brown head and thorax, and is about 1/32 inch [0.9 mm] long. The second through fourth instars continue to have this same general color pattern, except that the orange color on the abdomen gradually changes to a purple-gray with two black spots. The fourth instar increases to more than 2 mm long. The fifth instar is very different because the wing pads are easily visible and the general color is now black. The abdomen is blue-black with some darker black spots, and the total body length is about 3 mm.

Adults: The adults are approximately 1/8 × 1/32 inch [3.5 × 0.75 mm]. The males are usually slightly smaller than the females. The head, pronotum, and abdomen are gray-black in color and covered with fine hairs. The wings are white with a black spot, the corium, located in the middle front edge. The legs often have a dark burnt-orange tint. Individuals in a population, or in some cases, most of a local population, may have short wings, called brachypterous, which reach only halfway down the abdomen.

Life Cycle and Habits: The hairy chinch bug adults overwinter in the thatch and bases of clumps of grass in the turf. However, the common chinch bug prefers to fly to tall grasses of fields to find overwintering sites. These individuals then migrate in search of grain crops in the spring, but may establish in turf instead. The adults become active when the daytime temperatures reach 70°F [21.1°C]. The females feed for a short period of time and mate when males are encountered. Eventually the females begin to lay eggs by inserting them into the folds of grass blades or into the thatch. This usually occurs in

May from New York to Illinois. A single female may lay up to 200 eggs over 60 to 80 days. The eggs take about 20 to 30 days to hatch at temperatures below 70°F [21.1°C], but can hatch in as little as a week when above 80°F [26.7°C]. The young nymphs begin to feed by inserting their mouthparts in grass stems, usually while under a leaf sheath. The nymphs grow slowly at the beginning of the season because of cool temperatures, but speed their development by July. Generally the first generation matures by mid-July. At this time, considerable numbers of adults and larger nymphs can be seen walking about on sidewalks or crawling up the sides of light colored buildings. If a good hot, dry spring is available, turf injury by the first generation can be evident by June. Generally, the major damage is visible in July and August when the spring generation adults are feeding, and their second generation nymphs are becoming active. During the hot summer months, the new females lay eggs rapidly, and their young may mature by the end of August or the first of September. The second generation adults may lay a few eggs for a partial third generation if the season has been long. However, most of these late nymphs do not mature before winter temperatures drop. When cool temperatures arrive, the mature chinch bugs seek out protected areas for hibernation.

Control Options: These two chinch bug types are some of the oldest known American insect pests. The first records of damage to crops are from the 1780s. Because chinch bugs are major crop pests, a tremendous number of control strategies have been developed. Only those useful in turfgrass management have been selected. Chinch bugs are relatively easy to control when they are detected early.

Option 1: *Cultural Control—Watering the Turf*—Since this pest requires hot dry conditions for optimum survival and reproduction, irrigation during the spring and early summer may increase the incidence of pathogen spread, especially the lethal fungus, *Beauveria* spp. The adult chinch bugs can withstand water because of the protective hairs on the body, but the nymphs readily get wet and can drown.

Option 2: *Cultural Control—Use Resistant Turfgrasses*—The hairy chinch bug seems to prefer perennial ryegrasses and fine fescues, especially if these are in the sun and have greater than 0.5 inch [13 mm] of thatch. Bentgrass is also attacked, but this turf is rarely used in lawns. Bluegrass lawns with 50% or more ryegrass and/or fine fescue are the most likely to be attacked. In field tests, Yorktown, Yorktown II, and Citation perennial ryegrasses are the most susceptible to chinch bug buildup, while Score, Pennfine, and Manhattan are avoided. Jamestown and Banner fine fescues are more commonly attacked than FL-1, Mom Frr 25, and Mom Frr 33. In general, perennial ryegrasses, fine fescues, and tall fescues with endophytes are highly resistant to this pest.

Option 3: *Cultural Control—Recovery From Damage*—Turf with light to moderate damage will recover rather quickly if lightly fertilized and watered regularly. Heavily infested lawns may have significant plant mortality because of the toxic effect of chinch bug saliva, and reseeding will be necessary.

Option 4: *Biological Control*—Several researchers have been trying to develop a usable formulation of the *Beauveria* fungus, but at present no practical material is available. Several egg parasites and an adult parasite are known, but these do not seem to build up populations rapidly enough to control this pest. Currently, no work is being undertaken to augment these parasites. Several predators, especially the bigeyed bugs, *Geocoris* spp., are noted to kill large numbers of chinch bugs. Bigeyed bugs are often mistaken for chinch bugs because of their similarity in size and shape. Bigeyed bugs usually do not build up large populations until after considerable turf damage has occurred. Use of the insect parasitic nematodes *Steinernema* spp. and *Heterorhabtitis* spp. have given inconsistent results when used against these chinch bugs.

Option 5: *Chemical Control—Use Preventive Sprays*—In turf areas where chinch bugs have been a perennial problem, early insecticide sprays have been used to reduce the beginning spring population. This works well if applications are made in May after the adults have finished spring migrations and the young nymphs are just becoming active. It is highly recommended that preventive sprays be used only if sampling has been done to determine that chinch bugs are indeed present, especially if an unusually hot, dry spring has occurred.

Option 6: *Chemical Control—Target Spraying*—Chinch bugs are rather easy to detect in turf, and targeted insecticide sprays can be applied to reduce populations which appear to be building to damaging levels.

Sampling

Several sampling schemes have been developed for assessing chinch bug populations in turf. The simplest method is to visually inspect the turf by spreading the canopy. Chinch bug nymphs tend to hide in the deeper thatch, and careful inspection is necessary. Unfortunately, eggs and small chinch bugs are easily missed using this technique. A more reliable method is to use the flotation technique, counting the number of adults and nymphs present over a 10 minute span. Populations of 25 to 30 per ft^2 [270–325/m^2] warrant control, especially if these numbers are encountered in June and July. More complicated sampling methods use repeated sampling over a long period of time, relating the population numbers to temperature and humidity parameters and predict future populations.

Insecticides and Application

Most insecticides, when applied in liquid form, should not be watered in for chinch bug control, especially with high volume spray equipment. This is because the chinch bugs are surface and thatch residents. Watering in might wash the insecticide into the deep thatch layers without challenging the surface insects. Some of the granulars require some irrigation in order to activate the insecticide (release it from the granule). Be sure to check the instructions for current information on watering. Currently, bendiocarb (=Turcam, Ficam®), carbaryl (=Sevin®), chlorpyrifos (=Dursban®, Pageant®), cyfluthrin (=Tempo®), diazinon (not on golf courses or sod farms), ethion, ethoprop (=Mocap®), fluvalinate (=Mavrik®, Yardex®), isazophos (=Triumph®), and isofenphos (=Oftanol®) are registered for chinch bug control.

SOUTHERN CHINCH BUG

Species: *Blissus insularis* Barber. [Phylum Arthropoda: Class Insecta: Order Hemiptera: Family Lygaeidae] (Plate 3-4).

Distribution: This pest can be found from southern North Carolina to the Florida Keys and west to central Texas. This pest has apparently been spread with sod to other countries. It is also found in Hawaii.

Hosts: This insect is a major pest of St. Augustinegrass, though it will occasionally attack centipedegrass, zoysiagrass, bahiagrass, and bermudagrass.

Damage Symptoms: Usually irregular patches of St. Augustinegrass or bermudagrass turn yellow and then brown. These patches may expand or new patches may begin to form in the lawn. As the southern chinch bug populations build up they may kill extensive areas in a lawn and begin moving in mass across sidewalks and driveways. Though they do not often crawl to the surface of St. Augustinegrass, they may congregate on the tips of other grasses and weeds in the area.

Description of Stages: These pests are true bugs and have a gradual life cycle with egg, nymphal, and adult stages. This species is difficult to separate from other *Blissus* spp., and an expert is generally needed if specific identification is critical. Generally, if this bug is in St. Augustinegrass or bermudagrass, it will be the southern chinch bug.

Eggs: The eggs are elongate oval, approximately 0.03 × 0.01 inches [0.75 × 0.23 mm], and are squarely cut off at the top. The blunt end has four small tubercles visible through a dissecting microscope. The eggs are first white but change to bright orange before hatching.

Nymphs: There are 5 to 6 nymphal instars which change considerably in color and markings. The first instars are bright orange with a cream colored stripe running across the abdomen. The head and thorax become brownish with age. The second through fourth instars continue to have this same general color pattern, except that the orange color of the abdomen gradually changes to a dusky-gray with small black spots. The fourth instar increases to more than 3/32 inch [2 mm] long. The fifth instars are quite different because the wing pads have expanded and are easily visible. Occasionally an additional instar, the sixth, will be formed, especially in cooler weather. The abdomen becomes blue-black with some darker black spots, and the total body length is about 1/8 inch [3 mm].

Adults: The adults are approximately 1/8 × 1/32 inch [3.1 × 0.85 mm]. The males are usually slightly smaller than the females. The head, pronotum, and abdomen are gray-black in color to dark chestnut-brown and covered with fine yellow to white hairs. The wings are white with a black spot, the corium, located in the middle front edge. The legs are chestnut-brown. Individuals in a population may have short, nonfunctional wings which reach only halfway down the abdomen.

Life Cycle and Habits: In the southern range of this pest, adults and a few individuals of all stages overwinter in St. Augustinegrass turf, especially in the thatch. In the northernmost part of the range, only adults overwinter. These insects may become active anytime the temperatures rise above 65°F [18.3°C], but reproduction generally does not occur until April or May. Females prefer to deposit their eggs by forcing them between the leaf sheath and stem. Each female may lay 45 to 100 eggs over several weeks. Some eggs are deposited in the thatch. The eggs hatch in 8 to 9 days at 83°F [28.3°C] and 24 to 25 days at 70°F [21.1°C]. The young nymphs immediately begin feeding under the protection of the leaf sheath. Nymphs hatching elsewhere crawl into available spaces under leaf sheaths. Several nymphs are often congregated together. The nymphs go through 5 and occasionally 6 instars in 40 to 50 days during warm weather (above 80°F [26.7°C]). At cooler temperatures, below 70°F [21.1°C], the nymphs may take 2 to 3 months to mature. In most areas of the Gulf states, 3 to 5 overlapping generations occur each season. The first couple of generations are fairly well defined because of the initial start from the winter season. The first major adult peak usually occurs in June, and the second peak is in August. Subsequent peaks may be in October and December. The greatest amount of damage occurs during the dry season in the summer.

Control Options: The southern chinch bug is very difficult to control because of management practices in St. Augustinegrass and bermudagrass turf. Populations which are resistant to organophosphate insecticides have been found.

Option 1: *Cultural Control—Watering the Turf*—See: Chinch Bug.

Option 2: *Biological Control*—See: Chinch Bug.

Option 3 : *Cultural Control—Modify Agronomic Management*—Generally, St. Augustinegrass and bermudagrass are overfertilized, and this leads to rapid growth and immense buildup of thatch. In fact, many such turf areas are merely growing on their own thatch layer. This environment is very beneficial to southern chinch bug survival. Overly thatchy lawns should be dethatched, verticut, or top dressed to improve the general conditions. Reducing fertilizers and increasing irrigation can greatly reduce populations of this pest.

Option 4 : *Cultural Control—Use Resistant Turfgrasses*—Common St. Augustinegrass is highly susceptible to the southern chinch bug, while most bermudagrasses are fairly resistant. The St. Augustinegrass variety Floratam has been the standard in Florida for resistance, but in recent years, the chinch bug has seemed to develop the ability to attack this variety. New varieties are under development and Floralawn seems to be a good alternative. In the bermudagrasses, Tifton 292 seems to be a highly resistant variety. Check with suppliers and local county agents for locally adapted turfs with resistance to chinch bugs.

Option 5: *Chemical Control—Use Preventive Sprays*—In turf areas where chinch bugs have been a perennial problem, early insecticide sprays have been used to reduce the beginning population. Unfortunately, this technique is believed to be one of the major reasons for development of insecticide resistance.

Option 6: *Chemical Control—Target Spraying*—Chinch bugs are rather easy to detect in turf, and targeted insecticide sprays can be applied to reduce populations that appear to be building to damaging levels.

Sampling

See: Chinch Bug and Hairy Chinch Bug.

Insecticides and Application

Most insecticides, when applied in liquid form, should not be watered in for chinch bug control, especially when high volume spray equipment is used. This is because the chinch bugs are surface and thatch residents. Some of the granulars require some irrigation in order to activate the insecticide (release it from the granule). Be sure to check the instructions for current information on watering. Bendiocarb (=Turcam®, Ficam®), carbaryl (=Sevin®), chlorpyrifos (=Dursban®), cyfluthrin (=Tempo®), diazinon (not on golf courses or sod farms), ethoprop (=Mocap®), fluvalinate (=Mavrik®), isazophos (=Triumph®) (lawns and golf course greens, tees and aprons only), and isofenphos (=Oftanol®) are registered for chinch bug control. See: Chinch Bug and Hairy Chinch Bug.

Option 7: *Chemical Control—Treatment of Resistant Chinch Bugs*—Populations of southern chinch bugs have been found which are resistant to various insecticides, especially organophosphates. These pests are currently best controlled with synthetic pyrethroids and carbamates. Check with local state agencies to confirm resistance.

Option 8: *Cultural Control—Recovery from Damage*—See: Chinch Bug.

TWOLINED SPITTLEBUG

Species: *Prosapia bicincta* (Say). [Phylum Arthropoda: Class Insecta: Order Homoptera: Family Cercopidae] (Plate 3-5).

Distribution: Native of North America, most common in Gulf states into Central America. Found as far north as Maryland across to Kansas.

Hosts: Attacks southern turfgrasses such as bermudagrass, St. Augustinegrass, bahiagrass, and centipedegrass. Adults feed on herbaceous perennials and have been noted to damage holly.

Damage Symptoms: Large numbers of spittlemasses are unsightly and cause concerns by messing shoes and bare feet. Nymphs cause patches of turf to yellow, and the adults cause the leaves to turn brown and die. This gives the appearance of sparse, blighted turf. The adults may also cause damage to flowers or some ornamentals, especially Burford holly.

Description of Stages: This pest has a typical gradual life cycle.

Eggs: The eggs are elongate oval and taper to a rounded point at one end. The 0.04 × 0.01 inch [1.09 × 0.27 mm] eggs are first pale orange and change to an orange-red in a few days.

Nymphs: Freshly hatched nymphs are pale yellow with a small orange spot on each side of the abdomen, and are about 5/64 inch [2 mm] long. The nymphs molt four times, during which time the orange spots enlarge to cover the entire abdomen. The final, fifth instar nymph has well-developed wing pads with 2 transverse orange bands, and is about 5/16 inch [8 mm] long. The nymphs are usually covered by the spittle mass.

Adults: The boat-shaped adults are about 3/8 inch [10 mm] long, and have a dark brown to black color with two distinct reddish orange bands on the wings. The adults have orange-red bodies, do not produce spittlemasses, and can jump and fly short distances.

Life Cycle and Habits: The eggs overwinter in the turf and are usually deposited at the base of grass plants. Sometimes the eggs are inserted between a lower leaf sheath and stem, but more commonly they are inserted into the surrounding thatch. The eggs may be laid singly or in small groups. Though eggs are the normal overwintering stage, occasional adults may be found during the winter months in Florida. Eggs hatch in early spring when the turf is coming out of dormancy. The young nymphs must find a suitable feeding site within an hour or two or they die. After probing several places, the nymphs finally insert their mouthparts into suitable turfgrass tissue and they produce a spittle mass. This is actually an excretion from the anus. As feeding and growth continues, the nymphs may move about and sometimes several nymphs may share a single spittle mass. If too much moisture is present in the turf, the nymphs may move to the tips of the grass blades to form spittle masses. Depending upon temperatures, the nymphs take 34 to 60 days to mature. First peak adult emergences occur in late May and early June in Florida to July in South Carolina. Depending on temperatures and moisture, adults may be continuously present until October, or a second peak population of adults may be found in late August through September. Newly emerged adult females release a sex pheromone which attracts interested males. After mating, the females begin ovipositing in about a week. The females lay an average of 50 eggs over a two week period. During summer temperatures, the eggs take 14 to 23 days to hatch, but eggs laid in September or October do not hatch until the next spring. Nymphal feeding may only cause yellowing in the turf, but the adults cause a toxic reaction in which the affected stem turns brown and dies. Adults also may cause brown spots on flowers and ornamentals because of feeding. The adults fly for a period shortly after dark and are attracted to lights.

Control Options: This pest rarely damages well maintained turfgrasses (except centipedegrass) but is a nuisance because of the spittle masses.

Option 1: *Cultural Control—Control Irrigation*—Since the nymphs cannot survive dry conditions, especially when small, reducing irrigation in the spring and early summer may reduce populations. Using those turf management practices which reduce thatch buildup also makes it difficult for young nymphs to survive.

Option 2: *Natural Controls*—The fungus *Entomophthora grylli* attacks the adults, usually late in August through September. This fungus is probably encouraged by moist warm conditions, so irrigation on warm evenings may help in its spread.

Option 3: *Chemical Controls—Insecticide Applications*—If spittle masses are a nuisance problem or actual damage is occurring, applications of contact pesticides are generally effective. A reapplication may be needed in midsummer as adults migrate in from surrounding turf. Ornamental plants, especially hollies, may also need protection if large numbers of adults are active. Only carbaryl (=Sevin®) is currently registered for spittlebug control in turfgrass.

RHODESGRASS MEALYBUG (=RHODESGRASS SCALE)

Species: *Antonina graminis* (Maskell). [Phylum Arthropoda: Class Insecta: Order Homoptera: Family Pseudococcidae] (Plate 3-7).

Distribution: Recorded from southern South Carolina to southern California; found worldwide in tropical and subtropical regions of Africa, Australia, Central America, India, Japan, Pacific Islands, and South China.

Hosts: Attacks over 70 species of grasses, including rhodesgrass, bermudagrass, and St. Augustinegrass.

Damage Symptoms: This mealybug does not commonly kill turf unless stressful conditions are found. High cut bermudagrass during drought is more prone to severe damage. This pest produces considerable honeydew, and ants or bees may frequent turf which is heavily infested.

Description of Stages: Though this pest is a type of mealybug, it acts like true scales by being immobile once settled.

Eggs: Eggs are elongate oval and cream colored. The eggs are contained inside the remains of the female and her waxy cover.

Crawlers: These are the first instar nymphs which are the only mobile forms. The crawlers are flat, oval, cream colored insects with a median stripe tinged with purple. Short legs, six-segmented antennae, and two waxy tail filaments are evident.

Sessile Nymphs: The crawlers settle on the grass crown or at nodes, insert their mouthparts, and begin secreting a waxy coat. After the first molt the new sessile nymph takes on a saclike form without legs. Only the threadlike mouthpart and anal filament emerge from the body. A second molt occurs while the waxy cover grows larger and the anal excretory tube elongates.

Adults: Only females are known and these reproduce asexually. The adult body is also saclike, broadly oval, dark purplish brown and 1/16–1/8 inch [1.5–3 mm] long. The fluffy waxy covering turns yellow with age, and opening at the anterior and posterior ends expose parts of the body. A very long, 1/8–3/8 inch [3–10 mm], waxy, tubular filament arises from the anus through which honeydew is excreted.

Life Cycle and Habits: This pest continues its life cycle year-round but is slowed by winter temperatures. Reproduction is considerably reduced during the winter, but activity increases in the spring as the grass begins rapid growth. Peak populations are reached by July, and populations again are reduced through July and August as summer stress is brought about by dryness. In September and October the populations again resurge until peaking in early November. During the spring, females lay an average of 150 eggs over 50 days. As the crawlers hatch they remain under the waxy cover of the female for several hours before emerging. These crawlers tend to first walk to the tops of plants but eventually settle down by wedging themselves beneath a leaf sheath, usually at a node. Here the mouthparts are inserted and the excretory tube and waxy cover are started. In about 10 days, the first molt takes place and the walking legs disappear and the antennae are much smaller. The excretory tube continues to elongate and the waxy cover becomes thicker and larger. Two more molts take place under the waxy cover and maturity is reached in 25 to 30 days. During the summer, a generation averages two months but in winter this may take 3.5 to 4 months to complete. This scale apparently travels by transportation of sod or grass cuttings. The crawlers may climb onto the legs of animals and "hitch a ride." High temperatures, especially near 100°F [37.8°C], reduce scale development and may actually kill individuals. Exposure of the scale to 28°F [-2.2°C] for 24 hours is fatal. Thus, winter cold limits northern movement and survival of this pest. This scale increases turf mortality during periods of drought.

Control Options: This pest rarely kills turf unless it is poorly managed. Do not cut the turf too short.

Option 1: *Cultural Control—Water and Fertilize*—Frequent irrigation, fertilization, and mowing no shorter than two inches helps prevent damage by this pest even when high populations are present.

Option 2: *Biological Control—Conserve Parasites—* Two parasitic wasps have been successful in controlling rhodesgrass mealybug populations. A Hawaiian parasite, *Anagyrus antoninae* Timberlake, and an Indian parasite, *Dusmetia sangwani* Rao, have been established in the southern United States. Apparently, *Dusmetia* seems better adapted and can survive lower temperatures than *Anagyrus*. Both of these parasites, once established, should be conserved by avoiding insecticide sprays in some areas known to have rhodesgrass mealybugs and the parasites. This area will serve as an inoculum.

Option 3: *Chemical Control—Use of Contact Insecticides—*Many contact pesticides do not effectively reach the protected stages, especially the eggs and resting crawlers. Repeated applications may be necessary if this strategy is used. At present, only diazinon (not for use on golf courses or sod farms) is registered for management of this pest.

Option 4: *Chemical Control—Use Systemic Insecticides—*Systemic insecticides are more efficient in killing young feeding mealybugs. Some eggs may escape harm if hatching occurs after the insecticide's residual disappears. Thus, a second application may be necessary for killing freshly settled crawlers. No currently registered systemic insecticides have rhodesgrass mealybug on their labels.

BERMUDAGRASS SCALE

Species: *Odonaspis ruthae* Kotinsky. [Phylum Arthropoda: Class Insecta: Order Homoptera: Family Diaspididae] (Plate 3-8).

Distribution: This scale is found worldwide in tropical and subtropical regions. In the United States, this scale attacks bermudagrass from California to Florida. It is also known in Hawaii.

Hosts: This scale is most frequently reported on bermudagrass, though it has been found on centipedegrass, bahiagrass, St. Augustinegrass and tall fescue.

Damage Symptoms: Bermudagrass first appears to grow slowly and has a yellow color. This often looks like drought stress. Heavy infestations may dramatically thin and kill patches of bermudagrass. This type of damage is more evident during periods of hot, dry weather. Where bermudagrass enters winter dormancy, this scale can cause a delay in the spring green up.

Description of Stages: This is a typical armored scale that lays eggs and has winged male stages.

Eggs: The elongate oval eggs are pink to light burgundy-colored and are located within the female shell.

Crawlers: The eggs hatch into a first instar nymph called a crawler. This stage is pink and the body is very flat and oval. They have short legs, antennae, and eye spots.

Settled Nymphs: Settled crawlers begin to produce an oval, waxy test (shell) which is first straw yellow and then covered with white waxy secretions.

Adults: Adult females have shells, or tests, which are egg-shaped or oval in outline and 3/64–1/16 inch [1.0–1.75 mm] long. Often, there is a straw yellow area, the exuvium, near the larger end. The actual female body, inside the test, is oval and pinkish in color. Male scale tests are about one-half the size of females. The males are able to emerge from their test and are small gnatlike insects with one pair of wings. Their yellowish-pink bodies are about 0.02 inch [0.5 mm] long and have 2 to 3 long white waxy threads arising from the tip of the abdomen.

Life Cycle and Habits: Little is known about the actual time periods needed for development of this scale. Most studies have attempted to assess development by counting the numbers of different stages at various times of the year. From these studies, it appears that the bermudagrass scale may have two overlapping generations per year in the southern states. In Georgia and Florida, eggs are laid and crawlers are most active in the spring rainy season. Settled crawlers and adults can be found during much of the season, though little growth occurs during winter dormancy or during summer drought periods. It is suspected that this scale has continuous generations in warmer climates where the bermudagrass does not go dormant.

The eggs are laid and retained inside the female scale test. As the eggs hatch over several weeks, the tiny crawlers move along the stolons and lower grass stems. They prefer to settle under old leaf sheaths at the bases of crowns, but they may be found anywhere on the stolons and lower stems. Often, large numbers of settled crawlers and young adults can be found on stems and stolons in the soil or thatch layer. Rarely are the scales exposed on upper plant parts. As the settled scales grow, they begin to cover the body with loose waxy filaments which eventually give way to the formation of a solid, waxy shell-like test. At maturity, the scales often extend slightly from under the old leaf sheaths that originally hid the body. Populations can be so large at nodes and crowns that the scales seem to be stacked on top of each other.

Control Options: At present, no biological controls are known for this pest. Undoubtedly, small predators take their toll on the crawlers, but this has not been confirmed. Insecticide treatments have not been very successful in controlling this pest and none are currently registered for this purpose. Since eggs and crawlers may be present over extended periods, timed applications for crawlers are difficult to make. Damage assessment and control thresholds are not developed for this pest. However, when 30–40% of the turf shows yellowing from this pest, control procedures are probably warranted.

Sampling

When yellowed turf is encountered, the scale should be sampled for by digging out several affected stolons with attached aboveground stems. Inspect the nodes and bases of the stems for the oval, white scales. If the scale is confirmed, a sampling program should be followed.

A rating using a 3×4 foot plastic rectangle strung on one ft^2 grids (12 ft^2 total) should be used. This hoop is tossed onto the turf and ten of the ft^2 grids are rated for yellowing from scale activity. Each ft^2 with yellowed turf is counted as one. Fairways and roughs should be sampled every 50 yards, and four hoop samples should be taken for each green and apron. Tees should have two hoop samples. When presence of the scale activity is noted, samples should be taken every two months. If the activity is increasing and exceeds 4 ft^2 grids per 10 sampled [>40%], chemical controls are probably warranted. Below 4 ft^2 grids per 10 sampled [<40%], cultural controls should be used. Lawns can be generally rated for color and density.

Option 1: *Cultural Control—Water and Fertilize Turf*—Bermudagrass that is well fertilized and watered generally can outgrow this pest. However, damage can begin to appear if irrigation must be discontinued in the summer.

Option 2: *Chemical Control—Traditional Insecticides*—No insecticides are specifically registered for this pest. Literature reports indicate that diazinon reduced populations. Diazinon is no longer registered for golf course or sod farm use. Chlorpyrifos (= Dursban®) has traditionally had similar effects on scales as diazinon, and some golf course managers report success with this product. It should be applied when female scale tests are full of eggs. It must also be washed into the thatch with irrigation immediately after application in order to reach the target pests.

BILLBUGS—INTRODUCTION

Species: Billbugs are merely weevils—beetles with a beak-like snout, which have the thorax and head almost as long as the wing covers. Billbug larvae look something like white grubs but lack legs and have a more pointed abdomen. Most of the billbug species, of which there are more than 60 species, breed in grasses (Graminae) and sedges (Cyperaceae). Fortunately, only four species have been recorded as common or periodic pests of North American turfgrasses. The bluegrass billbug, *Sphenophorus parvulus* Gyllenhal, and hunting billbug *S. venatus vestitus* Chittenden, are the most common pests. The Phoenician (= Phoenix) billbug, *S. phoeniciensis* Chittenden, the Denver or Rocky Mountain billbug, *S. cicatristriatus* Fahraeus, are periodic pests in limited geographic areas. [Phylum Arthropoda: Class Insecta: Order Coleoptera: Family Curculionidae] (Figures 3.56 to 3.61).

Distribution: The bluegrass billbug attacks cool season turfs from Washington State and Utah across to the East Coast. The hunting billbug prefers warm-

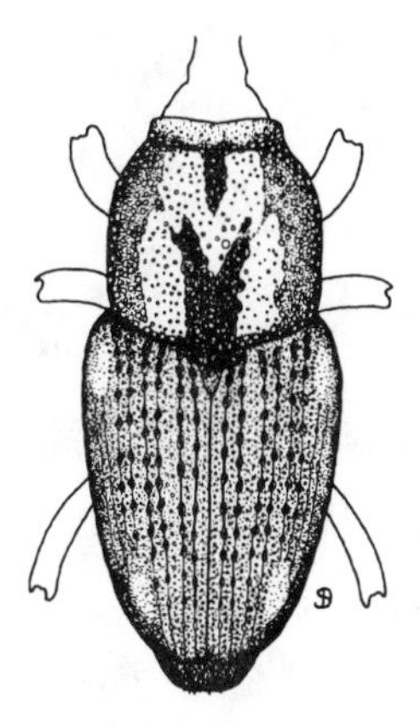

Figure 3.56 Phoenician billbug, *Sphenophorus phoeniciensis* Chittenden, dorsal view showing pattern of pits and furrows.

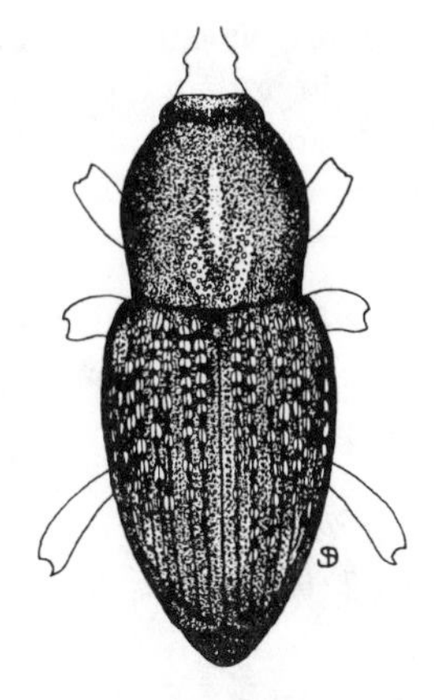

Figure 3.57 Denver billbug, *Sphenophorus cicatristriatus* Fahraeus, dorsal view showing pattern of pits and furrows.

Figure 3.58 A typical billbug, *Sphenophorus* sp., larva. (USDA)

season and transition zone turfgrass areas. The Phoenician billbug attacks turfgrasses in southern California, and the Denver billbug has damaged lawns in Colorado, New Mexico, and Wyoming.

Hosts: The bluegrass and Denver billbugs prefer Kentucky bluegrass and perennial ryegrass, while the hunting billbug and Phoenician billbug prefer bermudagrass and zoysiagrass.

Damage Symptoms: Billbug larvae bore into turfgrass stems, which causes the stems to turn straw colored as they die. This activity alone is not significant. However, larger larvae bore into the grass crown, killing the entire plant. They also may feed on roots and stolons. Infestations begin to appear as scattered dead stems. Eventually, 2–3 inch [5.0–7.5 cm] diameter spots die out; this is often misidentified as dollar spot disease. Heavy infestations may kill extensive turf areas.

Life Cycles: All the billbugs except for the hunting appear to have a single generation per year. The hunting billbug may breed continuously in southern states. Most billbugs spend the winter as adults, though a significant number of hunting and Denver billbugs overwinter as larvae. A spring feeding period is followed by egg laying, and larvae develop over the summer.

Identification of Species: The larvae and pupae are extremely difficult to identify, so try to find adult specimens. The adults are best identified by the patterns of pits and furrows found on the pronotum and wing covers. A 10× hand lens will be needed. The four common species can be identified using the pattern of pits and furrows on the back.

Control Options: Billbugs appear to be perennial pests once they have become established in an area. Often, whole neighborhoods with the same turfgrass type are attacked. Adult billbugs are notorious walkers. Because of this migratory habit, turf areas treated one year will often be reinfested from surrounding, untreated turf.

Option 1: *Cultural Control—Use Resistant Turfgrass*—Several cultivars of Kentucky bluegrass, bermudagrass, and zoysiagrass have demonstrated resistance or tolerance to billbug attack. Perennial ryegrasses, fine fescues, and tall fescues with endophytes are highly resistant to billbug attack. Contact seed suppliers for cultivars suited for your region. Billbugs generally do not seem to severely attack southern grasses such as centipedegrass and St. Augustinegrass.

Option 2: *Cultural Control—Water and Fertilizer Management*—Unlike white grubs, billbug eggs are not laid in soil where they are subjected to drying. However, low to moderate billbug larval infestations can often be masked by using water and fertilizers in the summer. Adequate water may reduce plant stresses due to root and crown damage. This is not a reliable method and should be used only if spring adult controls were inadequate.

Option 3: *Biological Controls*—Few natural predators or parasites have been found attacking billbugs. However, adults are occasionally attacked by the fungus, *Beauveria*. Keeping the turf environment moist in the spring migration period may help spread this fungus. The parasitic nematode, *Steinernema carpocapsae*, has been used successfully to control billbug larvae and adults when applied at 1.0 billion infective juveniles per acre.

Option 4: *Chemical Control—Spring Pesticide Applications*—The best strategy for control of most billbug species has been early applications of pesticides targeted to kill the adults emerging from hibernation sites. For the bluegrass billbug, this time is when the soil surface (at the 1-inch [2.5 cm] depth) reaches 67 to 69°F [19.4–20.6°C]. Pitfall traps can also be used to monitor spring migration for timing of pesticide use. Denver, hunting, and Phoenician billbug adults may migrate later in the spring. See: Bluegrass Billbug, Chemical Control—Spring Adults.

Option 5: *Chemical Control—Larval Controls*—Currently, no systemic insecticides seem to be effective for control of young billbug larvae burrowing in the stems. However, most of the soil insecticides will kill billbug larvae when they drop to the soil for root feeding. Larval activity may be monitored by taking square foot samples or soil cores. See: Bluegrass Billbug, Chemical Control—Summer Larvae.

BLUEGRASS BILLBUG

Species: *Sphenophorus parvulus* Gyllenhal. [Phylum Arthropoda: Class Insecta: Order Coleoptera: Family Curculionidae] (Figures 3.59 and 3.60; Plates 3-12 and 3-13).

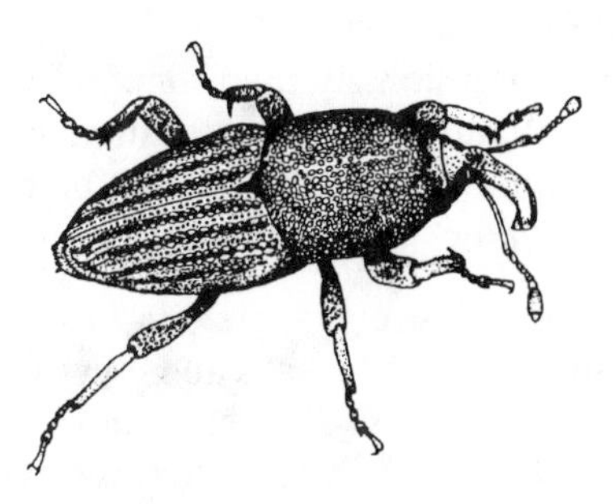

Figure 3.59 Bluegrass billbug, *Sphenophorus parvulus* Gyllenhal, adult.

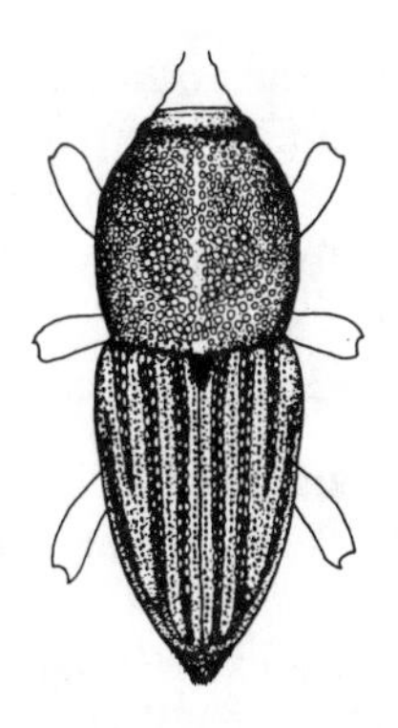

Figure 3.60 Bluegrass billbug, *Sphenophorus parvulus* Gyllenhal, dorsal view showing pattern of pits and furrows.

Distribution: This pest is most common in northern North America but may be found in southern states. It prefers cool season turf growing regions.

Hosts: Kentucky bluegrass is the preferred host, but this pest has been known to infest perennial ryegrass, red fescue and tall fescue.

Damage Symptoms: Light infestations in well kept turf results in small dead spots which look like the turf disease, dollar spot. Sometimes the damage looks like irregular mottling, browning, in the turf. Moderate infestations or chronic infestations result in sparse turf with some irregular dead patches. Heavy infestations result in complete destruction of the turf, usually in August. These infestations are sometimes confused with greenbug attacks. Billbug damaged turf turns a whitish straw color rather than yellow. Soil under damaged turf is solid, not spongy as in white grub attacks. Turf can be pulled out to reveal stems which have been hollowed out and are packed with sawdust-like fecal material.

Description of Stages: Billbugs have complete life cycles with a single generation per year and rarely a partial second generation.

Eggs: The pearly-white, kidney-shaped eggs are inserted into a small hole chewed in grass stems and are approximately $^1/_{16} \times {}^1/_{48}$ inch [1.6 × 0.6 mm].

Larvae: The typical weevil grubs are a robust C-shape with a slightly tapering abdomen. These larvae have no discernible legs, and have light yellowish brown head capsules when born. The head capsules gradually darken as the mature larvae reach about $^1/_4$ inch [6 mm] long.

Pupae: The pupae are $^5/_{16}$ inch [7–8 mm] long, and are first creamy-white. These change to a reddish brown just before the adult emergence. The pupae have the conspicuous billbug snout and pronotum.

Adults: The adults are $^5/_{16}$ inch [7–8 mm] long with a black body. When freshly emerged the adults are reddish brown. Fresh adults have a light brown coating over the body which can be rubbed off to reveal the shiny black shell. The pronotum is covered with small uniform punctures and the wing covers have distinct longitudinal furrows with pits. Occasionally the pronotal punctures fade along the top midline of the pronotum. Any other pattern indicates that another species may be present. If in doubt, this species should be confirmed by a specialist as several species of billbugs may be present in turf.

Life Cycle and Habits: In most of its range this pest overwinters in the adult stage. Adults have been found overwintering in thatch, cracks and crevices in the soil, worm holes, and in leaf litter near turf. The hibernating adults become more active in late April to mid-May when the soil surface temperatures rise above 65°F [18.3°C]. The adults wander about in search of suitable grasses and crops on which to feed. After feeding for a short period of time the female begins to insert 1 to 3 eggs in a feeding hole made in grass stems. The females may continue laying eggs into August but most eggs are laid by mid-July. Laboratory-kept females have been known to lay over 200 eggs, usually 2 to 5 per day. The eggs hatch in 6 days, depending on the temperature, and the young larvae begin to tunnel up and down the stem. If a stem is hollowed out while the larva is small, an exit hole may be formed and the larva will drop out and bore into another stem. Eventually the larva becomes too large to fit inside the grass stems and it drops to the ground to begin feeding externally on the grass crowns and roots. This is the point at which significant damage to the turf is noticed, especially if little rainfall or irrigation has occurred at this time. After 35 to 55 days the larva is full grown and pupates in a cell made of soil near the surface. The pupa gradually darkens and the reddish-brown, tineral adult emerges in 8 to 10 days. The new adults appear to be most common in late August through September. These do some minor feeding and seek out suitable sites for overwintering. Some adults have been observed trying to fly, but no great distances were covered. There is some evidence that adults that emerge early in August may begin laying eggs for a partial second generation. These larvae often do not develop rapidly enough to produce additional adults.

Control Options: See: Billbugs—Introduction.

Option 1: *Cultural Control—Use Resistant Turf Varieties*—Kentucky bluegrass varieties 'Touchdown,' 'Merion,' 'Nugget,' 'Adelphi,' 'Baron,' 'Cheri,' and 'Newport' are often susceptible to billbug attack. The varieties 'Park,' 'Arista,' 'NuDwarf,' 'Delta,' 'Kenblue,' and 'South Dakota Certified' are often resistant or tolerant to attack. Most perennial ryegrasses, especially those with endophytes, are resistant to billbugs as are the fescues. Occasionally, nonendophyte-protected ryegrasses and fescues are attacked when bluegrass nearby is heavily infested. It is strongly recommended that if roughs or other turf areas must be renovated after billbug damage, use bluegrass which has resistance or use a blend of turfgrasses containing resistant varieties or species.

Option 2: *Biological Control—Fungal Diseases*—Billbug adults and larvae seem to be susceptible to the entomophagous fungus, *Beauveria*. However, this fungus rarely attacks enough billbugs to have a significant effect on the population. No commercial preparations of *Beauveria* are currently available for use on billbugs.

Option 3: *Biological Control—Parasitic Nematodes*—The entomopathogenic nematodes, *Steinernema carpocapsae*, *S. glaseri*, and several *Heterorhabditis*, have been used to infect billbug larvae in the laboratory. In field trials, *S. carpocapsae* has provided satisfactory control of adults and larvae. For best results, apply the nematodes in early morning or evening, out of direct sunlight, and irrigate in immediately.

Option 4: *Chemical Control—Spring Adults*—This is the most commonly used strategy. Contact and/or stomach poisons are applied at the time that the adults come out of hibernation and are migrating around in search of oviposition sites. This may be from late April to mid-May. Recent research indicates that the adults become active when the soil surface temperature approaches 67 to 69°F [18.3–20°C].

Sampling

There are several methods for determining spring activity. The simplest is to use pitfall traps. Small plastic cups placed inside holes made by using a 4.5-inch [11.4 cm] cup cutter are ideal. These can be placed along the turf near or in flower beds so that they are out of the way. Adults can be easily counted by inspecting these traps 2 to 3 times a week. Suction or vacuum samplers have been relatively successful in recovering adults. Be sure to use these samplers during sunny days, as the billbugs may hide in cracks and crevices during cool overcast days. One of the most common methods of sampling is to watch driveways and sidewalks for migrating adults. This works well on hot sunny days but may miss the first activity period by a couple of weeks.

Degree-Day Timing

A degree-day model using the average method of calculation, a March 1 starting date, and a threshold temperature of 50°F [10°C] predicts that the first adult activity should occur between 280 and 352 DD_{base50} [155–195 $DD_{base10C}$]and the 30% first activity (the time that the last surface insecticide would be effective) should occur between 560 and 624 DD_{base50} [311–347 $DD_{base10C}$].

Insecticides and Application

Only a few insecticides with relatively long residual periods have been shown to be effective when used as an adulticide where migration may occur. If the entire area that is infested can be treated at one time, most of the turf insecticides will control the adults for an entire season. Chlorpyrifos (=Dursban®), cyfluthrin (=Tempo®), diazinon (not on golf courses or sod farms), ethoprop (=Mocap®), fonofos (=Crusade®, Mainstay®), isazophos (=Triumph®), and isofenphos (=Oftanol®) are registered for adult billbug control.

Options 5: *Chemical Control–Summer Larvae*–Since the billbug larvae drop out of the grass stems after several weeks of feeding, they should be susceptible to the normal soil insecticides. However, experience indicates that these larvae may do considerable damage before exiting to the soil, and many larvae may remain in the crowns and thicker stemmed rhizomes. Thus, when this strategy and timing is used, considerable damage often results unless irrigation and fertilization are also used.

Sampling

Early detection of summer larvae is difficult. However, by mid-June damaged tufts of turf can be pulled out to reveal the sawdust-like frass characteristic of this pest. By late June and into July the larvae are usually large enough to see in the soil and thatch by cutting open a flap of the turf and laying it back.

Degree-Day Timing

The larvae begin to emerge from the stems and are thus exposed to insecticides used between 925 and 1035 DD_{base50} [513–575 $DD_{base10C}$]. They can be controlled from this time until significant visual damage occurs between 1330 and 1485 DD_{base50} [739–825 $DD_{base10C}$].

Insecticides and Application

Most of the general grub insecticides (except for trichlorfon) seem to work well against billbug larvae. Apply when the first billbug grubs are detected at the crowns of grass plants. Irrigate in the application so as to carry the insecticide into the thatch zone. Bendiocarb (=Dycarb®, Ficam®, Turcam®), carbaryl (=Sevin®), diazinon (not on golf courses or sod farms), ethoprop (=Mocap®), fonofos (=Crusade®, Mainstay®), isazophos (=Triumph®), and isofenphos (=Oftanol®) are currently registered for billbug larval control.

Option 6: *Cultural Control–Masking of Damage*–With light to moderate billbug infestations, much of the damage can be masked with adequate irrigation and fertilization. The critical period for this irrigation and feeding is when the bluegrass is preparing itself for summer dormancy periods. If the parent plant is killed by a tunneling larva before the spring formed daughter plants have completely established roots, the new plants will also die. However, this lethal stress can be reduced if water is supplied at this time. Obviously, this strategy is a considerable gamble, especially if water is in short supply or the irrigation system fails for a week or two.

HUNTING BILLBUG

Species: *Sphenophorus venatus vestitus* Chittenden. [Phylum Arthropoda: Class Insecta: Order Coleoptera: Family Curculionidae] (Figure 3.61).

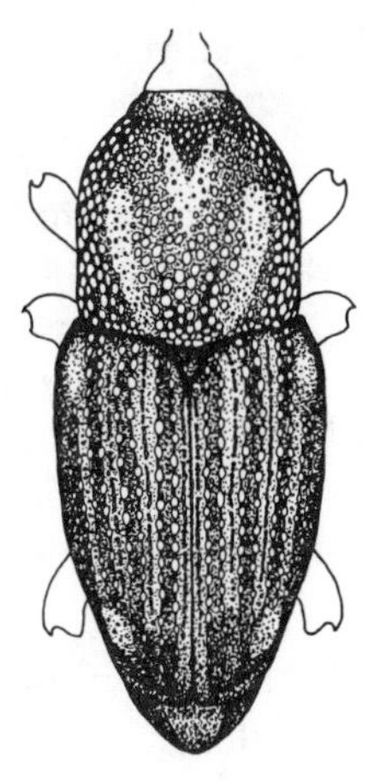

Figure 3.61 Hunting billbug, *Sphenophorus venatus vestitus* Chittenden, dorsal view showing pattern of pits and furrows.

Distribution: Several subspecies of *S. venatus* are found all across North America. *S. venatus vestitus* is most commonly found from Maryland across to Kansas and south.

Hosts: Attacks zoysiagrass and occasionally damages bermudagrass. May be found in bahiagrass, centipedegrass, and St. Augustinegrass.

Damage Symptoms: Patches of turf turn yellow and then brown. General infestations look similar to the results of summer drought or diseases. Because many roots have been damaged, the turf reacts more rapidly to dry conditions. Billbug damage can be confirmed by pulling up some of the damaged turf and looking for hollowed, sawdust filled stems.

Description of Stages: All stages are typical for weevils.

Eggs: The elongate, bean-shaped white eggs are inserted into stems of the preferred grasses.

Larvae: First instars are about 1/16 inch [1.5 mm] long, and have the typical weevil larval form. Mature larvae have tan to brown head capsules with black mandibles. Full grown larvae are 1/4–3/8 inch [7 to 10 mm] long.

Pupae: The pupae are first cream colored, changing to a reddish brown just before the adult emerges. The pupae have the distinct snout of the adult weevil.

Adults: The adults are generally larger and more robust than the bluegrass billbug. Adults range from 1/4–7/16 inch [6 to 11 mm], and often have a coating of soil which adheres to the body surface. Clean specimens have numerous visible punctures on the pronotum with a distinct Y-shaped, smooth, raised area behind the head. This area is enclosed by a shiny parenthesis-like mark on each side.

Life Cycle and Habits: Little is known about the biology of this pest. In the northern part of its range, this pest overwinters as dormant adults in the soil.

In southern states the adults may be found walking and feeding all year whenever temperatures are high enough for activity. Generally, most of the eggs are inserted into grass stems or leaf sheaths in the spring when zoysia and bermudagrasses are well out of winter dormancy. The eggs take 3 to 10 days to hatch and the young larvae mine down the inner surface of the leaf sheaths and bore into stems. As the larvae increase in size they drop into the ground and feed on roots and stolons. Mature larvae pupate after 3 to 5 weeks of feeding. The pupae take 3 to 7 days to mature. Because of the extended period of oviposition larvae may be found from May into October. However, in the northern zoysia growing regions most damage occurs in August. Damage in Gulf states may occur earlier, depending on the population size and weather. Draughty conditions seem to aggravate damage symptoms.

Control Options: See Billbugs—Introduction and Bluegrass Billbug

Option 1: *Cultural Control—Reduce Transport of Infested Turf*—The hunting billbug has been a severe pest of zoysiagrass and bermudagrass sod farms. In these situations the adults or larvae may be transported to new sites when lawns are sodded. Be sure to obtain certified pest free sod.

ANNUAL BLUEGRASS WEEVIL

Species: *Listronotus* (= *Hyperodes*) sp. near *anthracinus* (Dietz). Some confusion exists concerning the taxonomic position of this pest. [Phylum Arthropoda: Class Insecta: Order Coleoptera: Family Curculionidae] (Plate 3-11).

Distribution: *Listronotus anthracinus* has been collected only east of the Mississippi river, but the *Listronotus* causing turf damage has been localized in Connecticut, the Long Island area, Massachusetts, New York, Delaware, and Pennsylvania.

Hosts: Only annual bluegrass, *Poa annua* L., is damaged.

Damage Symptoms: Highly maintained *Poa annua* in greens, aprons, tees, fairways, and tennis courts is attacked. Light infestations cause a slight yellowing and browning of the turf. Moderate infestations cause small irregular patches of dead turf, and heavy attack kills turf in large areas. Damage begins to become obvious in late May or early June and is occasionally attributed to other causes. The hollowed grass stem is a sure diagnostic character.

Description of Stages: This pest has a complete life cycle (egg, larvae, pupa, and adults), with the larvae being typical for weevils, a C-shaped legless grub.

Eggs: The eggs are oblong, about 1/32 inch [1 mm] long, and change from yellow to smokey black when about to hatch.

Larvae: The larvae are C-shaped, legless, and have a creamy-white body. The head capsule is light brown in young larvae but becomes darker in older larvae. Newly hatched larvae are about $^{1}/_{32}$ inch [1 mm] long and grow to about $^{3}/_{16}$ inch [4.5 mm] when mature.

Pupae: Pupae are about $^{1}/_{8}$ inch [3.5 mm] long and have the beak, wing pads, and legs evident. They are first creamy white and turn red-brown before the adult emerges.

Adults: Adults are $^{1}/_{8}$–$^{5}/_{32}$ inch [3.5 to 4.0 mm] long, and have the combined length of the snout and prothorax distinctly shorter than the length of the wing covers. This is different from billbugs that have combined snout and prothoracic lengths equal to the wing cover length. This *Listronotus* is black with the wing covers coated with fine yellowish hairs, yellowish scales, and scattered spots of grayish-white scales. Newly emerged adults (callow adults) are orange-brown in color and require several days before becoming fully pigmented.

Life Cycle and Habits: This pest overwinters in the adult stage in protected areas near places *Poa annua* is cultured. They prefer to hide in tufts of bunch grasses and under leaf litter around bushes and trees. These adults become active in early spring, usually in April, and seek out actively growing *P. annua*. Here, the females chew small holes through the lower leaf sheaths and deposit from one to nine eggs; usually two eggs are laid per stem. The eggs take 4 to 5 days to hatch and the young grubs chew their way into the stems. Once inside, the larva feeds up and down the stem leaving tightly packed sawdust-like frass behind. This early feeding may not kill the stem but the leaves turn yellow and gradually wither. The larvae may move from stem to stem, hollowing out the centers until they are too large to fit inside. At this time the larvae feed externally, especially at the crown. The five larval instars are passed in 3 to 4 weeks and the fully developed larva digs about $^{3}/_{16}$–$^{1}/_{2}$ inch [5 to 10 mm] into the soil to form a earthen pupal cell. This cell is formed by the circular wiggling of the larva that packs the soil. The cell formation takes 3 to 4 days and the pupa remains in the cell for 5 to 7 days before molting into the adult form. The freshly emerged adult is light colored, soft, and remains in the pupal cell for 5 to 6 days to harden before emerging. These light colored adults are called callow adults. The entire cycle takes 25 to 45 days, depending upon food and temperature. The mature adults hide during the daytime but come to the tips of the grass stems to feed at night. This feeding continues through July and early August. The adults then seem to disappear, apparently going to their overwintering sites. There is some evidence that a few of these summer feeding weevils lay some eggs in August and September, but most wait until the following spring.

Control Options: Few natural controls appear to exist and this pest appears only where *Poa annua* is grown in specialized habitats.

Option 1: *Cultural Control—Destroy Overwintering Sites*—Removing leaf litter and cutting turf low in the fall may reduce populations. This has not proved to be an effective control.

Option 2: *Chemical Control—Early Spring Application of Insecticides*—Applications of soil insecticides between the time that *Forsythia* is in full bloom until flowering dogwood is in full bract appears to be the best time. This application is made to kill the adults before too many eggs have been laid. Chlorpyrifos (=Dursban®), cyfluthrin (=Tempo®), diazinon (not on golf courses or sod farms), isazofos (=Triumph®) and isofenphos (=Oftanol®) are registered for annual bluegrass weevil control.

CRANBERRY GIRDLER

Species: *Chrysoteuchia topiaria* (Zeller). [Phylum Arthropoda: Class Insecta: Order Lepidoptera: Family Pyralidae]. (Though technically a sod webworm, the cranberry girdler is more subterranean in habit and rarely feeds on turf leaves. Also called the subterranean sod webworm.)

Distribution: Found across North America but more common in the cool-season and transition turfgrass zones.

Hosts: Prefers cool-season grasses such as Kentucky bluegrass, bentgrass, and fine fescues.

Damage Symptoms: The larvae feed at the crown of grasses, which causes small circular areas of dead turf. Heavy populations can cause general turf death similar to grub damage.

Description of Stages: The stages are typical of moth complete life cycles, though the larvae do not have the dark spots diagnostic of most sod webworms.

Eggs: They are somewhat oval with broadly rounded ends and are approximately 0.017 × 0.013 inch [0.43 × 0.33 mm]. Only slightly raised longitudinal ridges and no cross ridges are present, as in other sod webworm eggs.

Larvae: This webworm's larvae are not typical for sod webworms because they lack the distinct dark spots over the body. Cranberry girdler larvae are dirty-white in color, with a tan head capsule. The mature larvae are 5/8–3/4 inch [16 to 20 mm] long.

Pupae: The pupae are located in a tough silken case attached to the base of plants, or in the soil. The pupae are typically light chestnut brown in color and are 5/16 × 3/32 inch [7–9 × 2.5 mm].

Adults: The adult moths have a wingspan of 9/16 to 13/16 inch [15 to 20 mm], and are quite colorful for sod webworms. The forewings have a series of longitudinal brown and cream stripes following the wing veins, a silver chevron across the wing tip followed with ocher, a series of 3 black spots, and a tip fringe of silver scales.

Life Cycle and Habits: Adult cranberry girdlers are active fliers from late June to mid-August. The adults emerge in the evening or at night to expand and harden their wings. The females begin to release a sex attractant that calls in receptive males for mating. Mated females generally do not begin laying eggs until the following day. On the second evening, fertilized females locate suitable habitat to deposit eggs. Unlike other sod webworms that lay eggs while flying, these females usually drop their eggs into the turf while resting on a plant. The females lay about 40% of their eggs on the second night and are usually finished within a week. Each female may lay 450 to 500 eggs. At 72°F [22.2°C] , eggs take 9 to 11 days to hatch. The young larvae prefer to set up residence in the thatch or upper soil surface next to grass crowns. As the larvae grow, they may actually bore into grass crowns but they are known to feed on roots, stems, and leaves. Webbed tunnels lined with fecal material are usually evident by August and September. The larvae mature rapidly, molting 7 to 8 times before cool temperatures cause the larvae to begin winter diapause. Mature or nearly mature larvae spin a tough silk case, called a hibernaculum, in the soil or thatch in October for overwintering protection. By November over 90% of the larvae will be prepupae enclosed inside the hibernaculae. In the following May, the larvae that did not mature in the fall resume feeding and may molt another time. Fully mature larvae do not feed, but pupate along with the rest of the population in May and early June. Depending on the temperature, pupae take 2 to 4 weeks to mature.

Control Options: See: Sod Webworms—Introduction. Since this species often bores into turfgrass crowns, it is usually sheltered from contact and stomach insecticides. Also, spring sod webworm treatments may not kill mature larvae which do not have to feed before pupating.

Option 1: *Natural Control*—Under normal conditions, birds (especially starlings, blackbirds, and killdeers) may reduce the population more than 80% by late fall. If considerable bird activity is noticed, check for white grubs. If no white grubs are present, water and fertilize, while not disturbing the bird activity.

Option 2: *Chemical Control—Use Pheromone Traps to Time Controls*—The sex pheromone, a combination of 1.0 mg (Z)-11-hexadecenal and 0.05 mg (Z)-9-hexadecenal, is commercially available for use in sticky traps. These traps can be placed out in June and July to monitor adult activity. Controls can be applied 7 to 10 days after peak emergence.

BURROWING SOD WEBWORMS

Species: Several species of *Acrolophus* are found across North America in association with turfgrasses. *A. plumifrontellus* (Clemens), *A. popaenellus* (Clemens), and *A. arcanellus* (Clemens) are the more common species encountered. [Phylum Arthropoda: Class Insecta: Order Lepidoptera: Family Acrolophidae] (Figures 3.62 to 3.64).

Figure 3.62 A burrowing sod webworm, *Acrolophus popeanellus* (Clemens), adult male. (after Rept. Ill State Entomol.)

Figure 3.63 A burrowing sod webworm, *Acrolophus arcanellus* (Clemens), adult male. (after Rept. Ill. State Entomol.)

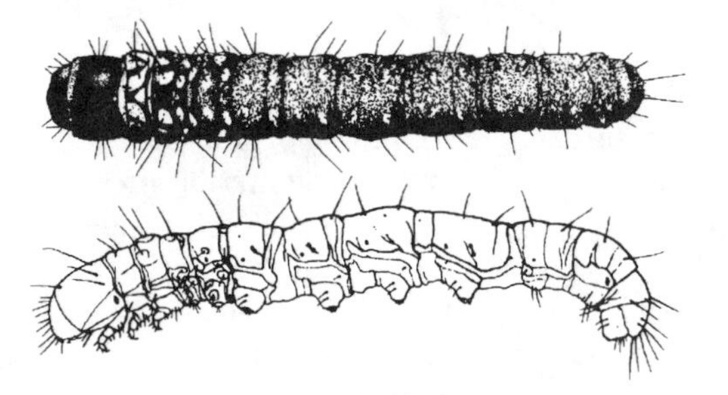

Figure 3.64 A burrowing sod webworm, *Acrolophus* sp., larva. Dorsal view (upper) and lateral view (lower) (After Rept. Ill State Entomol.)

Distribution: The above-named species are generally present east of the Rocky Mountains. Most activity is noticed in the transition and southern turfgrass zones.

Hosts: Little host specificity is known. Infestations have been recorded from Kentucky bluegrass, tall fescue, and bermudagrass turf.

Damage Symptoms: Few turfgrass areas have been significantly damaged. Damage often looks like sod webworm or cutworm activity. Occasionally the larvae will build silken tubes over the surface of the turf. These are unsightly and are often pulled out during mowing. These tubes may litter a lawn after the caterpillars pupate where they look like empty cigarette papers.

Description of Stages: Almost nothing is known about the immatures of burrowing webworms. Adults are often collected at lights and are relatively easy to identify.

Eggs: The eggs of *A. popaenellus* are roughly spherical and approximately 0.02 inch [0.5 mm] in diameter. They are cream colored when laid, but turn dark gray-brown after a few hours. Each egg has 18 to 20 longitudinal raised ridges.

Larvae: The larvae have a velvety grayish or brownish-white body with a distinctly chestnut brown head capsule and prothoracic collar. Under a 10X hand lens, the head has several raised ridges. The hooks (crochets) on the prolegs are arranged in a pear-shaped row of large hooks surrounded with many smaller hooks. Mature larvae are 1.0 × 1/8 inch [25–30 × 3 mm].

Pupae: The chestnut brown pupae are 9/16–13/16 inch [14 to 20 mm] long, with the large palps visible on the surface next to the smaller antennae.

Adults: The adults are mottled brown or reddish-brown moths with a wing span of 1.0–1 3/8 inch [25 to 35 mm]. The palps of most species are very large, hairy, and curved over the head to touch the thorax. Occasionally the palps extend across the thorax. These very active moths often have the wing scales rubbed off if they have been out for some time.

Life Cycle and Habits: Very little is known about this primitive group of moths. They are mainly tropical and subtropical in distribution, with species living on bromeliads and orchids. Many species in North America apparently feed on the roots and stems of grasses. What little is known is from observations of attacks on corn roots on no-till or minimum till corn fields. The adult moths are active from mid-June through July, though an occasional individual may be found through August. These adults fly at dusk until an hour or two after dark. They move very swiftly and upon landing, they quickly crawl and wiggle to the ground surface. The eggs are not attached to plant material but are dropped over the ground. The eggs take about 7 to 14 days to hatch and the young larvae establish themselves next to a plant stem. They soon begin to burrow vertically into the soil, where they construct a silk-lined tunnel. These burrows may extend from 4–24 inches [10 to 60 cm] into the ground, depending on the species and soil conditions. The silk tubes may be extended above the ground level and along grass stems. In very close cut turf, these tubes are occasionally built across the turf surface. The larvae spend the winter in their burrows, but resume activity in the spring. At maturity, the tunnels may be enlarged from 5/32–1/4 inch [4–6 mm] in diameter. The pupa is located in the tunnel near the ground surface.

Control Options: Burrowing sod webworms are rarely common enough to warrant controls in home lawns. However, their webbing tunnels may cause concern when they become entrapped in lawn mower blades after the adults have emerged. Occasionally, people will notice the silk tunnels which have been constructed onto the turf surface. Burrowing webworms can be controlled like most true sod webworms. See: Sod Webworms—Introduction.

EUROPEAN CRANE FLY OR LEATHER JACKET

Species: *Tipula paludosa* (Meigen) [Phylum Arthropoda: Class Insecta: Order Diptera: Family Tipulidae] (See Figure 3.65) [Other *Tipula* species may be found in cool season turf across North America. These can be treated like the European crane fly.]

Figure 3.65 A typical crane fly, *Tipula* sp., larva (leatherjacket). (after Rept. Ill State Entomol.)

Distribution: European crane fly occurs in British Columbia, Nova Scotia, Washington, and possibly Oregon; England and Europe.

Hosts: All cool-season turfgrasses.

Damage Symptoms: Bare areas, sparse growth, lodging of seed stalks and tall grasses.

Description of Stages: These flies have a complete life cycle.

Eggs: Black, 3/64 inch [1 mm] long, elongate-oval eggs with one side flattened and the other pointed are laid in turf.

Larvae: The eggs hatch into maggots, which are white and wormlike. As these larvae grow, they molt into gray to grayish-brown larvae which develop a tough skin and are commonly called "leather jackets." The larvae have four instars, and at maturity they may exceed 1.0 inch [25 mm] in length. The head capsule may be exposed during feeding or moving but is often withdrawn if disturbed. The larvae have two typical spiracular plates (breathing holes) at the end of the anal segment. These plates are surrounded by six fleshy anal lobes.

Pupae: The translucent, brownish pupae have the legs, wind pads, and antennae glued down, and are first present just below the soil surface. These 1-inch [25 mm] long pupae wiggle to the surface at emergence time.

Adults: These large crane flies have a 13/16 inch [20 mm] long, slender body with very long legs, almost 4 inches [10 cm] from the front to back leg. Adults are brownish-tan with smokey-brown wings.

Life Cycle and Habits: The European crane fly seems to require mild, winter temperatures, cool summers, and average annual rainfall of at least 24 inches [60 cm]. Adult flies, which look like giant mosquitoes, emerge from lawns, pastures, and roadsides from late August to mid-September. The adults may gather in large numbers on the sides of homes and other structures. These adults, however, cannot bite or sting. The adults mate and females begin to lay eggs within 24 hours after emerging. The eggs swell in a few days by absorbing soil moisture. In about two weeks the eggs hatch into small, brownish maggots which begin feeding by using their rasping mouthparts on plant roots, rhizomes, and foliage. By winter the larva has molted twice and reached the third instar. This instar feeds slowly during winter temperatures, but the fourth instar is reached in April and May. Turf damage is most evident in March, April, and May. The leather jackets stay underground during the day but come to the surface to feed on damp, warm nights. Feeding damage usually stops by late May into June. The larvae then rest in the upper soil and molt into the pupal stage in late July into mid-August. The pupae remain just below the soil surface for 11 to 12 days in August. When adults are ready to emerge, the pupae wriggle to the surface of the turf, from late August through September.

Control Options: Because this pest was imported from Europe, probably in the 1960s, few natural biological control agents are to be found. Even where natural controls are present in Europe, periodic outbreaks with damage occur.

Option 1: *Cultural Control—Habitat Modifications*—Crane fly eggs and first instar larvae are very susceptible to desiccation. Refrain from watering turf in September. Ryegrasses seem to be less preferred as a food host.

Option 2: *Biological Control—Parasitic Nematodes*—The commercially available nematode, *Steinernema carpocapsae*, has been effective when used at the rate of 22.2 billion infective larvae per acre; lower rates of 1 to 2 billion reduce crane fly larvae by about 50%. Other strains or species are being developed.

Option 3: *Chemical Control—Insecticide Sprays*—Good control has been obtained by applying insecticide sprays to infested turf between April 1 to April 15. Carbaryl (=Sevin®), chlorpyrifos (=Dursban®), diazinon (not on golf courses or sod farms), trichlorfon (=Dylox®, Proxol®) and isofenphos (=Oftanol®) are registered for this purpose.

FRIT FLY

Species: *Oscinella frit* (Linnaeus) though other stem boring fly larvae are known to attack grasses. [Phylum Arthropoda: Class Insecta: Order Diptera: Family Chloropidae] (Figure 3.66).

Distribution: This pest is a native of Europe, where it commonly attacks field crops. It has been found in virtually every state across North America.

Hosts: Has a wide host range in the grass family. Occasionally a pest in bentgrass and rarely a pest in Kentucky bluegrass and ryegrasses in the United States.

Damage Symptoms: Individual grass stems are killed by the boring activity of the larvae. Usually the seed head stems are destroyed, but when high populations are present, entire areas can be killed. This pest is most common in bentgrass areas where seed stems are allowed to form.

Figure 3.66 Frit fly, *Oscinella frit* (Linnaeus), adult. (USDA)

Description of Stages: Like all flies, this pest has a complete life cycle with egg, larval, pupal, and adult stages.

Eggs: The translucent shiny cream-colored eggs are slightly bean-shaped and about 1/64 inch [0.4 mm] long.

Larvae: The maggotlike larvae have no legs and no head capsule. However, the anterior end is pointed and has a pair of tiny black hooks used for rasping food. The mature larva is about 1/8 inch [3 mm] long.

Pupae: The pupae are light reddish-brown, oval, and about 5/64 inch [2 mm] long. Under magnification, only ringlike segments can be seen.

Adults: The adult flies are rather undistinguished black flies with yellowish or white markings. They are about the size of the small flies that gather around decaying fruit, approximately 3/32 inch [2–2.5 mm] long.

Life Cycle and Habits: The larvae generally overwinter in a small tunnel eaten out of a grass stem. In the spring, when the grass resumes growth, the maggot continues to feed by rasping inside the stem. As the maggot tunnels downward, it may pass nodes, killing the stem from that point outward. The maggot matures in several weeks and forms a pupa inside the stem or in the duff surrounding the grass plant. The adult flies emerge in a week or two and fly in search of new grass plants. The flies can often be seen in considerable numbers resting on the tips of grass blades in the morning or evening sun. They also tend to alight on lighter colored surfaces such as golf balls or equipment placed on the turf. The new adults insert eggs in the space between a leaf and stem and these eggs take about a week to hatch in warmer weather. Several generations can occur over a season and activity seems to be greatest when there is cool moist weather.

Control Options: Reasonable control of this pest has been obtained in Europe by using pesticides in the spring when the adults are active or by using systemics when the larvae are tunneling in the stem. At present, only diazinon (not for golf courses or sod farms) is registered for control of this pest in the United States.

THATCH AND ROOT INFESTING INSECT AND MITE PESTS

The insects that inhabit the thatch and soil zones can be some of the most devastating pests because they destroy turf roots, crowns, and underground rhizomes and stolons (Figure 3.67). These critical parts of the turf plant can tolerate light to moderate damage but when heavy damage occurs, the plant dies.

This group of pests is also the most difficult to control because of their isolation from control materials. Pesticides and biological controls have to travel through the turf canopy, then thatch which may be compacted and

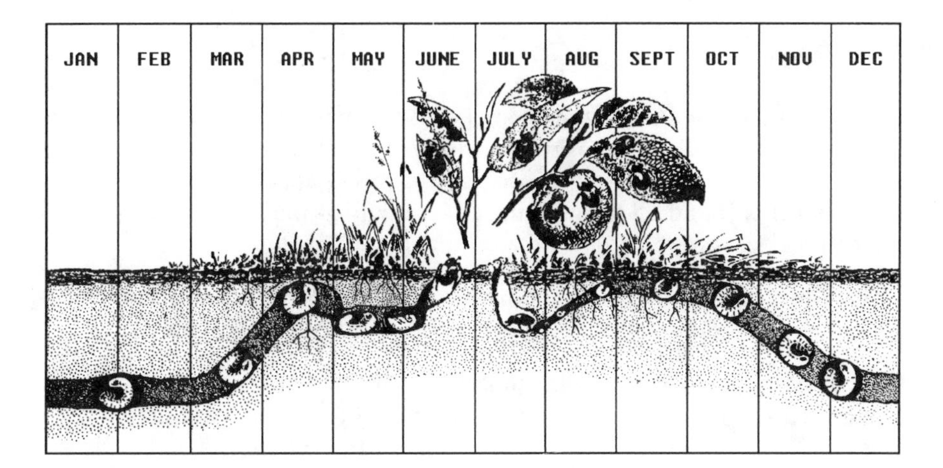

Figure 3.67 Diagram of typical Japanese beetle life cycle in turf environment. (redrawn from USDA)

impervious to water, and finally the high organic matter in the upper inches of soil. Most pesticides tend to adsorb onto the organic matter and surfaces of living plant tissues. Tiny biological controls such as entomopathogenic nematodes and bacteria are fractions of a millimeter long and quite susceptible to ultraviolet light radiation as well as immediate death from drying.

Management of thatch thickness and texture is important to managing pests that live in the thatch/soil zone. Turf managers should carefully review techniques of core aeration, top dressing, and verticutting, as well as turf cultivar selection and usage of fertilizers and irrigation.

White grubs and mole crickets are the most common and serious pests in this group. White grubs can literally graze off all the roots, so that the turf lifts up like a loose carpet. Mole crickets appear to be more selective in their feeding, but their eating of turf roots results in the death and thinning of turf. There are many other insects in this zone, most of which are beneficial or inconsequential. However, some scales (ground pearls and others), root aphids, mealybugs, and wireworms can cause noticeable damage from time to time. Unfortunately, little is known about the life histories of these occasional pests.

MOLE CRICKETS – INTRODUCTION

Species: Four species of mole crickets may be found in North American turfgrass. The tawny mole cricket (previously misidentified as the Changa or Puerto Rican mole cricket), *Scapteriscus vicinus* Scudder, and the southern mole cricket, *S. borellii* Giglio-Tos (= *S. acletus* Rehn and Hebard), are the most damaging species. The short-winged mole cricket, *S. abbreviatus* Scudder, occurs occasionally in pest levels and the native mole cricket, *Gryllotalpa* *hexadactyla* Perty, is present in small numbers. [Phylum Arthropoda: Class Insecta: Order Orthoptera: Family Gryllotalpidae].

Distribution: Turf damaging mole crickets are found south of a line running from mid-North Carolina through mid-Louisiana and into southeastern Texas. Most consistent damage occurs in Georgia and Florida. The native mole cricket is found all through the eastern half of the United States and is more of an occasional curiosity than a turf pest.

Hosts: Most damage occurs to southern grasses such as centipedegrass, bahiagrass, bermudagrass, and St. Augustinegrass.

Damage Symptoms: These pests tunnel through soil, like their mammal counterpart. This tunneling breaks up the soil around turf roots and the turfgrass often dies due to desiccation. The trails themselves are considered unsightly and interfere with ball roll on greens. During mating and overwintering the adults often throw up mounds of soil around their permanent burrows. At this time, the adults are not feeding extensively enough to kill large patches of turf. Real damage occurs in the summer months when the nymphs are actively feeding on the turfgrass roots. Heavy infestations during this period may result in large dead patches and exposed soil. St. Augustinegrass does not show as severe a response to mole crickets, possibly because of its different growth habits.

Life Cycles and Habits: These pests have gradual life cycles with eggs, nymphal, and adult stages. The nymphs look like the adults but are smaller and do not have wings. Most species mature by fall but do not lay eggs until the following spring.

Identification of Species: The three turf infesting mole crickets can be identified using the following key (Figure 3.68):

1a. Four dactyls, "claws", on the foretibia. Hindfemur longer than pronotum .... species of *Grylotalpa* and *Neocurtillus*.

b. Two dactyls on the foretibia. Hindfemur shorter than pronotum, (Species of *Scapteriscus*) 2

2a. Tibial dactyls widely separated, by more than 0.30 mm, gap appearing U-shaped 3

b. Tibial dactyls narrowly separated, by less than 0.30 mm, gap appearing V-shaped .... *S. vicinus* Scudder, tawny mole cricket

3a. Forewing short, extending no more than one-third the abdomen length. Hindwings hidden .... *S. abbreviatus* Scudder, short-winged mole cricket

b. Forewing reaching end of abdomen. Hindwings extending beyond tip of abdomen .... *S. borellii* Giglio-Tos, southern mole cricket

Control Options: These turfgrass pests are some of the most difficult to control. Their ability to tunnel, fly, and eat a wide range of foods, contributes to the problems. They can tunnel downward and escape most pesticide applications, especially during dry spells. Spring migrations can occur for several

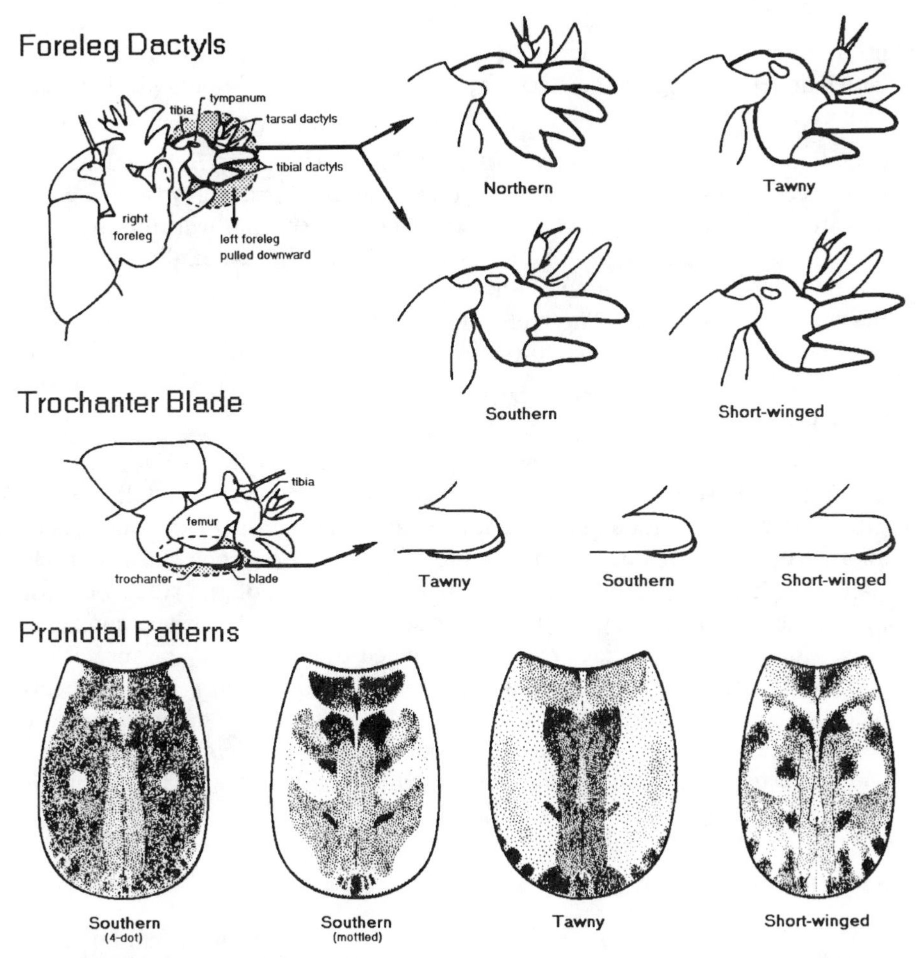

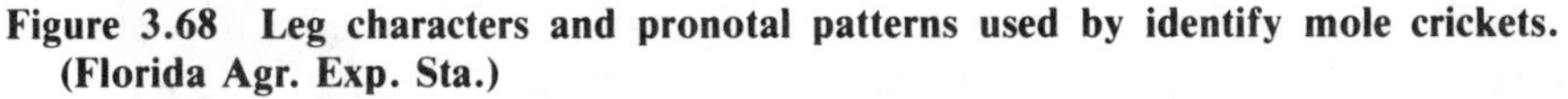

Figure 3.68 Leg characters and pronotal patterns used by identify mole crickets. (Florida Agr. Exp. Sta.)

months and adults may travel up to 6 miles from pastures and road rights-of-way. Reinfestations occur constantly.

Option 1: *Biological Control—Encouragement of Natural Agents*—Mole crickets are cannibalistic and many young nymphs perish in this manner. Predacious insects, birds, and mammals prey on mole crickets but may dig up turf in the effort. The parasitic wasp, *Larra bicolor* Fab. (Hymenoptera: Sphecidae), has been introduced, but it has not been effective. Apparently, it cannot easily maintain itself without mild winter temperatures and nectar producing flowers. Obviously, this is not a very satisfactory strategy because extensive damage can occur before these control agents have any real effect.

Option 2: *Biological Control—Parasitic Nematodes*—Several insect parasitic nematodes, species of *Steinernema* and *Heterorhabditis*, have been used in laboratory experiments to cause mortality in mole crickets. Field tests have not resulted in satisfactory levels of control, but new species of nematodes and better understanding of their ecology may result in usable biologicals in the near future.

Option 3: *Chemical Control—Use of Food/Attractant Baits*—Baits are most effective if applied to destroy young nymphs after a heavy rain or good irrigation. The nymphs will come to the surface to feed and will find the bait. Adult mole crickets are less liable to pick up the bait, especially the spring reproducing ones. Baits should be applied while the soil is moist, but not when the grass foliage is wet. The bait granules often stick to the foliage of wet plants and do not filter down to the soil surface where they are needed. Also, do not water for several days after applying a bait, as this may leach out the pesticide or degrade the bait granule.

Option 4: *Chemical Control—Targeted Insecticides for Spring Adults*—Control of mole crickets at this time is very difficult because the adults are not often actively feeding, and many of the females are migrating from one area to another in search of mates and suitable egg laying areas. Once in a suitable spot, females may burrow deep into the soil, well out of the reach of most insecticides. If this strategy is used, it must be understood that any control is only temporary. Insecticide treatments applied one week may be ineffective two weeks later when new migrants have reinvaded the turf. Thus, retreatments or a repeat treatment program must be followed if no mole crickets are to be tolerated. This is obviously an expensive and time-consuming process which is usually not economically warranted, considering the small amount of damage eliminated.

Sampling for Spring Adults

This stage is difficult to sample. Usually only tunneling and mound making are the obvious signs of activity. The traditional method of using a disclosing irritant may not reveal the total mole cricket population. This is because the adults may be hiding 1–2 feet [30–60 cm] down in the soil and may not be reached by the irritant. Disclosing solutions work best on adults when applied after rains or heavy irrigation. Use a rating grid to determine where activity is most prevelent. These areas should be sampled for nymphs in May and June.

Insecticides and Application

None of the registered insecticides give outstanding results in a single application for spring mole cricket control. Most of the rapid acting insecticides provide some dead mole crickets on the surface for easy observation, but these treatments do not prohibit reinfestations. All of the materials work best when applied after irrigation or rain, followed with immediate watering in. Acephate (=Orthene®) and baits containing carbaryl (=Sevin®) or chlorpyrifos (=Dursban®) have produced some control during this period.

Option 5 : *Chemical Control—Targeted Insecticides for Summer*

Nymphs—The majority of mole cricket nymphs are present from June into September, though nymphs are found in some areas year-round. During this summer period it is easier to use insecticides to reduce the general population and reduce the potential for damage. The young nymphs are usually at the surface or within a few inches of the surface, and are constantly feeding. Thus, they are more likely to come into contact with any insecticides present.

Sampling for Summer Nymphs

Summer nymph populations are often missed when the nymphs are small. These tiny insects make minute trails on the surface of bare soil but are usually completely hidden by turf cover. Because of this, a disclosing solution must be used to detect their presence. By August, the nymphs have matured considerably and their activity is easy to visually detect. This late detection should be avoided because larger nymphs are more difficult to kill and considerable damage may have occurred.

Insecticides and Application

Most of the registered insecticides work best against this stage of the mole cricket. As with spring applied insecticides, these generally work best when applied after a rain or after irrigation. Generally, baits are more effective during this period of time (see above). Acephate (=Orthene®), bendiocarb (=Dycarb®, Ficam®, Turcam®), carbaryl (=Sevin®), chlorpyrifos (=Dursban®), cyfluthrin (=Tempo®), diazinon (not on golf courses or sod farms), ethoprop (=Mocap®), fonofos (=Crusade®, Mainstay®), isazophos (=Triumph®), and isofenphos (=Oftanol®) are registered for mole cricket control.

TAWNY MOLE CRICKET

Species: *Scapteriscus vicinus* Scudder (misidentified as *S. didactylus*, the Changa or Puerto Rican mole cricket) [Phylum Arthropoda: Class Insecta: Order Orthoptera: Family Gryllotalpidae] (Figure 3.69; Plate 3-3).

Distribution: Introduced into Georgia around 1899; now found throughout Florida and south of a line running from the southern tip of South Carolina, across Georgia and to the southern tip of Alabama; a native of coastal South America from northern Argentina through Brazil and in Columbia, Panama, and Costa Rica.

Hosts: Prefers to live in bahiagrass, but also damages close-cut bermudagrass. Occasionally damages St. Augustinegrass, centipedegrass, and zoysiagrass.

Damage Symptoms: Produces typical mole cricket damage to turf. This species prefers to feed on roots, but also damages turf by tunneling and throwing up mounds of soil around the permanent burrows. Loosened and uprooted

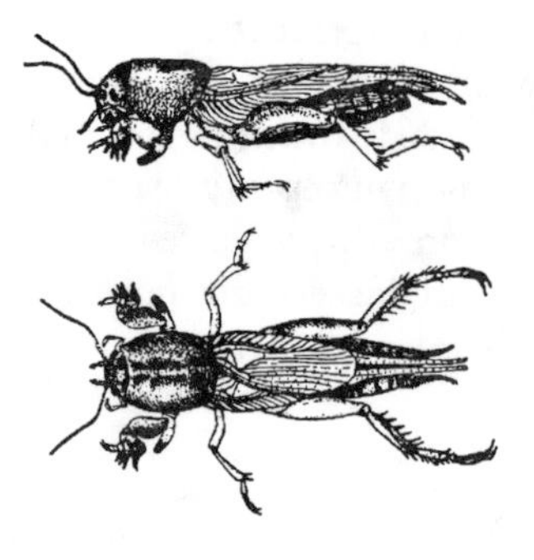

Figure 3.69 Tawny mole cricket, *Scapteriscus vicinus* Scudder, adult. (USDA)

bahiagrass usually withers and dies, leaving trails of dead turf. Heavy populations can completely kill turf, leaving bare patches of soil.

Description of Stages: This pest has a typical gradual life cycle.

Eggs: Round translucent white eggs are placed in clusters in chambers 3–10 inches [7.5–25 cm] below the soil surface.

Nymphs: This stages looks like miniature adults. Early instars do not have visible wing pads and later instars begin developing visible wing pads. The nymphs molt 6 to 8 times; the exact number is not known. Even at this early age, this species can be identified by the V-shaped space between the tibial dactyls.

Adults: Fully grown adults are about 1¼ × ⅜ inch [30–34 × 8–10 mm]. They are a light tawny brown and have the obviously modified front legs and enlarged thorax typical of mole crickets. The forewings are shorter than the abdomen, and males have the darker rasp at the forewing base. The V-shaped space between the tibial dactyls is species characteristic.

Life Cycle and Habits: Females are attracted to the calls of males in the spring when temperature and moisture allow. Egg laying may begin in March, but 75% of the eggs are laid between May 1 and June 15. During the mating and oviposition periods the adults are extremely active, emerging from the soil at dusk to crawl about, tunnel, and fly. Flight in mass commonly occurs after rainfall, and adults are attracted to lights. Mated females dig down a few inches into suitable soil and construct an egg chamber in which they lay an average of 35 eggs. Females may construct three to five chambers and lay 100 to 150 eggs total. The eggs hatch in about 20 days and the young nymphs dig upward to seek out food. These nymphs prefer to feed on plant roots but will eat small insects, and can be cannibalistic. Larger nymphs will often attack smaller nymphs and eggs. Only 10% to 15% of the adults and nymphs have been found with insect parts in the gut. The nymphs continue to feed, molt, and grow through the summer months. By the end of October, 85% of the nymphs have reached adulthood. The remaining nymphs continue development very slowly and most mature by the following spring. Adult and nymphal behavior are highly regulated by temperature and soil moisture. Most feeding occurs at night, especially after rain showers or irrigation during warm weather. These insects may tunnel 20 feet per night in moist soil. During the day, individuals tend to return to a permanent burrow which is deeper in the ground. The crickets may remain in these permanent burrows for considerable periods during cool winter temperatures or dry spells. Male mole crickets locate preferred habitats in the spring and call with their toadlike trill (3.3 kHz with 60 cycles/sec at 77°F [25°C]) from the entrance of their burrows. This occurs for about an hour shortly after sunset. Mole crickets can fly more than six miles in a night and may fly more than once. This accounts for the great difficulty in eliminating populations in any given area.

Control Options: See: Mole Crickets—Introduction.

SOUTHERN MOLE CRICKET

Species: *Scapteriscus borellii* Giglio-Tos (= *S. acletus* Rehn and Hebard) [Phylum Arthropoda: Class Insecta: Order Orthoptera: Family Gryllotalpidae].

Distribution: Introduced into Georgia at the turn of the century; now found south of a line running from mid-North Carolina through mid-Louisiana and into eastern Texas; probably a native of South America, where similar mole crickets have been found in northern Argentina, Uruguay, Paraguay, Bolivia, and Brazil.

Hosts: Apparently this species prefers to prey on other insects and even each other. However, considerable feeding on plants occurs; 41% were found to have plant material in the gut. Often found in bahiagrass, bermudagrass, and occasionally St. Augustinegrass. Rarely damages centipedegrass or zoysiagrass.

Damage Symptoms: Produces typical mole cricket damage, though most of the damage is due to tunneling. Nymphs and adults tunnel and throw up mounds of soil. Uprooted turf soon wilts and dies from desiccation.

Description of Stages: This pest has a typical gradual life cycle.

Eggs: Round translucent white eggs are placed in clusters of up to 35 in chambers underground.

Nymphs: The nymphs look like small adults but do not have functional wings. The wings develop as pads which enlarge with each molt. This species can be identified by the U-shaped space between the tibial dactyls. Some populations have mottled patterns on the pronotum, while other populations have a dark background with four small pale spots.

Adults: Adults are about 1¼ × ⅜ inch [32 × 9 mm]. They are reddish to dark brown and have either four distinct pale spots on the prothorax, or molting. The forewings are shorter than the abdomen, and the hindwings just extend beyond the tip of the abdomen. The U-shaped area between the tibial dactyls is diagnostic of the species. Males also have the darker rasp and file on the forewing base.

Life Cycle and Habits: This species has a cycle much like the tawny mole cricket. Egg laying occurs from mid-March into June, but some adults continue laying eggs into September. The nymphs mature rapidly through the summer but only about 25% reach adulthood by winter. Some males call during the fall and mating at this time may occur. The nymphs complete development slowly during the winter, with more adults appearing from February through April with a peak in May. Nymphs and adults of this species have been found during much of the year. Most activity is found after rain or irrigation, and a spring flight period occurs during mating and oviposition. Males produce a bell-like trill (2.8 kHz with 135 cycles/sec) for an hour or longer just after sunset. The southern mole cricket is more of an omnivore than the tawny mole cricket. Only 5% have been found with plant material alone in the gut, while 59% had only animal material in the gut and the rest had a combination. Thus, it is suspected that turfgrass damage caused by the southern mole cricket is due to its tunneling and soil mounding rather than actual root feeding.

Control Options: See: Mole Crickets—Introduction.

SHORT-WINGED MOLE CRICKET

Species: *Scapteriscus abbreviatus* Scudder [Phylum Arthropods: Class Insecta: Order Orthoptera: Family Gryllotalpidae] (Figure 3.70).

Distribution: Florida and Georgia in the United States; native to Argentina, Paraguay, and Brazil; also introduced into Puerto Rico, Virgin Islands, Cuba, Nassau, and Haiti.

Hosts: Seems to prefer plant material for food.

Damage Symptoms: Produces typical mole cricket damage but the damage is often attributed to the nymphs of one of the other species of *Scapteriscus*. This pest does not fly and thus tunnels to disperse.

Description of Stages: This pest has a typical gradual life cycle.

Eggs: Similar to other mole crickets.

Nymphs: Similar to other mole crickets. Difficult to distinguish from the southern mole cricket, which also has a V-shaped space between the tibial dactyls. However, the ocelli are smaller in the short-winged mole cricket.

Adults: Similar to other mole crickets except that the hindwings extend no more than one-third the length of the abdomen. The hindwings are very small and the ocelli are smaller than in *S. borellii*.

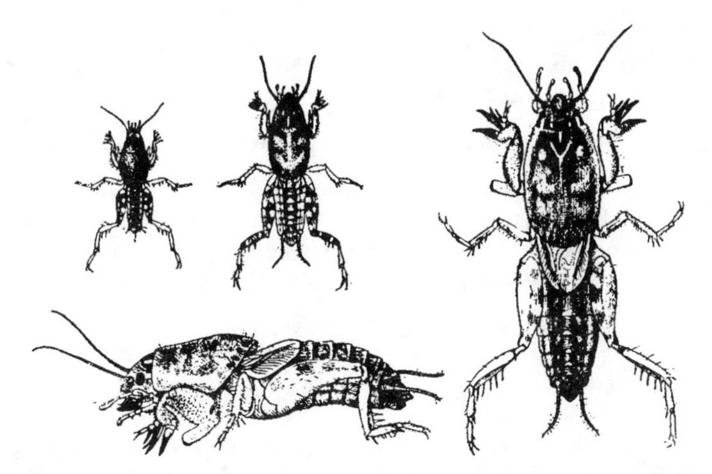

Figure 3.70 Short-winged mole cricket, *Scapteriscus abbreviatus* Scudder. (top, left to right) 2nd instar nymph, 4th instar nymph, adult; (bottom left) adult, side view. (USDA)

Life Cycle and Habits: This species is probably the least understood of the turf infesting mole crickets. It was apparently more of a pest before the tawny mole cricket was imported. However, in recent years more short-winged mole crickets have been found in turf, and its spread seems to be increasing because of transporting soil. This species cannot fly and must naturally spread by crawling or tunneling. The males do not make a mating call chirp but apparently produce low chirping sounds if a prospective mate is found. This pest seems to have a cycle similar to the southern mole cricket; that is, most reach adulthood by the fall.

Control Options: See: Mole Crickets—Introduction.

GROUND PEARLS

Species: Two species of *Margarodes* are involved; usually *M. meridionalis* Morrill is the species found attacking turfgrasses [Phylum Arthropoda: Class Insecta: Order Homoptera: Family Margarodidae] (Plate 3-9).

Distribution: Generally found attacking warm season turfgrasses from Southern California across to North Carolina.

Hosts: Attacks bermudagrass, centipedegrass, zoysiagrass, and St. Augustinegrass. Most commonly damages bermudagrass and centipedegrass.

Damage Symptoms: Irregular patches of the turf appear unthrifty, and over a year or two thin out or die. Turf often turns yellowish. Damage is most common during dry spells.

Description of Stages: Very little is known about this group of pests and descriptions are scant. Apparently a modified gradual life cycle such as is found in some mealybugs and scales is followed.

Eggs: Light pinkish eggs are deposited in a white waxy sac by the females.

Nymphs: The first nymphal instar, called the crawler, is about 0.008 inch [0.2 mm] long. This stage attaches to a grass root and begins to cover itself with a hard coating of yellowish to light purple wax. This is the ground pearl. The encysted nymphs range from 0.02–5/64 inch [0.5–2.0 mm] in diameter.

Adults: The adult females are pinkish sac-like forms, about 1/16 inch [1.6 mm] long, and have well-developed front legs and shorter second and third legs. The males are rarely seen, but are tiny white to pinkish gnat-like insects.

Life Cycle and Habits: Because these insects are essentially subterranean for most of their lives, little is known about their exact behavior. Generally it is thought that there is only one generation per year. Mature females occur in late spring, during which time they emerge from their hard waxy cysts and crawl to the soil surface, where they mate with the tiny winged males. Once mated, the females dig 2–3 inches [5–7 cm] deep into the soil and form a new waxy coat in which they lay about 100 eggs. The eggs hatch into crawlers which disperse in the soil in search of grass roots. When suitable grass roots are found, the crawlers attach themselves and secrete a hard waxy coating which becomes the ground pearl stage. The nymphs continue to develop inside this cyst and overwinter attached to the root.

Control Options: Since these pests are protected during most of their life cycle inside the waxy cyst, no insecticides have been found to be effective.

Option 1: *Cultural Control—Maintain Healthy Turf*—Watering during drought and good fertilization have been the most effective methods used to counter the attacks of ground pearls.

WHITE GRUBS—INTRODUCTION

Species: White grubs (grubworms or simply, grubs) are the C-shaped larvae of a large group of beetles called scarabs. Many species of scarabs are found in North America, and several of these commonly attack turfgrasses. The most important species are: Japanese beetle, *Popillia japonica* Newman; May or June beetles, *Phyllophaga* spp.; northern masked chafer, *Cyclocephala borealis* Arrow; southern masked chafer, *C. lurida* Burmeister; southwestern masked chafer, *C. pasadenae* Casey; western masked chafer, *Cyclocephala hirta* LeConte; black turfgrass ataenius, *Ataenius spretulus* (Haldeman); and the green June beetle, *Cotinis nitida* (Linnaeus). Other, more localized, white grub pests are: European chafer, *Rhizotrogus majalis* (Razoumowsky); the Asiatic garden beetle, *Maladera castanea* (Arrow); and Oriental beetle,

Exomala orientalis (Waterhouse). [Phylum Arthropoda: Class Insecta: Order Coleoptera: Family Scarabaeidae] (Plate 3–10).

Distribution: White grubs are perennial pests of all the cool-season and transition zone turfgrasses. Occasionally white grubs will attack southern grasses when the turf is highly maintained and cut short. May/June beetles, masked chafers, green June beetles, and the black turfgrass ataenius can be found across most of North America. The Japanese beetle is present in eastern states, though occasional infestations are introduced in plains and western states. The European chafer, Asiatic garden beetle, and Oriental beetles are fairly restricted to the northeastern states.

Hosts: All species of turfgrass may be infested.

Damage Symptoms: White grubs eat organic matter, including the roots of plants. Therefore, damage first appears to be drought stress. Heavily infested turf first appears off-color gray-green, and wilts rapidly in the hot sun. Continued feeding will cause the turf to die in large irregular patches. The tunneling of the larvae causes the turf to feel spongy underfoot, and the turf can be rolled back like a loose carpet. Grub populations may not cause observable turf injury, but predatory mammals such as skunks, raccoons, opossums, armadillos, and moles or birds may dig in search of a meal.

Life Cycles and Habits: Scarabs have a complete life cycle with eggs, larvae, pupae, and adults. Japanese beetles, masked chafers, European chafers, Asiatic garden beetles, Oriental beetles, and green June beetles have annual life cycles. The May/June beetles usually take 2 to 3 years to develop in northern states but some southern species have annual cycles. The black turfgrass ataenius has 2 to 3 generations per summer. Most turf scarabs overwinter as larvae but the black turfgrass ataenius and May/June beetle adults overwinter.

Identification of Species: The adults are easily identified to genus but the grubs are the stage usually found in the turf. The grubs are identified by the form, shape, and arrangement of bristles (the raster) on the last abdominal segments. A 10× or 15× hand lens is usually adequate for identification and the common white grub groups can be identified using a raster pictorial key (Figures 3.71 through 3.78).

Control Options: White grubs seem to be periodic pests, attacking turf areas irregularly from year to year. The major factor influencing development of damaging numbers of grubs is soil moisture and rainfall. In general, in years with normal or above normal rainfall, grub populations increase. Well maintained turf next to ornamental plants favored by the adults seems to be more commonly attacked. However, masked and European chafer adults do not feed, and these pests build up in well watered and maintained turf. Black turfgrass ataenius and green June beetle adults seem to be highly attracted to turf with decaying thatch layers.

Option 1: *Cultural Control—Host Plant Modifications*—Certain species of

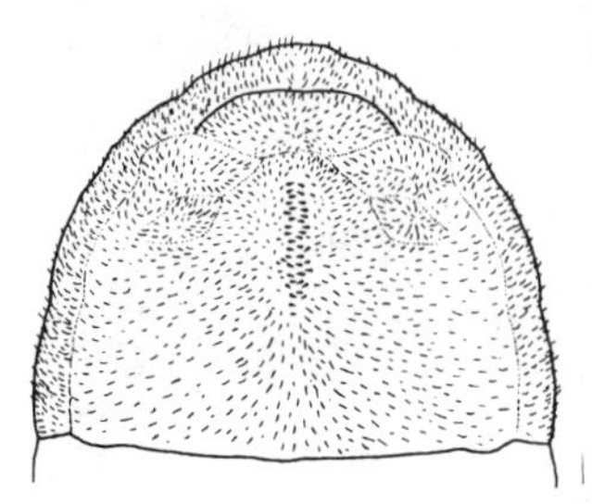

Figure 3.71 Green June beetle, *Cotinis nitida* (Linnaeus), raster pattern.

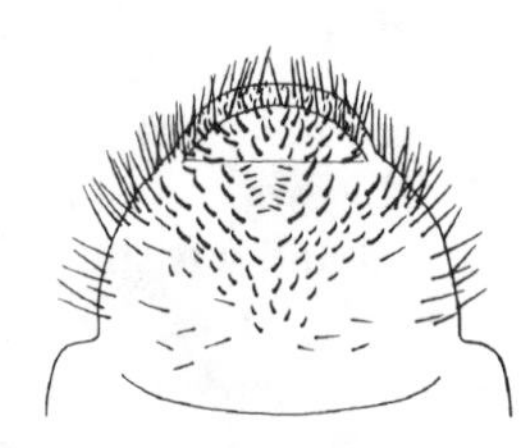

Figure 3.72 Japanese beetle, *Popillia japonica* Newman, rastern pattern. (redrawn from Ritcher)

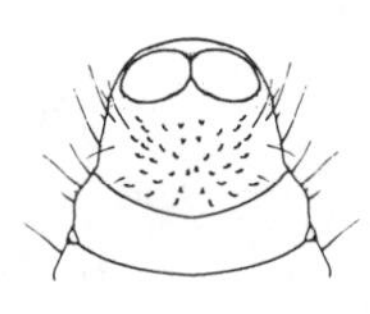

Figure 3.73 Black turfgrass ataenius, *Ataenius spretulus* (Haldeman), raster pattern and anal pads. (redrawn from Wegner)

scarab adults prefer specific host plants. Where Japanese beetles are common, do not plant roses, grapes, and lindens along high maintenance turf areas. May/June beetles prefer oaks, and the green June beetles feed on ripening fruit such as peaches. The fine and tall fescues are not as severely attacked as Kentucky bluegrass and perennial ryegrass.

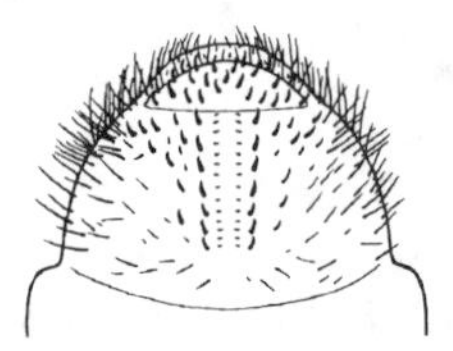

Figure 3.74 Oriental beetle, *Exomala orientalis* (Waterhouse), raster pattern. (redrawn from Ritcher)

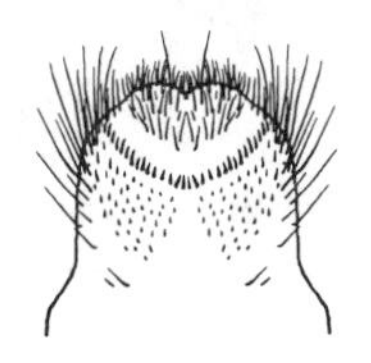

Figure 3.75 Asiatic garden beetle, *Maladera castanea* (Arrow), raster pattern. (redrawn from USDA)

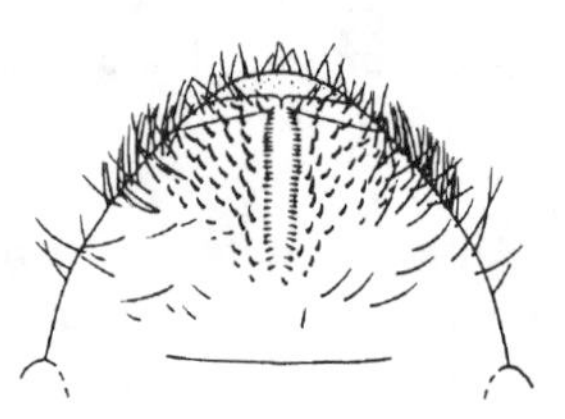

Figure 3.76 Typical May/June beetle, *Phyllophaga* sp., raster pattern. (redrawn from Ritcher)

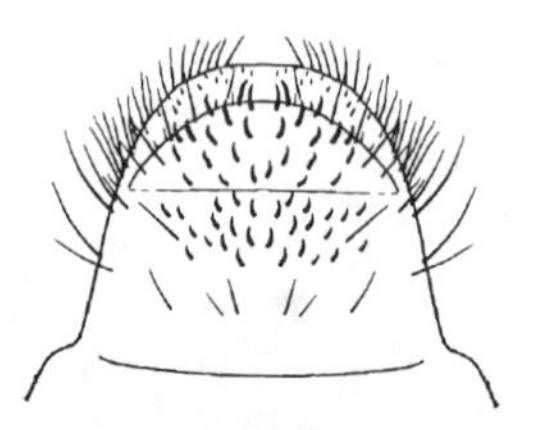

Figure 3.77 Masked chafer, *Cyclocephala* sp., raster pattern. (redrawn from Ritcher)

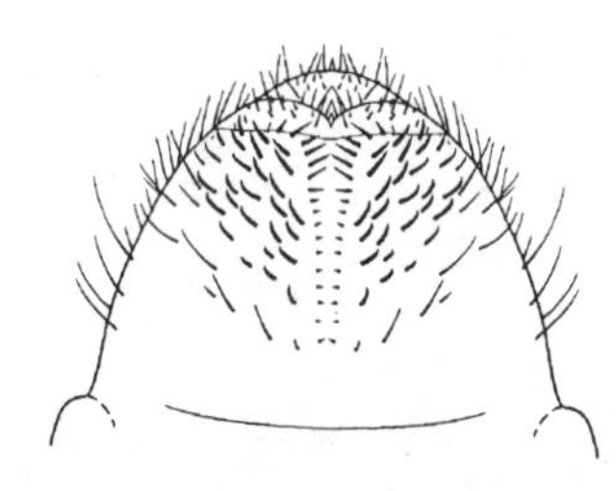

Figure 3.78 European chafer, *Rhizotrogus majalis* (Razoumowsky), raster pattern. (redrawn from Ritcher)

Option 2: *Cultural Control—Water Management*—Practically all white grub species require moist soil for their eggs to hatch. The young larvae are also very susceptible to desiccation. In areas where turf can stand some moisture stress, do not water in July and early August when white grub eggs and young larvae are present. On the other hand, moderate grub infestations can be outgrown if adequate water and fertilizer is applied in August through September and again in May when the grubs are feeding. This latter strategy is not preferred because mammals may dig up the turf, or irrigation bans may occur.

Option 3: *Natural Controls—Parasites*—Several parasitic wasps, *Tiphia* spp. and scoliids, attack white grubs and may effectively reduce populations in certain areas. Masked chafers and green June beetles are the species most commonly attacked. However, these parasitic wasps may take 2 to 3 years to build up effective populations, during which time turf damage may occur. Several parasites have been imported for control of the Japanese beetle, but most are poorly established and restricted to only a few states.

Option 4: *Biological Control—Milky Diseases*—Several strains of the bacterium, *Bacillus popilliae*, have been found which attack white grubs. However, the commercial preparation of this bacterium is extracted from Japanese beetle grubs and is most active for this species. This bacterium is picked up by feeding grubs, and causes the body fluids to turn a milky white before grub death. Fresh bacterial preparations should be used and 3 to 5 years are needed to provide lasting controls. Unfortunately, recent studies in Kentucky and Ohio indicate that the currently available products have not performed well in their soils.

Option 5: *Biological Control—Parasitic Nematodes*—Insect parasitic nematodes in the genera *Steinernema* (= *Neoaplectana*) and *Heterorhabtitis* have been shown to be effective against white grubs. Field trials of *S. carpocapsae* strains have generally resulted in less than 50% control, though *H. heliothidis* strains have achieved 80% control or better. At present, available strains do not appear to be effective from one season to the next.

Option 6: *Chemical Control—Preventive Pesticide Applications*—Since white grub occurrence is rather sporadic, applying pesticides for control of anticipated grub populations is **not** recommended. However, in areas where adult activity has been observed or perennial infestations have occurred, preventive applications may be warranted. Currently, isofenphos (=Oftanol®) and isazofos (= Triumph®) are the only registered products which seem to have extended activity.

Option 7: *Chemical Control—Early Reactive Pesticide Applications*—Most of the modern soil insecticides have short active residual periods (three weeks or less) and must be used when the grubs are actively feeding. No registered insecticide is 100% effective; they usually kill 75% to 90% of the grubs present in any given area. This is why reapplications may be necessary when grub populations get very high. Timing of treatments is critical for success. You should apply the pesticide early, when the grubs are small and actively feeding, yet late enough to catch all of the population. In general, reducing thatch and using good irrigation after making a pesticide application will increase control.

Option 8: *Chemical Control—Late Season Reactive Pesticide Applications*—Occasionally turfgrass damaging populations of white grubs may go undetected until September or October. By this time the annual white grubs are usually third instars and may be 70–80 times the body weight of a newly hatched grub. These mature grubs are voracious feeders, but are ready to dig down into the soil when cold weather arrives. Chemical control of these large grubs is difficult, at best. If a late season insecticide application is needed, diazinon (not on golf courses or sod farms), isazofos (=Triumph®) and trichlorfon (=Dylox®, Proxol®) have been the most successful. Be sure to irrigate well after the application in order to keep the grubs near the soil/thatch interface and to wash in the pesticide.

Option 9: *Chemical Control—Spring Pesticide Applications*—As with the late fall pesticide applications, spring treatments are often ineffective. Though the grubs feed during the spring, they are quite large and the span of time for treatment is short. If a spring application is deemed necessary, check to make sure that the grubs are actively feeding at the soil/thatch level. Diazinon (not on golf courses or sod farms), isazofos (=Triumph®) and trichlorfon (=Dylox®, Proxol®) have been the most effective pesticides at this time.

Maximizing Control of White Grubs with Insecticides

Adult Sampling. Adult activity of May/June beetles, masked chafers, European chafers, Oriental beetles, and Asiatic garden beetles can be monitored using light traps. Useful predictive data can be obtained by monitoring beetle captures one to two times a week. Simply plot the number of beetles collected over the date sampled on graph paper. If the number of beetles collected drops for 7 to 10 days in a row, you can assume that the peak emergence and oviposition time has passed. Most species have eggs that hatch within 14 to 21 days. Therefore, grub insecticides can be applied 3 to 4 weeks after the peak adult activity was noted, in order to target the young grubs feeding at the soil/thatch interface.

Grub Sampling. White grub populations should be assessed when the grubs become large enough to be easily seen (August for the annual grubs and early June for black turfgrass ataenius). Assess by taking square foot samples several places over the turf area. Populations of annual grub species which are less than six grubs per ft^2 can usually be masked by water and fertilizers. Populations between 10 and 15 per ft^2 can cause significant turf damage later in the fall, September and October. Of course, populations occasionally reach 40 to 60 grubs per ft^2 and these levels can cause damage by late August.

Time spent doing grub sampling can usually be reduced by sampling only in the most likely turfgrass habitats. Most of the annual white grubs seem to prefer grass in sunny areas. The night flying grubs are often attracted to street lights and may lay large numbers of eggs under or near these lights. Black turfgrass ataenius adults prefer to lay their eggs in the compacted, moist, and decaying thatch. The green June beetle prefers sunny, thatchy turf or areas which have had manure applied as a fertilizer. Japanese beetle adults usually attack high quality turf near favorite food trees and shrubs.

Insecticides and Application

An appropriate registered pesticide should be selected according to the current needs and situation. If irrigation is available, liquid applications are very effective; granular insecticides are often more effective where irrigation is not possible. Bendiocarb (=Dycarb®, Ficam®, Turcam®), carbaryl (=Sevin®), chlorpyrifos (=Dursban®), diazinon (not on golf courses or sod farms), ethoprop (=Mocap®), fonofos (=Crusade®, Mainstay®), isazophos (=Triumph®), isofenphos (=Oftanol®) and trichlorfon (=Dylox®, Proxol®) are registered for management of white grubs. Table 3.1 contains published data on the performance of currently registered insecticides.

Recent studies have established that 95% to 99% of any pesticide used for grub control ends up in the thatch. Pre- or postirrigation does not seem to change this binding. If the thatch layer is one inch thick or more, the grubs probably will not contact challenging doses of the insecticides.

Several of the grub insecticides seem to be working less effectively in some geographic areas. Grub resistance was suspected, but the actual problem has been documented to be microbial degradation (=enhanced degradation). Modern synthetic pesticides generally degrade rapidly. However, many are subject to additional degradation by bacteria and fungi which use the compounds as food sources. These microbes tend to build up if a pesticide is used continuously. To reduce the chances of creating enhanced microbial degradation problems, use a pesticide only once, when needed, and alternate pesticides.

Table 3.1. Efficacy of White Grub Insecticides—1976–1989[a]

Insecticide	Rate lb.ai./a.	Avg % Control	# Tests	Range % Control	% of Tests Below 70%
Bendiocarb	2.0	75.4	16	25–100	19
	4.0	83.5	23	38–99	17
Carbaryl	8.0	77.3	19	13–95	31
Chlorpyrifos	4.0	48.3	25	0–96	64
Diazinon	4.0	67.9	14	47–99	42
	5.5	74.5	40	25–100	30
Ethoprop	5.0	72.1	24	48–97	42
Fonofos	4.0	65.4	10	53–93	30
Isazophos	2.0	85.2	38	46–100	16
Isofenphos	2.0	80.5	55	40–100	22
Trichlorfon	8.0	76.0	46	0–98	20

[a] Data from Insecticide and Acaricide Tests, Entomological Society of America.

In general, irrigating after an insecticide application is made will improve performance for soil insect control. It is also generally recommended that grass clippings be returned to the lawn for one to two mowings after a grub insecticide application. Do not wait more than 30 days to recheck the grub infestation, especially if the original population was high. If the grub population has not been reduced below six grubs per square foot, consider reapplication of another pesticide. Remember, the smaller the grubs the easier they are to kill with insecticides.

BLACK TURFGRASS ATAENIUS

Species: *Ataenius spretulus* (Haldeman). [Phylum Arthropoda: Class Insecta: Order Coleoptera: Family Scarabaeidae] (Figure 3.73; Plate 3–10).

Distribution: The species is found in all states east of the Rocky Mountains as well as California. Similar species in this genus as well as *Aphodius* are know to attack turfgrasses worldwide.

Hosts: Commonly attacks bentgrass, annual bluegrass, and Kentucky bluegrass.

Damage Symptoms: This grub usually damages cool season turfgrasses on golf courses. Turf first wilts and does not respond long after irrigation. The turf is easily peeled back because of the lack of roots. Death of turf in irregular patches occurs in June or August.

Description of Stages: The larvae look like typical white grubs but the adults are more elongate than other turf infesting scarabs. Several species of *Ataenius* and *Aphodius* require an expert for accurate species identification.

Eggs: The pearly white eggs are approximately 0.02 × 0.03 inch [0.52 × 0.72 mm], and are laid in clusters of 11–12 eggs. The eggs expand slightly when moist.

Larvae: The C-shaped white grubs are very small but third instars can be separated from other grubs using a 10X hand lens. The tip of the abdomen has two distinct anal pads and the few raster bristles are scattered at random. Smaller grubs have these characters but a microscope may be needed to see them. Full grown larvae are approximately 5/16 inch [8 mm] long.

Pupae: The small, 5/16 inch [6–8 mm] long pupae are first white but become reddish with maturity.

Adults: The small, shiny black beetles are 1/8 to 7/32 × 1/16 to 3/32 inch [3.6 to 5.5 × 1.7 to 2.4 mm]. The prothorax has small pits scattered over the surface, and the wing covers have distinct longitudinal grooves. Newly emerged adults are reddish to dark chestnut brown, but these become black in a few days.

Life Cycle and Habits: There are many species of *Ataenius* in the United States and most are dung feeders. However, a couple of species, especially *A. spretulus* and *A. strigatus*, feed on decaying humus as well as living plant roots. *A. spretulus* was first recorded as damaging golf greens in the 1930s, but major outbreaks of this pest began to occur in the 1960s and 1970s on golf course greens, tees, and aprons. Damage may extend into the fairways. Optimum habitat for this grub seems to be short cut turf with a moist compacted thatch layer. Adult black turfgrass ataenius beetles overwinter 1-2 inches [2.5-5 cm] in the soil under leaf litter and plant material along the edges of fairways and in wooded roughs. The adults emerge in early spring, usually when spring crocus and red buds are in bloom. These adults warm in the spring sun and fly to turf areas where they dig into the thatch. Upon finding a suitable oviposition site, usually in April and early May, the females lay clusters of 11 to 12 eggs in the thatch just above the soil. Most of the eggs are present from mid-May to early June, during which time they take about a week to hatch. The tiny white grubs feed on the organic material in the thatch, including grass roots. Grub populations of 200 to 300 per ft^2 [2000-3000/m^2] are common and grub populations of 50 per ft^2 [540/m^2] can severely damage the turf. The first generation larvae take about four weeks to mature. These mature grubs dig into the soil and make a compact pupal chamber. The pupa takes a little more than a week to mature and becomes reddish brown before the adult emerges. Young adults are also reddish brown and these "callow" adults may emerge and craw or fly about before becoming completely black in a few days. The first generation adults are active in July, laying a new batch of eggs. The second generation of grubs severely damage the turf in August when rainfall is scarce. Because the summer adults may lay eggs over an extended period of time, a few larvae can be found in September and October. If the larvae do not mature by winter, they die. The second generation adults emerge in late August through September and these seek overwintering sites, usually near the edges of wooded lots.

Control Options: See: White Grubs—Introduction. This grub has two generations per year and insecticides for the grubs may be needed for either generation.

Option 1: *Chemical Control—Timing Using Indicator Plant Phenology—* Control of the first generation of grubs is best achieved if pesticides are applied when black locust (*Robinia pseudoacacia*) trees or Vanhoutte spirea (*Spiraea* x *vanhouttei*) shrubs are in full bloom. The second generation of larvae can be controlled when the Rose-of-Sharon (*Hibiscus syriacus*) begins to bloom.

Option 2: *Biological Control—Milky Diseases—Ataenius* and *Aphodius* beetles have their own strains of these diseases. None are commercially available. Infected grubs will not feed and these often remain alive, though not feeding, after an insecticide application. Do not assume that the insecticide did not work if milky *Ataenius* are present.

Option 3: *Chemical Control—Preventive Pesticide Applications*—Though this technique is not recommended for control of most annual grubs, applying pesticides to kill the adults of BTA is an effect management option. Applications of chlorpyrifos (=Dursban®) or fonofos (=Crusade®, Mainstay®) at the time that adults are migrating have been very effective at reducing resultant larval population to insignificant levels.

Option 4: *Chemical Control—Late Reactive Pesticide Applications*—Occasionally turfgrass damaging populations of *Ataenius* may go undetected until the turf begins to wilt dramatically. By this time the white grubs are usually third instars and may be ready to pupate. Chemical control of these mature grubs is difficult, at best. If a late reactive insecticide application is needed, isazophos (=Triumph®) and trichlorfon (=Dylox®, Proxol®) have been the most successful. Be sure to irrigate well after the application in order to keep the grubs near the soil/thatch interface and to wash in the pesticide.

ASIATIC GARDEN BEETLE

Species: *Maladera castanea* (Arrow). [Phylum Arthropoda: Class Insecta: Order Coleoptera: Family Scarabaeidae] (Figure 3.75; Plate 3-10).

Distribution: Introduced from Japan in the 1920s. Most common in northeastern United States from New England states across to Ohio and down into South Carolina.

Hosts: Larvae occasionally attack turf but seem to prefer a variety of roots from weeds, flowers, and vegetables. The adults feed on over 100 species of plants, preferring asters, dahlias, mums, roses, and a variety of trees and vegetables.

Damage Symptoms: The grubs cause typical damage to turf, wilting and irregular patches of dead turf. However, since this species likes roots of other plants, the grubs may be grouped around weedy areas. Grubs may also be congregated next to flower beds where plants preferred by the adults are located. The adults strip foliage off plants, leaving a ragged appearance. They do not skeletonize like Japanese beetles. Flowers often have the petals eaten off.

Description of Stages: The stages are typical of a scarab having a single generation per year. Since this beetle is rather small, most of the stages are likewise diminutive.

Eggs: The eggs are laid in clusters of 3–15 and are loosely held together by a gelatinous material. The individual eggs are oval and about 3/64 inch [1 mm] long. After absorbing water the eggs become spherical.

Larvae: Newly hatched larvae are about 1/16 inch [1.4 mm] long and have light brown head capsules. Full grown larvae are 5/8 inch [15–18 mm] long when stretched out. These grubs can be identified by the longitudinal anal slit and transverse curved row of brown spines making up the rasters.

Pupae: The pupae rest in the last larval skin and are about 3/8 inch [8–10 mm] long. At first they are white and gradually turn tan.

Adults: The adults are 3/8 inch [7–10 mm] long and broadly wedge shaped. They are a chestnut brown and often have a slight iridescent sheen to the elytra when alive. The abdomen protrudes slightly from under the wing covers, and the undersurface of the thorax has an irregular covering of short yellow hairs. The hind legs are distinctly larger and broader than the others.

Life Cycle and Habits: The adult beetles may be active from late June to the end of October, but most of the adults are found from mid-July to mid-August. The adults emerge at night and activity fly when temperatures are above 70°F [21.1°C]. When the temperature drops below this, they tend to walk up the plants or grasses to feed rather than fly. The adults are strongly attracted to lights. During the day, the beetles hide in the soil around favored food plants. After feeding several nights the females begin laying eggs in small clusters. The females tend to search out turf and pastures for egg laying, and generally deposit the eggs 1–2 inches [2.5–5 cm] deep in the soil. The females lay eggs over several weeks and average 60 eggs. The eggs normally hatch in 10 days during summer temperatures. The young larvae dig to the soil surface, where they feed on roots and decomposing organic material. Most first instar larvae are found in August and early September. The second instars are found in September, and many do not reach third instars until the following spring. About half the population overwinters as second instars, and the remainder are partially developed third instars. As cool October temperatures arrive, the larvae burrow down 8–17 inches [20–43 cm] to pass the winter. The larvae return to the soil surface in the spring and all seem to mature by mid-June, at which time they pupate 1.5–4 inches [4–10 cm] in the soil in compacted earthen cells. The pupal stage is relatively short, lasting 8 to 15 days. The adult remains in the old pupal skin, changing from white to the mature chestnut brown for a few days before digging to the surface.

Control Options: See: White Grubs—Introduction— Generally, because of its small size, grub populations below 20 ft^2 [215/m^2] are not severely damaging to turf if water and fertilizer is available. Damaging populations tend to build up when several years of rainy summer weather have occurred.

EUROPEAN CHAFER

Species: *Rhizotrogus* (=*Amphimallon*) *majalis* (Razoumowsky) [Phylum Arthropoda: Class Insecta: Order Coleoptera: Family Scarabaeidae] (Figure 3.78; Plate 3-10).

Distribution: This European pest was first detected in Newark, NY in 1940. Since then the pest has spread into Connecticut and upper New York, west to Michigan, and south from West Virginia across Maryland.

Hosts: The grubs feed on a wide variety of plant roots and organic matter in the soil. They are known to feed on the roots and thatch of all cool-season turfgrasses.

Damage Symptoms: Typical grub damage of thin turf, wilting, and death in irregular patches can be found in the fall and early spring.

Description of Stages: This pest has stages typical of an annual white grub.

Eggs: The freshly laid eggs are an oval shape, approximately 0.03 × 0.02 inches [0.73 × 0.49 mm], and a shiny, milky white color. After absorbing water, the eggs become dull gray and swell to 3/32 × 7/64 inch [2.0 × 2.7 mm].

Larvae: The first instar larvae are about 5/32 inch [4 mm] long, and these grow to approximately 11/16 inch [17 mm] when fully grown third instars. All three instars are typical C-shaped white grubs, but these can be identified by the raster which has two parallel rows of bristles that diverge laterally at the anus. These grubs are slightly smaller than the June beetles. *Phyllophaga* grubs have two parallel rows of bristles on the raster which do not diverge at the anus.

Pupae: The pupa looks like most scarab pupae but is slightly larger than that of the Japanese beetle and smaller than the *Phyllophaga*.

Adults: The adults look much like some of the light colored June beetles. However, the European chafer is 17/32 inch [13–14 mm] long, shorter than most June beetles, and the wing covers have distinct longitudinal grooves (striatae). The most distinctive characteristic is the absence of a tooth on the tarsal claw of the middle leg. The *Phyllophaga* have a distinct tooth.

Life Cycle and Habits: The adults emerge from the pupal cells in mid-June and continue mating and oviposition until late July. Most activity occurs from the last week of June through the second week of July. The adults emerge at sundown and fly to nearby trees and shrubs silhouetted against the sky. Here, large numbers fly, with a considerable buzzing noise, for 20 to 35 minutes. When the sky is truly dark the adults settle on the foliage and begin copulation. Copulation continues in mass until daybreak, when the adults return to the soil. Cool or rainy nights greatly reduce flight and mating activities. Apparently adults may return several times for mating, but eventually females dig into the soil to lay eggs. Each female lays 15 to 20 eggs in 2 to 5 days. The eggs are usually laid singly in compacted cells of soil, 2–6 inches [5–15 cm] deep. The eggs swell as they absorb soil moisture and hatch in about two weeks. The first instar may remain deep in the soil if surface soil moistures are low. Eventually the young larvae move to the surface and feed on plant roots. If food is sufficient the first instar matures in about three weeks and the second instar takes about four more weeks to mature. The third instars feed for a period in the fall before moving down for the winter. This pest moves up, down, and sideways depending upon soil moisture and food availability. This grub may feed longer than many other species in the fall before moving down. This species is also one of the first to return to the surface in the spring, often in March. Pupation occurs in mid-May 2-6 inches [5-15 cm] in the soil.

Control Options: See: White Grubs—Introduction. This pest is controlled like most white grub species, except it seems to be less susceptible to moisture stress.

GREEN JUNE BEETLE

Species: *Cotinus nitida* (Linnaeus) [Phylum Arthropoda: Class Insecta: Order Coleoptera: Family Scarabaeidae] (Figure 3.71; Plate 3-10).

Distribution: This native of North America is commonly found from southern Pennsylvania across to Oklahoma and south. It's most commonly a turf pest in the transition zones of Tennessee and Kentucky to the Carolinas.

Hosts: This pest has grubs which feed on the roots of many species of turfgrasses and field crops. It seems to prefer areas with high organic matter, especially heavily manured areas.

Damage Symptoms: The larvae occasionally attack turf sufficiently to kill it in irregular areas. More commonly, the larvae and adults make burrows in the turf, throwing up mounds of soil. This activity resembles the soil mounds made by ants. The larvae also have the alarming habit of crawling about on their backs after rains. These migrating grubs creep across sidewalks, driveways, and often end up in swimming pools and garages.

Description of Stages: This grub has a life cycle typical of annual grubs.

Eggs: The eggs are planted in loose masses of 10 to 30. They are oval when laid, 1/16 × 5/64 inch [1.5 × 2.1 mm], but upon absorbing water become spherical, 7/64 inch [2.5 × 2.8 mm].

Larvae: The grubs are somewhat atypical for scarabs. They have the normal C-shape, but usually move by stretching out on their backs to creep along with an undulating motion. Because of this mode of movement the legs are considerably smaller than other white grubs. This type of movement is diagnostic. First instars are 1/4 inch [6-6.5 mm] long, second instars are 5/8 inch [15-17 mm] long, and mature third instars are 1 3/4 inch [45-48 mm] long.

Pupae: The pupa is formed in a distinct earthen cell and measures approximately 1.0 × 1/2 inch [25 × 13 mm]. They are first light yellowish brown but gain some of the metallic green sheen of the adult before emerging.

Adults: The adults are about 13/16 to 1.0 inch [20 to 25 mm] long, and generally are a striking velvety green color with orange-yellow margins. The lower surface is a shiny metallic green. The head has a distinctive flat horn.

Life Cycle and Habits: The green adults begin to emerge in late June but are most common in July and August. This species flies during the daytime and makes a considerable buzzing noise which alarms some people. The adults may congregate around seeping wounds of trees, and are very common on ripe grapes, figs, peaches, plums, melons, some vegetables, and ears of corn. Once a feeding site is established, several adults may be actively flying about or attempting to mate. Newly emerged females call males to the ground for mating by producing an attractive odor. Females ready to lay eggs dig a burrow into the ground, throwing up a small mound of soil. The females excavate a small cavity in the soil 2–5 inches [5–13 cm] down in which they lay 10 to 30 eggs. These eggs are packed into a ball of soil about the size of a walnut. Females may emerge and dig a new burrow elsewhere but more commonly they continue in the first burrow to dig 1 to 2 additional egg chambers. The eggs swell as they absorb moisture and the grubs hatch out in 2 to 3 weeks. The young grubs are very active and can move a considerable distance in a short period of time. Since the eggs are clustered, the larvae first appear to work in groups or colonies. The grubs feed on organic material, preferring manure and rotting plant remains. By the time the grubs are second instars they have constructed individual burrows which average 6–12 inches [15–30 cm] deep. At night the grubs deposit soil around the burrow entrance, making small mounds 2 to 3 inches [5–8 cm] in diameter which resemble ant mounds. The grubs may also emerge and crawl about on their backs during warm, wet evenings. This is especially common after a warm afternoon thunderstorm. As winter temperatures arrive, the grubs dig deeper and remain inactive at the bottom of the burrow. These grubs may become active at any time when the temperatures rise. Larval digging may occur anytime during the late summer, through winter and into spring. By late May and early June the grubs have matured and they construct an earthen cell glued together with a secretion. In about three weeks the adults break out of the pupal cell and dig to the surface.

Control Options: See: White Grubs in Turfgrass.

Option 1: *Cultural Control—Reduce Organic Fertilizers*—This pest prefers soils with very high organic content and is encouraged by the addition of animal manures. Do not apply manures to turf in the spring or early summer as these may be attractive to egg laying females.

Option 2: *Chemical Control—Apply Adequate Irrigation*—When pesticides are used, the grubs often escape contact because of their deep burrowing. Quickly water pesticides in with at least 1 inch of water so that the burrows are flooded.

JAPANESE BEETLE

Species: *Popillia japonica* Newman [Phylum Arthropoda: Class Insecta: Order Coleoptera: Family Scarabaeidae] (Figure 3.79).

Distribution: This import is generally found east of a line roughly running from Michigan, southern Wisconsin and Illinois, south to Alabama. Occasional introductions are made into western states such as California when the adult beetles or larvae are shipped in commerce. The original population was detected in New Jersey in 1916, having been introduced from Japan.

Hosts: The adult beetles are general herbivores and are known to feed on over 400 species of broadleafed plants, though only about 50 species are preferred. The grubs will also feed on a wide variety of plant roots including ornamental trees and shrubs, garden and truck crops, and turfgrasses. They seem to especially relish Kentucky bluegrass, perennial ryegrass, tall fescues and bentgrass.

Damage Symptoms: The adults are skeletonizers; that is, they eat the leaf tissue between the leaf veins but leave the veins behind. This leaves the appearance of lace and the remaining tissue withers and dies. Often the adults will attack flower buds and fruit. The grubs can kill small seedling plants but most commonly damage turf. The turf first appears off-color, as if under water stress. Irrigation of this turf causes a short, but not long-lasting response, or no response at all. The turf feels spongy under foot and can be easily pulled back like old carpet to reveal the grubs. Large populations of grubs kill the turf in irregular patches.

Description of Stages: The life stages of the Japanese beetle are typical of white grubs.

Eggs: The white oval eggs are usually about 1/16 × 3/64 inch [1.5 × 1.0 mm]. They are placed in the soil where they absorb moisture and become more roundish.

Larvae: The larvae are typical white grubs which can be separated from other soil dwelling white grubs by the presence of a V-shaped series of bristles on the raster. First instar larvae are about 1/16 inch [1.5 mm] long, while the mature third instars are about 1 1/4 inch [32 mm] long.

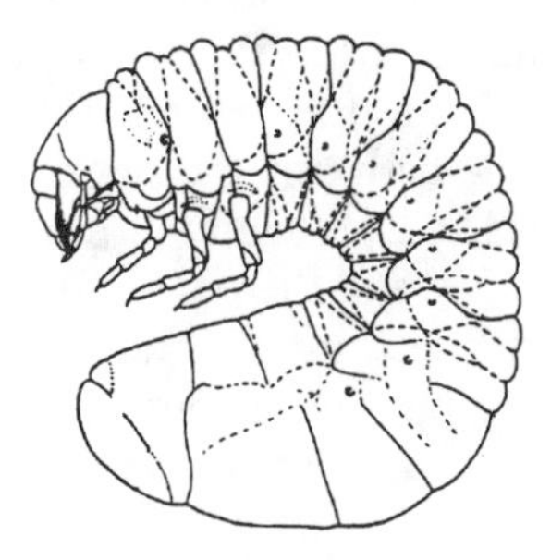

Figure 3.79 Side view of Japanese beetle grub. (redrawn from Ritcher)

Pupae: The pupae are cream colored first and become light reddish-brown with age. The average pupa is about 9/16 × 9/32 inch [14 × 7 mm].

Adults: The adults are a brilliant metallic green color, generally oval in outline, 5/16 to 7/16 × 3/16 to 9/32 inch [8–11 × 5–7 mm] wide. The wing covers are a coppery brown color and the abdomen has a row of five tufts of white hairs on each side. These white tufts are diagnostic. The males have a sharp tip on the foreleg tibia, while the female has a long rounded tip.

Life Cycle and Habits: Larvae which have matured by June pupate and the adult beetles emerge during the last week of June through mid-July. On warm sunny days the new beetles crawl onto low growing plants and rest for awhile before taking flight. The first beetles out of the ground seek out suitable food plants and begin to feed as soon as possible. These early arrivals begin to release a congregation pheromone (odor), or if they are females, an additional sex pheromone, which is attractive to adults which emerge later. This causes additional adults to gather in masses on the unfortunate plants first selected. In cool weather, the adults may feign death by dropping from the plants but normally they will take flight. Mating takes place on the food plants, or occasionally on the ground, and several matings by both males and females is common. After feeding for a day or two, the females leave feeding sites in the afternoon and burrow into the soil to lay eggs at a depth of 2–4 inches [5–10 cm]. Females may lay 1 to 5 eggs scattered in an area before leaving the soil the following morning or a day or two later. These females return to feed and mate. This cycle of feeding, mating and egg laying continues until the female has laid 40 to 60 eggs. Most of the eggs, 75% of a population, are generally laid by mid-August though adults may be found until the first frost of fall. If the soil is sufficiently moist, the eggs will swell in a few days and egg development takes only 8 to 9 days at 80–90°F [26.7–32.2°C]or as long as 30 days at 65°F [18.3°C]. The first instar larvae dig to the soil surface, where they feed on roots and organic material. If sufficient food and moisture are available, the first instars can complete development in 17 days at 78°F [25.6°C], or as long as 30 days at 68°F [20°C]. The second instars take 18 days to mature at 78°F [25.6°C] and 56 days at 68°F [20°C]. While this development is going on, the grubs tunnel considerably laterally in search of fresh roots. This leaves the turf in a very spongy condition. Generally, most of the grubs are in the third instar by early fall and are ready to dig into the soil to hibernate. The grubs move down 4–8 inches [10–20 cm] into the soil as cool temperatures arrive. At this depth the soil rarely gets below 25°F [-3.9°C] and the grubs survive with no difficulty. The grubs return to the surface in the spring as the soil temperature warms. Generally the grubs can be expected to be active at the surface when the surface soil temperatures are about 60°F [15.6°C], usually in mid-April. The grubs continue their development in the spring and the few second instars seem to mature in time to pupate along with the third instars. The mature grubs form a prepupa in mid-June. This form empties its gut and has a translucent appearance. The pupa is formed in the split skin of the prepupa in an earthen cell 1–3 inches [2.5–7.5 cm] in the soil.

Control Options: See: White Grubs in Turfgrass. Since this pest is important in agriculture, considerable effort has been placed on developing control strategies.

Option 1: *Cultural Control—Habitat Modification*—Since the eggs and young grubs are very susceptible to dry soils, do not irrigate during the time that the eggs and first instar larvae are developing. However, if natural rainfall occurs this tactic will not work. Do not plant highly desirable adult Japanese beetle food plants near the turf. This is especially possible on golf course fairways.

Option 2: *Biological Controls — Natural Parasites* — Several parasitic wasps, *Tiphia* spp., and a parasitic fly, *Hyperecteina aldrichi*, are know to attack the Japanese beetle. These parasites do not seem to reliably reduce the populations. However, the *Tiphia* appear to be more efficient in southern states.

Option 3: *Biological Control — Bacterial Milky Disease* — The bacterial milky diseases, *Bacillus popilliae* Dutky and *B. lentimorbus* Dutky, are slightly effective at controlling the grubs if applied correctly. The spore count must build up for 2 to 3 years to be very effective, and during this time you should not use an insecticide against the grubs which are needed to complete the bacterium cycle. Some recent data suggests that these bacterial diseases may not be performing well in certain areas. This may be due to reduced virulence or grub resistance. More commonly, different white grub species have displaced the Japanese beetle grub. However, in general, this pathogen has performed well in northeastern states for 10 to 30 years.

Option 4: *Mechanical Control—Trapping*—Several traps have been developed to capture the adults. These traps generally use a mixture of the aggregation and sex pheromones. Recent data indicates that these traps do not reduce grub populations, and in some cases may contribute to foliar plant damage.

NORTHERN MASKED CHAFER

Species: *Cyclocephala borealis* Arrow. [Phylum Arthropoda: Class Insecta: Order Coleoptera: Family Scarabaeidae] The western masked chafer, *C. hirta* LeConte, which is common in central California is similar to the northern masked chafer (Figures 3.77 and 3.80; Plate 3–10).

Distribution: A native of the Americas from Canada to South America. Commonly a pest in the cool season turf areas from New England across to Illinois.

Hosts: Commonly attacks Kentucky bluegrass and perenial ryegrass. Adults do not feed.

Damage Symptoms: Typical grub damage to turf may occur in the fall or spring. Heavy infestations cause the turf to appear off-color due to water

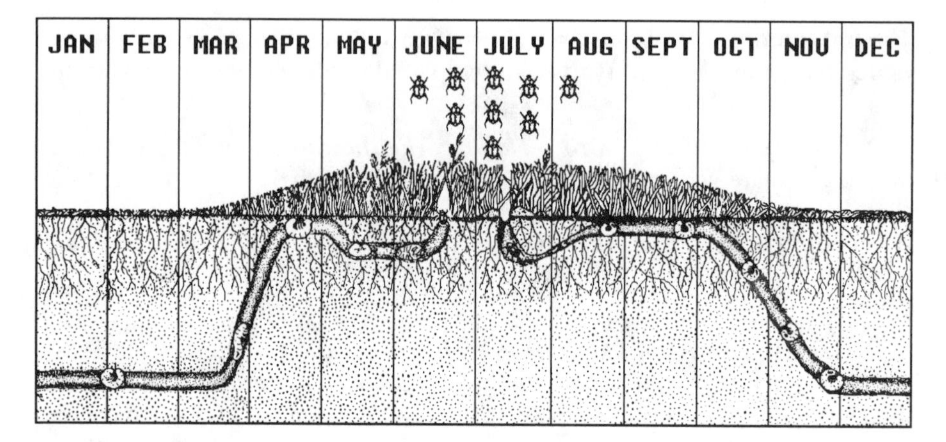

Figure 3.80 Diagram of typical masked chafer cycle in turf environment. (redrawn from USDA)

stress, but this turf dies in irregular patches. Infested turf feels spongy underfoot and is easily lifted because of the absence of roots.

Description of Stages: Typical annual scarab life cycle with a complete metamorphosis.

Eggs: The eggs are barrel-shaped, pearly white, and approximately $^{1}/_{16} \times {}^{3}/_{64}$ inch [1.68 × 1.3 mm] when freshly laid. After absorbing soil moisture, the eggs increase in diameter to $^{1}/_{16}$ inch [1.58 mm] and are more oval.

Larvae: Newly hatched first instar larvae are about $^{3}/_{16}$ inch [5 mm] long and grow to about $^{7}/_{8}$ inch [23 mm] when mature third instars. This larva looks very much like *C. lurida* and can be distinguished by characteristics of the mouthparts. The grub can be field identified as being a *Cyclocephala* by the irregular pattern of large bristles comprising the raster.

Pupae: The pupae are first creamy white in color and gradually turn reddish brown before adult emergence.

Adults: The adults are yellowish ocher-brown and have darker markings on the heads and eyes from which the name "masked" chafer is derived. The body is covered with fine hairs, which is a good field character for distinguishing this species from the southern masked chafer. Males are slightly larger, $^{15}/_{32} \times {}^{1}/_{4}$ inch [11.8 × 6.76 mm], and darker than females, $^{7}/_{16} \times {}^{1}/_{4}$ inch [11.0 × 6.66 mm]. Males have large front tarsal claws and a longer antennal club. This club is longer than the combined length of the other antennal segments and separates it from *C. lurida* males in which the club is shorter than the combined length of the other antennal segments.

Life Cycle and Habits: Adult beetles usually begin emergence in late June and are active into mid-July. Males come to the soil surface after dark before females emerge. This species apparently is active later at night than the closely related, *C. lurida*. Maximum activity occurs around midnight. Unmated females come to the soil surface releasing a sex pheromone which attracts the males. Many males often cluster around calling females, and the successful male clasps the female with his modified legs. Mated females and males fly at night and are strongly attracted to lights. The males tend to fly within two feet [60 cm] of the ground, while females seem to fly at higher altitudes. Neither males nor females feed on plant material. Mated females dig down 4-6 inches [10-15 cm] and lay an average of 11 to 12 eggs. If soil moistures are sufficient, the eggs swell within a few days and hatch in 20 to 22 days. The young larvae burrow to the soil surface in search of plant roots. The larvae also eat general organic material. The larvae grow rapidly when adequate moisture and roots are present and third instars are common by September. It's during this time of the season that most of the damage occurs. As the soil temperatures begin to drop in the fall, the larvae begin to dig downward to hibernate. Larvae may dig down 18 inches [45 cm] but most are within 12 inches [30 cm] of the surface. Winter mortality may be heavy, with over 50% dying. Grubs surviving the winter return to the soil surface in late April and May to feed. The larvae again move down slightly in late May and early June to pupate. The pupa is formed within the old exoskeleton, which splits down the center line.

Control Options: See: White Grubs—Introduction.

Option 1: *Cultural Control—Withhold Irrigation*—Since the eggs require moisture for development, restricting irrigation in July and early August may significantly reduce survival.

Option 2: *Biological Control*—Milky diseases caused by strains of *Bacillus popilliae* are known, but most commercial preparations do not contain these host-specific strains. Larvae are also commonly attacked by *Tiphia* wasps, but these wasps may require several seasons to build up sufficient numbers to adequately control grubs in an area.

SOUTHERN MASKED CHAFER

Species: *Cyclocephala lurida* Bland (= *C. immaculata*). [Phylum Arthropoda: Class Insecta: Order Coleoptera: Family Scarabaeidae] The southwestern masked chafer, *C. pasadenae* Casey, occurs in western Texas to southern California and is very similar to the southern masked chafer in biology. (Figure 3.77; Plate 3-10)

Distribution: A native of North America, but also has been collected from Central and South America. Commonly a pest from southern Pennsylvania across to Nebraska and south.

Hosts: Commonly attacks turfgrasses in the transition zones (Kentucky bluegrass and tall fescues) and in southern bermudagrass areas.

Damage Symptoms: Turf begins to show drought stress in late fall or spring and does not rapidly recover after rain or irrigation. Heavy infestations result in turf dying in irregular patches. Birds, skunks, raccoons, and opossums commonly dig up turf along the edges of the dead patches.

Description of Stages: Typical annual scarab life cycle with a complete metamorphosis.

Eggs: The eggs are oval when laid, and are about 1/16 inch [1.7 mm]. These pearly white eggs absorb moisture from surrounding soil and increase to 5/64 inch [2.1 mm] in diameter and have a nearly spherical shape.

Larvae: First instars are about 3/16 inch [4.5 mm] long at hatching and reach 7/8–1.0 inch [22–25 mm] when maturing as third instars. The mouthparts must be dissected to distinguish this species from *C. borealis*. The larvae have the irregular pattern of bristles on the raster which is typical of all *Cyclocephala* larvae.

Pupae: The 11/16 × 5/16 inch [17 × 8 mm] pupae are first creamy white and gradually change to reddish-brown just before the adult emerges.

Adults: The adults are a dull dark yellow-ocher and have darker brownish-black markings on the heads and eyes. The thorax and wing covers do not have conspicuous hair, which distinguishes this species from *C. borealis*. The adults are 7/16 to 9/16 × 7/32 to 9/32 inch [11–14 × 6–7 mm]. Males have an enlarged fifth tarsal segment on the forelegs which is used to grasp the female. However, the antennal club of males is shorter or equal to the combined length of the other antennal segments; *C. borealis* males have the club longer.

Life Cycle and Habits: Adult beetles usually begin emergence in mid-June and are active into mid-July. Males come to the soil surface after dark before females emerge. This species apparently is active earlier in the evening than its sibling species *C. borealis*. Males begin to emerge just before sunset and skim the ground surface in search of unmated females. Unmated females come to the soil surface, climb upon a grass blade and release a sex pheromone. Many males often cluster around calling females, and the successful male clasps the female with his modified legs. Mated females and males fly at night and are strongly attracted to lights. The males tend to fly within two feet [60 cm] of the ground, while females seem to fly at higher altitudes; most of the activity is finished by midnight. Neither males nor females feed on plant material but merely mate and disperse at night. Mated females dig down 4–6 inches [10–15 cm] and lay 11 to 14 eggs. If soil moistures are sufficient, the eggs swell within eight days and hatch in 14 to 18 days at 70 to 75°F [21.1–23.9°C]. The young larvae burrow to the soil surface in search of plant roots. The larvae also eat general organic material. The larvae grow rapidly when adequate moisture and roots are present. The second instars are reached in 20 to 24 days at 80°F [26.7°C], and third instars are common by September. It's during this time of the season that most of the damage occurs. As the soil temperatures begin to drop in the fall, the larvae begin to dig downward to hibernate. Larvae may dig down 12 inches [30 cm], but most are within 3–6 inches [7.5–15 cm], at least in southern states. Grubs surviving the winter return to the soil surface in late April and May to feed. The larvae again move down slightly in late May and early June to pupate. The pupa takes about 17 days to mature.

Control Options: See: White Grubs—Introduction and Northern Masked Chafer.

ORIENTAL BEETLE

Species: *Exomala* (=*Anomala*) *orientalis* (Waterhouse). [Phylum Arthropoda: Class Insecta: Order Coleoptera: Family Scarabaeidae] (Figures 3.74 and 3.81; Plate 3–10).

Distribution: This native of Japan was first detected in Connecticut in 1920. It has since moved, mainly in soil of nursery stock, to New York, Pennsylvania, Ohio, New Jersey, most of the New England states, and down to the Carolinas.

Hosts: Readily attacks the roots of cool-season turfgrasses. The adults feed little and are occasionally found on flowers, especially daisies.

Damage Symptoms: Typical grub damage consisting of wilting turf which turn to irregular dead patches. This pest prefers to feed on turf in sunny areas. Adult feeding damage is rarely detectable.

Description of Stages: This pest has a cycle normal for an annual white grub.

Eggs: The white eggs are first oval in shape, being $3/64 \times 1/16$ inch [1.2×1.5 mm], and swell to $1/16 \times 5/64$ inch [1.5×1.9 mm] after a few days in moist soil.

Larvae: The C-shaped white grubs are approximately the same size as Japanese beetle grubs. Mature grubs are about 1 inch [25 mm] long. The anus is transverse and the raster has two parallel rows of about 14 short stout spines. These might be confused with young *Phyllophaga* larvae. However, *Phyllophaga* larvae have a Y-shaped anal opening.

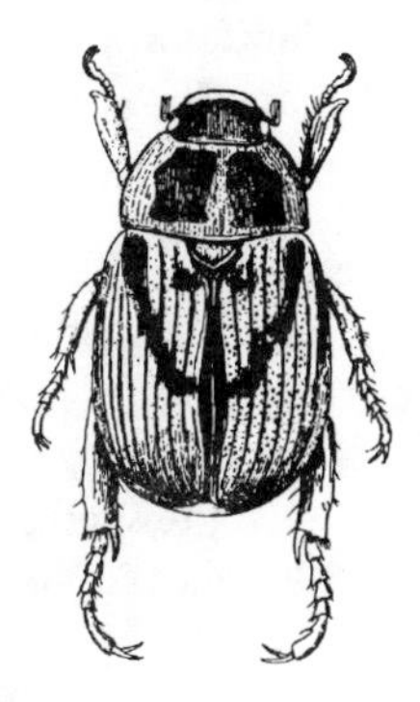

Figure 3.81 Oriental beetle, *Exomala orientalis* (Waterhouse), adult. (Conn. Exp. Sta.)

Pupae: The pupa is about 3/8 × 3/16 inch [10 × 5 mm]. They are first cream colored and turn to light brown. The tip of the abdomen has a thick fringe of hair-like setae.

Adults: The adults are 3/8 inch [9–10 mm] long and vary considerably in markings on the thorax and wing covers. Individuals may be entirely brownish-black to entirely straw colored except for a brown head and mark on the pronotum. Usually the head is solid dark brown, the pronotum is dark in the center outlined in straw color, and the wing covers have longitudinal grooves and are mottled with patterns of dark brown on straw.

Life Cycle and Habits: The adults begin to emerge in late June and some individuals may be found into September. Most of the adults are active in July. The beetles may fly short distances in the morning and evenings and are commonly found on flowers, where they are chewing on the petals. Some of the adults are attracted to lights but never in large numbers. A few days after mating, the females burrow into the soil to lay eggs. The eggs are laid in small groups between 3 and 9 inches [7.5–23 cm] in the soil. Females lay an average of 26 eggs. The eggs must be in moist soil so that water can be absorbed and development continued. At normal soil temperatures in July and August, the eggs take 18 to 24 days to hatch. The young grubs move to the soil surface to feed on roots and organic material. The second instar larvae are found in 3 to 4 weeks and these usually molt into third instars in another 3 to 4 weeks. The majority of the larvae overwinter as third instars but some overwinter in the second instar. Larvae burrow down late October and November and return to feed the following April and May. The pupae are present from mid to late June and usually take about two weeks to mature.

Control Options: See: White Grubs—Introduction.

MAY AND JUNE BEETLES, *PHYLLOPHAGA*

Species: Over 150 species of *Phyllophaga* are known in North America, but of these, only about 25 have been found attacking turfgrasses. An expert is needed to make species identification of the adults and larvae. [Phylum Arthropoda: Class Insecta: Order Coleoptera: Family Scarabaeidae] (Figures 2.79 and 3.82; Plate 3–10).

Distribution: Some species, such as *P. hirticola* (Knoch), *P. crenulata* (Froelich), *P. tristis* (Fabricius), and *P. ephilida* (Say), are generally distributed from the great plains east. *P. crinita* Burmeister is an important pest in Texas and Oklahoma, and can also be found in Louisiana to Georgia. *P. latifrons* (LeConte) commonly attacks St. Augustinegrass in Florida. Only a few species like *P. anxia* (LeConte) and *P. fervida* (Fabricius) are found all across North America.

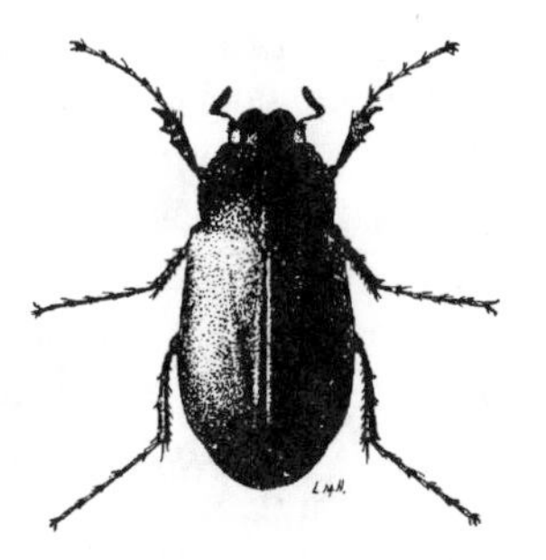

Figure 3.82 A May/June beetle, *Phyllophaga hirticula* (Knoch), adult. (after Rept. Ill State Entomol.)

Hosts: In general, *Phyllophaga* grubs will feed on the roots of many types of plants including turfgrasses. The adults feed on the foliage of many trees and shrubs.

Damage Symptoms: In recent years this group has been less severe as damaging pests in the northeastern states, probably because of competition from imported grub pests. However, in southern and western states *Phyllophaga* are often present in high populations. Damaged turf wilts as if under drought stress, and eventually this turf dies in irregular patches. Some southern grasses may not show these symptoms as readily because of moist sandy soil and deep roots. However, digging by mammals can be more of a problem.

Description of Stages: The life stages of *Phyllophaga* are typical of white grubs.

Eggs: The white oval eggs are usually about 5/64 inch [2 mm] long and are placed single in earthen cells. The eggs absorb moisture and become more roundish.

Larvae: *Phyllophaga* larvae are very difficult to separate to species, and an expert is needed to make a proper identification. However, almost all *Phyllophaga* larvae have a broadly Y-shaped anus and two parallel rows of bristles pointing towards each other on the raster. Full grown grubs are about 13/16–1 3/16 inch [20–30 mm] long.

Pupae: The pupae are white at first and become brownish with age. Most pupae are 19/32–1.0 inch [15–25 mm] long.

Adults: The different species all have the same general shape but may differ considerably in size and color. An expert is needed for species identification, and usually only males can be used for accurate determination. Adults may be a light ochre brown, as in *P. crinita,* to almost black, as in *P. anxia*. Adults are usually 19/32–1.0 inch [15–25 mm] long, depending on the species.

Life Cycle and Habits: Northern species take 3 to 5 years to complete development, usually three years in New York across to Nebraska and south. Species in the middle states and south may take 1 to 2 years to develop. *P. crinita* is a common Southwestern pest which takes only one year to develop. *P. latifrons* in Florida also has an annual cycle. The adults begin to emerge in April and

May in southern states, and usually May and June for the rest of the United States. This is why they are called "Maybugs" and "Junebugs." *P. crinita* often emerges in July and early August in Texas and Oklahoma. The adults emerge from the soil at dusk and fly to trees and shrubs for a meal of foliage. If the evening temperatures are high enough, the adults continue feeding until dawn, when they return to the soil for hiding. High populations can severely defoliate preferred plants such as oaks and maples. The adults are also highly attracted to night lights and often alarm people with their buzzing flight. After feeding several nights and mating, females burrow into moist soil to a depth of 2 to 6 inches [5–15 cm] where they lay 20 to 30 eggs individually packed into balls of soil. These eggs must absorb moisture from the surrounding soil to develop. After 20 to 40 days the eggs hatch and the young larvae burrow upward in search of plant roots. Each species seems to be able to feed on a variety of plant roots, though certain plants are preferred. Annual cycle *Phyllophaga* develop rapidly during the summer and fall, and continue to feed as long as the soil temperatures do not get too cold. Those species usually pupate in February and March. The *Phyllophaga* with two-year cycles generally reach the second instar by fall, and this stage overwinters. These grubs return to the soil surface the following spring and feed until late summer, when they pupate. The pupae usually transform into adults by late fall but these adults overwinter until the next spring. The three-year *Phyllophaga* remain as larvae for two years and do not pupate until the following summer. Again, the pupae may transform into the adult stage by early fall but these adults delay emergence until the following spring. In the far northern states, some *Phyllophaga* may need 3 to 4 summers to complete larval development before they pupate.

Control Options: See: White Grubs in Turfgrass. The northern *Phyllophaga* rarely reach damaging populations in turfgrass situations. The southern *Phyllophaga* which damage turfgrasses can be treated like annual white grubs except spring treatments are not beneficial, since most of the pests are in the pupal or adult stages.

NUISANCE INVERTEBRATE, INSECT, AND MITE PESTS

When considered as an ecosystem, the turfgrass habitat is a rich source of plant and general organic matter which can be used by general herbivores. We usually consider these herbivores "pests" when they cause noticeable damage to the look of the turf. These herbivores are merely one part of the complete ecosystem. Turf herbivores, their remains, and the organic matter found in the turf canopy, thatch, and upper soil levels provide abundant food resources for a variety of other animals. Still other animals use the insulating quality of the turf canopy and thatch and the soil beneath to build nests or resting places. When these other animals associated with the turf ecosystem cause concern or problems with the human turf managers and users, the animals become "nuisance" pests.

Each nuisance pest may be viewed in dramatically different ways, depending upon the use of the turf and personal views of the manager and users of the turf. Earthworms are generally considered to be highly beneficial because of their ability to assist in the decomposition of thatch, conditioning, and aeration of the soil. However, earthworms collect grass leaves and smaller stems in the fall and mix them with castings around their base burrows. During the winter months, these mounds become hardened and result in "lumpy" lawns which are not appreciated by most home owners. Earthworms on putting greens leave mounds of castings which can dull mowers and interfere with ball roll. Therefore, though earthworms are generally beneficial, they can be nuisance pests under certain conditions. Millipedes, centipedes, and many wasps live in the turf habitat. Though most of these animals will not hurt people, most people fear them because of a lack of knowledge and appreciation. These are nuisance pests which many users of turf demand to be controlled.

In most instances, turf managers need to try to educate the general users of turf about the benefits of the numerous organisms which are considered nuisance pests. On the other hand, fire ants, ticks, fleas, and chiggers may use the turf ecosystem while actively attacking humans and their pets. Because these pests do not actively feed on and damage turf, they are here considered nuisance pests.

EARTHWORMS

Species: Several species are common across the United States. The common earthworm or night crawler, *Lumbricus terrestris* Linnaeus, and the red earthworm, *L. rubellus* Hoffmeister, are common larger species. Smaller species belong to the genera *Allolobophora* and *Eisenia.* [Phylum Annelida: Class Oligochaeta: Family Lumbricidae] (Figure 3.5).

Distribution: Certain species are more common in local areas, but earthworms are present in all of North America.

Hosts: Earthworms feed on decaying organic material such as decaying thatch and leaf litter.

Damage Symptoms: Earthworms are considered beneficial because of their ability to decompose plant remains and incorporate these into the soil. However, on close-cut, well maintained turf, such as greens and tees or in grass tennis courts, worm castings may be a problem. Earthworms generally deposit soil and fecal material at the ground surface. These piles of soil/feces, called castings, can harden in clay soils to damage a smooth surface. This same activity in lawns can lead to "lumpy" ground under the turf canopy.

Description of Stages: Earthworms are elongate, slender invertebrates with many ring-like segments. The young and adults look alike and growth is accomplished by adding segments to the body. Depending on the species, earthworms can be 3/4–8 inches [2–20 cm] long. Earthworms can be distinguished from roundworms (nematodes) by the presence of short hairs called setae. These can be seen under a microscope if the worm is small, or felt by running the finger along the lower side of a mature worm. Earthworms move by contracting and expanding the body, while nematodes wiggle.

Eggs: The eggs are laid in a gelatinous cocoon which is 3/64–9/32 inch [1–7 mm] in diameter. These cocoons are translucent and blood vessels or the young worms can be seen developing inside. A cocoon may contain one or several eggs, but usually only one worm hatches.

Immatures & Adults: Both of these stages look alike. The sexually mature worms have a swollen band on the body, the clitellum, which is used to secrete the cocoon.

Life Cycle and Habits: Earthworms are active anytime during the year when temperatures are above freezing. During the warmer months, the earthworms come to the soil surface to pick up parts of dead plant material. They may pull grass clippings and dead leaves down their holes for food. The worms may also feed on organic material in the soil as they tunnel. While tunneling, the worms ingest great quantities of soil and this is often deposited, excreted, at the soil surface as castings. This tunneling and soil deposition is one of nature's ways of aerating and mixing soils. During rainy nights, the worms often leave the soil to crawl about in search of mates and food. If two sexually mature earthworms come together, they often mate. Earthworms are hermaphroditic (having male and female sex organs). During copulation, earthworms join by attaching to each other through mucus bands. Sperm from each worm travels down ducts to enter the female pores of their current mate. The worms do not fertilize themselves. After mating, the fertilized eggs are deposited into a bandlike mucus cocoon that slips past the egg ducts and off the tip of the worm. These cocoons are deposited in the soil and if moisture is sufficient, the eggs will mature in 2 to 3 weeks. The young worms emerge and immediately begin to feed in the soil. The young worms take 2 to 3 months to become sexually mature under ideal conditions. Some worms become inactive during dry weather and form a mucus lined chamber deep in the soil.

Control Options: Earthworms are beneficial animals and no controls are recommended. Though no pesticides are recommended for control of earthworms, certain insecticides (carbaryl, chlorpyrifos, fenvalerate, guthion, methomyl, nicotine and propoxur) and fungicides (benomyl and captan) are highly toxic to them. Repeated use of these products may decrease earthworm populations.

Option 1: *Cultural Control—Reduce Soil Mounding*—Soil mounding in lawns commonly occurs during the cool, fall and spring periods. These "lumpy" lawns can be reduced by rolling with a heavy lawn roller, especially after a rain or irrigation. Earthworm castings on golf course greens and tees can be picked up during mowing, rolled down, or raked down.

SLUGS AND SNAILS

Species: Several species of slugs and snails may be found in turf but most feed on other plants. Common slugs are: Gray garden slug, *Agriolimax reticulatus* (Muller); spotted garden slug, *Limax maximus* Linnaeus; and tawny garden slug, *L. flavus* Linnaeus. The brown garden snail, *Helix aspersa* Muller, is occasionally found crossing turf in search of more palatable food. [Phylum Mollusca: Class Gastropoda] (Figures 3.3 and 3.4).

Distribution: Slugs and snails can be found world wide in tropical to temperate zones.

Hosts: Each slug and snail has preferred food plants, though some are even carnivores. Slugs and snails in turfgrasses are usually after broadleaf weeds such as clovers and ground ivy. Some of these pests will rarely feed on southern grasses such as St. Augustinegrass.

Damage Symptoms: Because slugs and snails do not normally attack turfgrasses, they are considered nuisance pests. They leave slime trails on sidewalks and over grass blades. They are also no fun to step on with bare feet.

Description of Stages: Slugs and snails go through a gradual growth and the immatures look like the adults.

Eggs: The translucent white or amber-colored eggs are 3/16–7/32 inch [5 to 5.5 mm] in diameter and are laid in clusters of several dozen.

Immatures & Adults: Slugs are merely snails without a shell. Young slugs are 3/16–9/16 inch [5–15 mm] long and most adult slugs are 1 3/16–1 9/16 inch [30–45 mm] long, but some large species can get to be 2 3/4–3 1/2 inch [70–90 mm] long. They are usually grey or brown with spots or mottling. Snails may have flattened or elongate shells and tropical forms may be brightly colored. Both snails and slugs move on a flat foot which secretes a mucus trail.

Life Cycle and Habits: The gray garden slug will be used as an example. This species overwinters in the adult stage and sometimes in the egg stage, as long as freezing temperatures are avoided. In the spring, the adults become active and lay clusters of eggs in crevices in the soil or under boards, rocks or leaf litter. The eggs take about 20 days to hatch at 80°F [26.7°C] and twice as long at 50°F [10°C]. The young slugs feed on decaying plant material as well as leafy green plants. The slugs can become sexually mature in 10 to 15 weeks. Slugs and snails are generally creatures of the night that hide under litter, rocks, or boards during the day. Slugs and snails also prefer moist environment. Slugs may become dormant in dry summer periods by hiding in crack and crevices and losing up to 20% to 40% of their body moisture. Snails often crawl upon vertical surfaces and form a mucus plug over the shell opening which reduces water loss during dry periods.

Control Options: These pests generally do not have to be controlled in turf unless flower beds are nearby or weeds are present.

Option 1: *Cultural Control—Modify the Environment*—Removing plant residues from flower beds and controlling the weeds in lawns will help reduce slug populations. The practice of using mulches around trees and shrubs also contributes to slug survival. Try to create landscaped areas which allow the surface to dry rapidly after watering or rains.

Option 2: *Chemical Control—Use Molluscicides*—Several pesticides are registered for slug and snail control. These can be applied as general sprays or in bait form. Metaldehyde (=Bug-Geta®, Deadline®, Slug-Geta®) and methiocarb (=Grandslam®, Mesurol®) are usually available as baits.

SPIDERS AND TARANTULAS

Species: Several hundred species of spiders may be found in the turf. Most common groups are the wolf spiders, family Lycosidae, the grass and funnel-web spiders, family Agelenidae, and the sheet-web spiders, family Linyphiidae. The tarantulas found in the southwestern states are merely large, hairy spiders in the family Theraphosidae. Dangerous spiders such as the recluse spiders, *Loxoceles* spp., and the black widow, *Latrodectus mactans* (Fabricius), are not associated with turf. [Phylum Arthropoda: Class Arachnida: Order Araneae] (Figure 3.6).

Distribution: Different species of spiders are located across the world.

Hosts: Spiders are predators on other arthropods, and species can be found in all turfgrass habitats.

Damage Symptoms: Spiders do not damage turf but are considered beneficial in reducing other arthropod populations.

Description of Stages: Spiders undergo little change during their development. Young spiders look like the adults but may have slightly different color patterns. All spiders have four pairs of legs, no antennae and the cephalothorax narrowly joined to the abdomen.

Eggs: The eggs are usually round and deposited in clusters in a silken sac.

Immatures & Adults: It is beyond this work to describe all the forms of spiders found in the turf. Usually wolf spiders are medium to large spiders, 9/16–1 3/4 inch [15–45 mm] in diameter, with gray or brown and often with darker stripes on the thorax. Wolf spiders do not usually spin webs but may dig a burrow in the ground. Funnel-web spiders cover the turf with a sheet of webbing which ends in a funnel-like tunnel which leads into a burrow. The sheet-web spiders include the bowl and doily spiders. These tiny spiders, 5/32–3/8 inch [4–10 mm], construct a bowl-shaped web between grass blades and irregular webbing over the bowl. These are often visible in the morning dew. Tarantulas merely construct a burrow in the ground.

Life Cycle and Habits: Each spider has a unique life cycle but generally a single generation is completed per year. The tarantulas often take two years to mature. The wolf spiders are active predators and do not spin a web. The females often carry the egg case with them and the young ride on her back for awhile before foraging on their own. The tarantulas also are active predators but usually stay near their base burrow. The other spiders build different types of web nests which are used to trap their insect prey.

Control Options: No controls are recommended because spiders are beneficial. Occasionally wolf spiders will enter a house in search of food. In the southern states the tarantulas, because of their large size, cause concern but their bite is no worse than a bee sting. In fact, tarantulas tame quickly in captivity and make excellent pets. If spiders need to be controlled, spot treatments or mechanical crushing is sufficient.

CHIGGERS

Species: *Trombicula alfreddugesi* (Oudemans), *T. batatas* (Linnaeus), and *Acariscus masoni* (Ewing) are common species. Others in these genera may occasionally cause problems. [Phylum Arthropoda: Class Arachnida: Order Acarina: Family Trombiculidae] (Figure 3.83).

Distribution: *T. alfreddugesi* is found from Canada to South America, while *T. batatas* and *A. masoni* are more common in the southern states.

Hosts: These mites are parasitic on reptiles, amphibians, mammals, and birds in their larval stage. The nymphs and adults are apparently free living and do not take a blood meal. Chiggers can be found in any turf, but seem to prefer taller grasses and weeds.

Damage Symptoms: No damage is done to turf. The larvae cause itchy, red welts on the skin of humans when they attempt to feed.

Description of Stages: Chiggers have typical mite stages of eggs, larvae, nymphs, and adults.

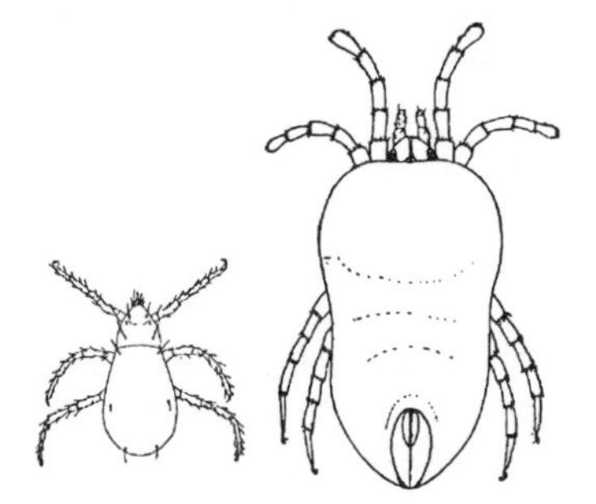

Figure 3.83 Chigger larva (left) and adult female (right). (USDA)

Eggs: The eggs are microscopic, roundish and reddish when about to hatch.

Larvae: The larval chigger is orange-yellow or light red and has only three pairs of legs. They are about 0.006 inch [0.15 mm] long and barely visible with the unaided eye.

Nymphs: The nymphs are 3/64 inch [1–1.2 mm] long, bright red, and covered with complex flattened hairs. These now have four pairs of legs. Under a microscope or hand lens, these mites have puffy or lobed cephalothoracic areas.

Adults: The adults look like the nymphs but are slightly larger.

Life Cycle and Habits: Apparently the adults and nymphs overwinter. These are general scavengers or predators. Eggs are laid in the spring and these hatch in a week. The young larvae climb onto animals or on the clothing of people. Here they seek out a suitable soft spot and often bury their head into the base of a hair follicle. They inject an anticoagulant saliva and begin to suck blood. The saliva is allergenic and causes severe itching and a red swelling. Often the mites are trapped in the hair follicle and cause even more inflammation. Successfully fed larvae drop off the host and enter into a resting state. They then molt into the nymph which has four pairs of legs and this, in turn, molts into the adult, usually without feeding. The complete life cycle may take 50 days, but usually only one cycle occurs per year.

Control Options: Chiggers are very difficult to control because of their ability to feed on so many hosts. Most people can adequately protect themselves by properly applying repellents which state that they are effective against chiggers. Most well maintained lawns do not get chigger populations. Chiggers prefer tall grasses and weedy areas where rodents frequent.

Option 1: *Cultural Control—Modify Habitats*—Since chiggers do best in areas where grasses are allow to grow tall and other shrubby weeds grow, keep these areas mowed and remove weedy growth where possible.

Option 2: *Chemical Control—Area Sprays*—Several insecticides are registered for chigger control. Apply these materials evenly over the lawn and surrounding plant material. Reapplications may be necessary if small rodents are present. Carbaryl (=Sevin®), chlorpyrifos (=Dursban®), cyfluthrin (=Tempo®), diazinon (not on golf courses or sod farms), and fluvalinate (=Mavrik®, Yardex®) are registered for this purpose.

TICKS

Species: The most common lawn tick is the American dog tick, *Dermacentor variabilis* (Say). Other ticks occasionally found in lawns and around homes are the brown dog tick, *Ripicephalus sanguineus* (Latreille), and the lone star tick,

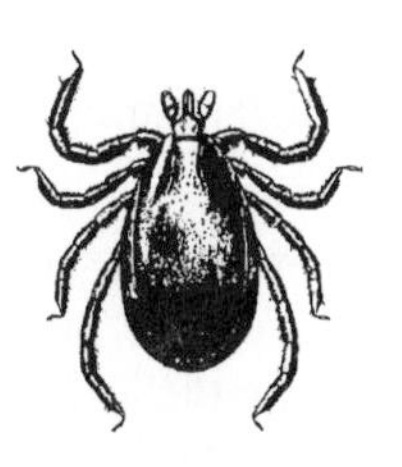

Figure 3.84 Blacklegged tick, *Ixodes scapularis* Say, male. (USDA)

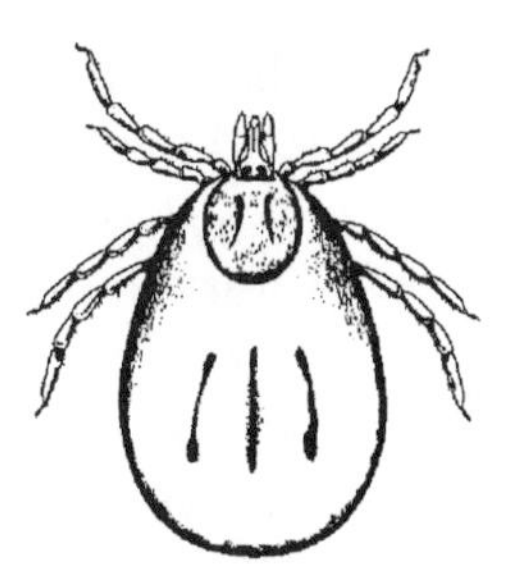

Figure 3.85 Blacklegged tick, *Ixodes scapularis* Say, engorged female. (USDA)

Ambylomma americanium (Linnaeus). The deer tick, correctly called the blacklegged tick, *Ixodes scapularis* Say [previously, *I. dammini*], is not usually associated with well maintained turf. [Phylum Arthropoda: Class Arachnida: Order Acari: Family Ixodidae] (Figures 3.8, 3.84, and 3.85).

Distribution: The American dog tick is most common in the states east of the Rocky mountains as well as California, Idaho, and Washington. The brown dog tick is cosmopolitan, but prefers the warmer states, and tropical and semitropical countries. The lone star tick is found from Texas and Oklahoma across to the Atlantic Coast and south into Central and South America. Blacklegged ticks are found in New England states.

Hosts: Ticks do not attack turfgrasses but are blood sucking parasites of animals. Tall grasses and weeds provide shelter for ticks and they may develop large populations in these areas. The brown dog tick prefers to feed on dog blood in all its stages. It rarely attacks other animals or humans. The American dog tick prefers small rodent hosts in the larval and nymphal stages but adults will feed on humans, dogs, and livestock. All the stages of the lone star tick will suck blood from humans and larger animals.

Damage Symptoms: No damage is done to turfgrasses. Ticks are known to vector several diseases including Rocky Mountain spotted fever (which can be present across the United States), viral encephalitis, and tularemia. The blacklegged tick transmits Lyme disease. The American dog tick and lone star tick also can cause tick paralysis, a neurotoxic reaction to the saliva of ticks. This paralysis is found in certain people, especially when these ticks attach to the skin at the base of the back of the head.

Description of Stages: Ticks lay eggs which hatch into six-legged larvae, called seed ticks. Seed ticks molt into normal eight-legged nymphs which molt into adults.

Eggs: The eggs are shiny and oval, about 1/64 inch [0.4 × 0.5 mm], yellowish to brown and laid in clusters of 1000 to 6000.

Larvae: The larvae or seed ticks are 0.02–0.04 inch [0.5 to 1.0 mm] long and have only three pairs of legs.

Nymphs: The nymphs have four pairs of legs and are smaller than the adults, being 1/16 to 3/32 inch [1.5 to 2.5 mm] long.

Adults: The adults are easily identified by looking at the shape of the mouthparts, anal openings, and markings on the tip of the abdomen and shield. Male ticks have a hard shield, the scutum covering the abdomen, while females have the scutum extending only halfway down the abdomen. The lone star tick is the easiest to identify because it usually has a white spot at the base of the scutum. The brown dog tick is uniformly reddish brown, and the American dog tick is reddish brown with white mottling. Unfed adult ticks are 1/8–1/4 inch [3–6 mm] long, but engorged females may be 3/8–5/8 inch [10–15 mm] long.

Life Cycle and Habits: These three ticks are similar in that they are 3-host ticks. That is, they have to feed three separate times on the blood of a host before the cycle can be completed. Some other ticks are 2-host ticks or 4-host ticks. The adults of these pest ticks grab ahold of the hair of a passing animal and climb to a suitable place to insert the mouthparts and begin feeding on blood. Males hunt for females attached to an animal and mate on the host. After feeding for several days, the fully engorged females drop off and seek a protected spot in leaf litter, a crack or crevice in the ground, or under furniture and the like in a house. The female may deposit 1000 to 5000 eggs over a 3-week period. The eggs are held together by a gelatinous material which reduces desiccation. The eggs hatch in about a month and the seed ticks seek out a host. Ticks look for hosts by climbing to the tips of tall grasses or the leaves of bushes and shrubs. Here they hold on with a couple of legs and hold the other legs outward. The legs have hook-like claws which grab ahold of the fur of a passing animal or the clothing of humans. After feeding for several days, the seed ticks drop off the host and find a place to rest and molt into the nymph. Seed ticks rarely survive winter temperatures. The nymphs must again find a suitable host and feed before dropping to molt into the adult stage. Most ticks, outside, overwinter in the adult stage. These adults become active when the spring temperatures get to be 70°F or higher. Usually only one generation per year is found, but the brown dog tick may breed continuously in tropical areas or in dwellings.

Control Options: Ticks are like fleas in that a well planned set of controls must be followed. Ticks on domestic pets should be treated by a veterinarian or under their care. Pet bedding and small rodent populations must be carefully treated to reduce larval and nymphal populations of ticks. Generally, lawn and garden treatments are most effective during the spring periods when the newly active adults are looking for a host. The brown dog tick may be established inside a home and a pest control operator will be needed for proper treatments. Often additional treatments will be necessary if pets are allowed to roam in weedy or wooded areas and if tick eggs hatch at a later date. Blacklegged ticks are best managed by keeping turf and weeds cut short enough to discourage mice and voles.

SOWBUGS AND PILLBUGS

Species: Several species may be present. Usually the pillbug is *Armadillium vulgare* (Latrelle) and sowbugs may be species of *Oniscus* and *Procellio*. [Phylum Arthropoda: Class Crustacea: Order Isopoda] (Figure 3.8).

Distribution: Sowbugs and pillbugs are found worldwide.

Hosts: Sowbugs and pillbugs rarely feed on turfgrasses but they may feed on the germinating seed of newly planted turf. Usually they are scavengers, feeding on decaying organic material or young tender parts, roots, and stems, of seedling plants.

Damage Symptoms: Broadleaf plants may have cotyledons eaten off or young leaves will have ragged edges. These are considered nuisance pests when they invade homes in search of cool moist places to hide.

Description of Stages: These crustaceans go thorough a gradual metamorphosis similar to some insects, i.e., the immatures look like the adults. Pillbugs are able to roll into a compact ball when disturbed. Sowbugs are more active and cannot roll into a ball.

Eggs: The small white eggs are retained in a pouch carried by the female underneath her body.

Immatures: The young isopods have less than seven pairs of legs and are carried in the brood pouch. After emerging from this pouch they look like the adults except they are smaller.

Adults: The adults are elongate oval and 3/16–3/8 inch [5–10 mm] long. They may be solid or mottled gray or brown, have a pair of conspicuous antennae, and seven pairs of legs.

Life Cycle and Habits: Depending upon the species and temperatures, 1 to 3 broods per year may be found. Generally, isopods are continuous breeders and all stages can be found any time of the year. When mature, the adults mate and a female may produce 25 to 200 eggs held in a brood pouch. The young stay in the pouch for 1 to 2 months, apparently obtaining food from their mother. After emerging, the young may take several months to a year to mature. Sowbugs and pillbugs may live 2 to 3 years. Most species molt in two stages, the front half of the exoskeleton is shed, then the last half. This is why some individuals will look like they are lighter colored on one half of the body.

Control Options: Since these arthropods require moist environments and organic material for food, modifying the habitat is one of the best strategies.

In general, these animals are considered beneficial because they help decompose waste plant material.

Option 1: *Cultural Control—Modify the Environment*—Sowbugs and pillbugs are generally a problem where mulches are used around ornamentals and flower beds. They can also feed on grass clippings if they are piled in an area or are allowed to pile up in the turf. Reduce mulching or encourage drying of the soil surface to help reduce populations.

Option 2: *Chemical Control—Use Pesticides*—Several insecticides will also kill this crustacean. Check for labels which include sowbugs or pillbugs.

CENTIPEDES

Species: Several species of different families may be in the turf or soil. Adult centipedes with 15 pairs of short legs are in the order Lithobiomorpha and usually in the genus *Lithobius*. Adults with 21–23 pairs of legs are in the order Scolopendromorpha. This family includes the large tropical species which can get to be over 150 mm. The third group are burrowing centipedes of the order Geophilomorpha, which have 29 or more pairs of legs and are very slender and elongate. [Phylum Arthropoda: Class Chilopoda] (Figure 3.12).

Distribution: Centipedes are found worldwide.

Hosts: These arthropods are predators and are found in turfgrasses where they are hunting other arthropod prey.

Damage Symptoms: These cause no damage to turf but are considered nuisance pests if they happen to crawl inside a house in search of food. All centipedes have poison fangs beneath the mouth opening. However, only the larger individuals, greater than $1\frac{1}{2}$ inch [40 mm], have the ability to pierce human skin. The poison is painful but not generally dangerous to people unless they are hypersensitive.

Description of Stages: Centipedes are easily separated from millipedes by their long antennae, rapid running ability, and the presence of only one pair of legs per trunk segment attached to the side of the body. Little is known about the life stages of centipedes.

Eggs: The spherical, white eggs are about $\frac{5}{64}$–$\frac{5}{32}$ inch [2–4 mm] in diameter and are generally laid singly. Some species have females which dig a burrow and protect an egg cluster.

Immatures and Adults: The immatures look much like the adults but are often lighter colored. Most above ground centipedes are $\frac{13}{16}$–$1\frac{1}{2}$ inch [20–40 mm] long, while the burrowing centipedes may be up to 3 inches [70 mm] long but only $\frac{1}{8}$ inch [3 mm] wide.

Life Cycle and Habits: Little is known about the life cycles and habits of centipedes. Most species overwinter as adults in protected places in the soil, under rocks, or under bark of dead trees. The females lay eggs during the spring and summer and these hatch in several weeks. The young mature over a season while actively preying on insects, spiders, and other arthropods.

Control Options: No controls are needed unless these arthropods are beginning to invade houses. Even then, simple crushing and removal are usually adequate. The greatest problem with these animals is the erroneous fear that centipede bites are extremely dangerous to people. See: Millipedes, controls.

MILLIPEDES

Species: Many species can be found. Usually 2–3 species are present in any turf habitat. The garden millipede, *Oxidus gracilis* Kotch, is a common species with a flattened dorsum. Other flattened species are usually in the genus *Polydesmus* while cylindrical species are usually in the genus *Callipus*. [Phylum Arthropoda: Class Diplopoda] (Figure 3.13).

Distribution: *Oxidus gracilis* is worldwide in distribution and other species may have regional territories.

Hosts: Millipedes are general scavengers, usually feeding on decaying animal and plant materials.

Damage Symptoms: Because these arthropods rarely attack living plant material they are considered nuisance pests. Occasionally they will feed on the roots and leaves of seedling plants.

Description of Stages: Millipedes can be separated from centipedes by the presence of paired legs on the trunk. Be careful not to count the legs on the first few segments because the second pair of legs may be absent or reduced for copulatory behavior. They have a simple life cycle in which the young look like the adults but have fewer pairs of legs.

Eggs: Millipede eggs are usually laid in clusters. They are white to brown and about 1/64 inch [0.4 mm] in diameter. Often they are sticky and soil readily adheres to them.

Immatures & Adults: Newly hatched millipedes have only three pairs of legs but can be separated from insects by not having a conspicuous abdomen without legs. As the young grow, segments and legs are added. Adults may be 9/16–1 3/8 inch [15–35 mm] long.

Life Cycle and Habits: Little is known about the life cycles of millipedes. Apparently most overwinter as immatures or adults. Adults lay eggs in the moist spring and fall months in clusters of 20 to 100. These clusters are usually in cavities in the soil or under rocks or pieces of wood. The eggs hatch in 1 to 3 weeks and the young take 21 to 25 weeks to mature. The adults can live for one to several years.

Control Options: These nuisance pests are considered beneficial unless they invade homes in search of cool, moist places to hide. Since they feed on decaying organic material, removal of this food will reduce populations.

Option 1: *Cultural Control—Modify Habitat*—Mulching around flower beds and ornamentals often allows millipede populations to increase. Allow these areas to periodically dry and remove any grass clippings or leaf litter from next to buildings.

Option 2: *Chemical Control—Use Pesticides*—Several insecticides will also kill this group of arthropods. Check insecticide labels for an appropriate product. General turf sprays for this group is not recommended. Treat around homes and buildings by applying a residual material next to doors and porches where millipedes may attempt to enter.

EARWIGS

Species: Several species may be present, usually the European earwig, *Forficula auricularia* Linnaeus, ringlegged earwig, *Euborellia annulipes* (Lucas), or the shore earwig, *Labidura riparia* (Dallas). [Phylum Arthropoda: Class Insecta: Order Dermaptera] (Figures 3.86 and 3.87).

Distribution: The European and ringlegged earwigs have been transported around the world and are found in the United States along the eastern and western coasts. The shore earwig is found in tropical and subtropical areas of the world and is restricted to the Gulf states across to Arizona and California.

Figure 3.86 European earwig, *Forficula auricularia* Linnaeus, adult. (redrawn from Gibson)

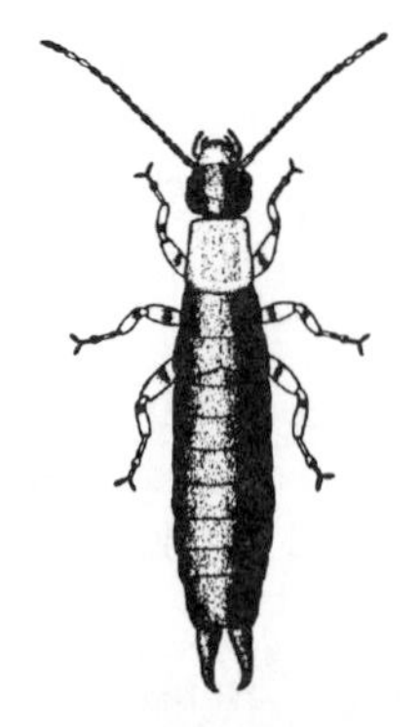

Figure 3.87 Ringlegged earwig, *Euborellia annulipes* (Lucas), adult. (Ohio Agr. Exp. Sta.)

Hosts: Earwigs are scavengers and predators and can be found in turfgrass areas where suitable food can be found.

Damage Symptoms: Earwigs rarely damage turf and are usually considered beneficial except when they become a nuisance pest. Earwigs usually prey on other insects or feed on freshly killed plant and animal materials. Their secretive behavior, hiding in cracks and crevices and inside flowers, causes some alarm when people discover them. Their forceps-like tails are used for defense and larger ones can lightly pinch the skin. These forceps are not poisonous though many earwigs release a foul odor if handled. In southern states, large numbers of these pests may enter homes and hide in flower pots, in clothes dropped on the floor, and in the crack between door and window sills. In spite of their namesake, earwigs do not enter the ears of humans in an attempt to eat brains! However, earwigs have been known to enter ears of campers at night, apparently in search of a dark place to hide.

Description of Stages: These insects have a gradual metamorphosis with egg, nymphal, and adult stages. The most characteristic feature of an earwig is its forceps-like cerci at the tip of the abdomen.

Eggs: The eggs are white, oval, and about $1/32 \times 3/64$ inch [0.8×1 mm]. They are placed in clusters in a chamber in the ground.

Nymphs: The nymphs look like the adults but usually have thin narrow forceps, lighter color, and no wings.

Adults: The adults are brown to black and are often marked with yellow or orange. They may be $5/8$–1.0 inch [16–25 mm] long and males generally have large curved forceps, while females usually have thinner ones. Some species have both winged and wingless adults in the same population.

Life Cycle and Habits: Female earwigs are protective of the eggs and young. Females of the European earwig usually lay 30 to 90 eggs in a brood chamber in the fall or early winter. They will clean the eggs of molds and fungi and attempt to protect them from predacious insects. The eggs usually hatch in 70 to 90 days and the young nymphs travel only short distances from their nest. By the time they are third instars, the nymphs are more venturesome and will forage considerable distances form their original nest. Often they get lost but seem to be welcome in the nests of other earwigs. Development takes 2 to 3 months in cool weather. In hot dry summer months the earwigs are less active and hide in cool, moist areas. Sometimes European earwig females will lay another batch of eggs in the spring. The ringlegged earwigs and shore earwigs continuously breed during the warmer months. They may produce 1 to 6 clutches of eggs per year. The ringlegged earwig nymphs take about 90 days to mature, and the shore earwigs take 40 to 60 days. The shore earwigs are not as protective of the nymphs as other species and the females often eat the young after the brood chamber is opened.

Control Options: Because earwigs are generally beneficial in their predatory activities, no controls are recommended for general turf treatments. However, when these insects invade homes, barrier and house invasion treatments may be necessary.

Option 1: *Cultural Control—Reduce Habitat Suitability*—Earwigs need other insects for food and require moisture. Reduce reliance on mulches and remove excess debris from around the home, i.e., boards, leaf litter, rocks, etc., can greatly reduce earwig populations. Irrigate turf areas only and try not to wet the soil next to structures. Make sure rain downspouts drain well away from the house.

Option 2: *Chemical Control—Barrier Insecticide Treatments*—Several residual insecticides are useful in reducing earwig invasion into homes. Treat the area around structures as well as door and window sills and under the eaves. Bendiocarb (=Dycar®, Ficam®, Turcam®), carbaryl (=Sevin®), chlorpyrifos (=Dursban®), cyfluthrin (=Tempo®), diazinon, fluvalinate (=Mavrik®, Yardex®), fonofos (=Crusade®, Mainstay®) are registered for control of earwigs in turf.

BIGEYED BUGS

Species: Several species of the genus, *Geocoris*. *Geocoris bullatus* (the large bigeyed bug), *G. uliginosus* and *G. punctipes* are commonly found in turf. [Phylum Arthropoda: Class Insecta: Order Hemiptera: Family Lygaeidae] (Figure 3.28).

Distribution: *G. bullatus* and *G. uliginosus* is widely distributed in the United States, from coast to coast and in southern Canada. *G. punctipes* is found south of a line from New Jersey west to southern Indiana and Colorado; also southwest through Arizona and California.

Hosts: These can be abundant predators of pests of turfgrasses as well as field crops. Tough primarily predators, bigeyed bugs often feed on plants but have not been shown to cause any damage.

Damage Symptoms: Bigeyed bugs in turf are often associated with chinch bug populations, and because of similarities in size and shape are blamed for damage. Bigeyed bugs are considered important beneficial predators.

Description of Stages: These bugs have a simple life cycle.

Eggs: Most species lay ovoid or subcylindrical eggs, slightly less than 3/64 inch [1 mm] long, and have a typical circle of 6–10 peglike projections on the anterior end. The eggs range from white to yellowish tan, but the eggs of *G. bullatus* have been stated to be light pink.

Nymphs: Five nymphal instars are found in all species. The nymphs are elongate oval, like most lygaeid bugs, and resemble nymphal chinch bugs except for the obvious eye characters and color. Bigeyed bugs have laterally bulging eyes which distinctly curve backward to overlap the front of the thorax. The head is about as wide as the body proper. *G. uliginosus* nymphs have blackish-brown head, thorax, and wing covers, while *G. bullatus* and *G. punctipes* nymphs are straw colored with irregular darker spots.

Adults: Bigeyed bug adults are about 1/8 inch [3–4 mm] long, oblong-oval bugs with heads much broader than long, and possessing conspicuous eyes which curve back around the front of the thorax. Adults run and fly rapidly if disturbed. *G. uliginosus* adults are almost entirely black with light areas along each side. Adults of *G. bullatus* and *G. punctipes* are mottled buff or straw-colored with numerous small punctures.

Life Cycle and Habits: Most bigeyed bugs have several overlapping generations per year. In northern states, *G. bullatus* usually overwinters as eggs, and *G. uliginosus* and *G. punctipes* seem to overwinter as adults. In general, these bugs appear to require warmer temperatures in order to have rapid development. Using *G. uliginosus* as an example, adults become active in the spring when daytime temperatures exceed 75°F [23.9°C]. Apparently, these active adults emerge from their overwintering hiding places such as cracks and crevices in the soil or in litter around plants. The adults feed on small arthropod prey such as mites, aphids, other bugs, moth larvae, and insect eggs. They also take moisture from plant tissues or surface dew. After finding suitable prey populations, females attach their eggs to plant stems and leaves. These eggs hatch in 11 to 23 days at 70 to 80°F [21.1–26.7°C]. Hatching may not occur if the average temperature drops below 60°F [15.6°C]. The nymphs also prey on small insects and mites and take about 30 to 40 days to mature when growing at 75°F [23.9°C]. However, considerably longer times are needed at lower temperatures. Adults live 40 to 80 days, depending on environmental variables and females can lay 100 to 300 eggs during this time. Thus, two to four generations can occur per season, depending on the temperature. Since these predators are not prone to exclusively inhabit turf, they do not seem to control surface feeding turfgrass pests. In fact, they often appear to arrive in chinch bug infested turf after damage is apparent. It is at this time that the bigeyed bugs are confused with chinch bugs because of their similarities in size and shape. Some evidence indicates that the bigeyed bugs may eventually decimate the chinch bug population but usually after considerable damage has occurred.

Control Options: Bigeyed bugs are considered beneficial predators, though they often pierce plant tissues with their sucking mouthparts in order to obtain liquid. At present, no system has been developed to encourage an early buildup of bigeyed bugs in order to control greenbugs or chinch bugs.

LEAFHOPPERS

Species: Many different species can be found feeding on the different species of turfgrasses. Species may be large and green like the sharpshooters, *Draecu-*

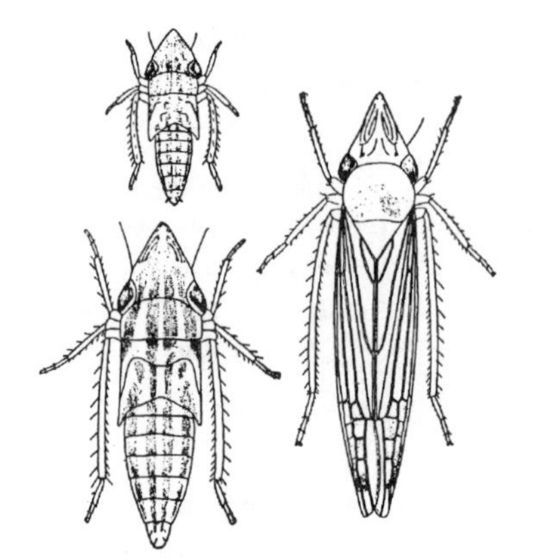

Figure 3.88 A sharpshooter leafhopper, *Draeculacephala mollipes* (Say). Young nymph (top left), mature nymph (lower left) and adult. (USDA)

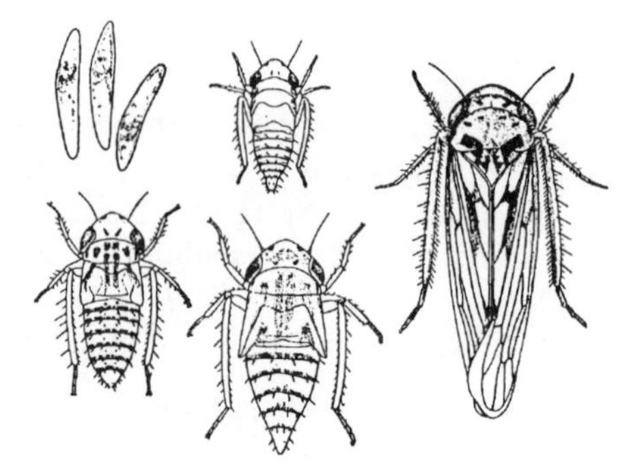

Figure 3.89 Gray lawn leafhopper, *Exitianus exitiosus* (Uhler). Eggs (top left), three nymphal instars, and adult (right). (USDA)

lacephala spp., to small, yellow or brown species like the gray lawn leafhopper, *Exitianus exitiosus* (Uhler), and the painted leafhopper, *Endria inimica* (Say). Other leafhoppers without common names are often in the genera: *Agallia*, *Forcipata*, *Latalus*, and *Polyamia*. [Phylum Arthropoda: Class Insecta: Order Hemiptera: Family Cicadellidae] (Figures 3.88 through 3.90).

Distribution: Leafhoppers can be found across the United States in turfgrass habitats.

Hosts: All species of turfgrasses have leafhopper species present.

Damage Symptoms: Leafhoppers are not considered to be important pests of turfgrasses, though thousands of individuals can be found in a square yard of turf. The slight damage observed is yellowing and spotting of leaves and stems where these insects pierce the tissues with their sucking mouthparts. Often homeowners become concerned with the large numbers which literally jump and fly in a cloud in front of their feet.

Description of Stages: These insects have a gradual metamorphosis with egg, nymphal and adult stages.

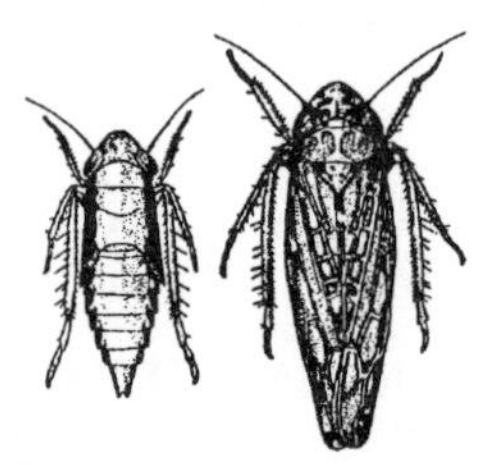

Figure 3.90 Lawn leafhopper, *Deltocephalus hospes* (Kirkaldy), nymph (left) and adult (right). (USDA)

Eggs: The elongate, white, slightly bean shaped eggs, are about 3/64 inch [1 mm] long. They are inserted into slits cut into stems or placed into leaf sheaths.

Nymphs: The wedge-shaped nymphs, with bristle-like antennae, have jumping hind legs but often they will move sideways or run to the other side of a leaf to avoid observation. The nymphs are often lighter colored than the adults.

Adults: The adults are also wedge-shaped and jump and fly rapidly if disturbed. An expert is needed to make species identifications.

Life Cycle and Habits: Most leafhoppers overwinter in the egg or adult stages, depending on the species. Those that overwinter as adults become active when the turf comes out of winter dormancy. These adults feed and mate, and females may insert 75–200 eggs into grass stems over several weeks. The eggs hatch in a week or two and the young nymphs feed at the base of leaves on along the stems. Most nymphs go through 5 instars in a couple of weeks to a month. The new adults begin another generation and the total population explodes by mid-summer. Depending on the leafhopper species and location, up to five generations per year may be found.

Control Options: These surface insects are not usually controlled because little damage is caused by their feeding. However, large populations are considered a nuisance and some turf discoloration may occur in hot, dry periods. Little information exists on the exact biology of leafhoppers in turf and control techniques.

Option 1 : *Chemical Control—General Cover Sprays*—General cover sprays are needed only when the leafhoppers are too much of a nuisance. Often repeat treatments will be needed as the adults can easily fly from one lawn to another. Acephate (=Orthene®), bendiocarb (=Dycarb®, Ficam®, Turcam®), carbaryl (=Sevin®), chlorpyrifos (=Dursban®), and fluvalinate (=Mavrik®, Yardex®) are registered for leafhopper control.

GROUND BEETLES

Species: Many species with several shapes and forms can be found in turfgrasses. [Phylum Arthropoda: Class Insecta: Order Coleoptera: Family Carabidae] (Figures 3.27, 3.91, and 3.92).

Distribution: Many species are found throughout the world.

Hosts: These are generally predaceous beetles, though some feed on plant materials.

Damage Symptoms: No damage is done to turfgrasses, and these are considered beneficial. However, homeowners often are upset when these fast running beetles enter a home.

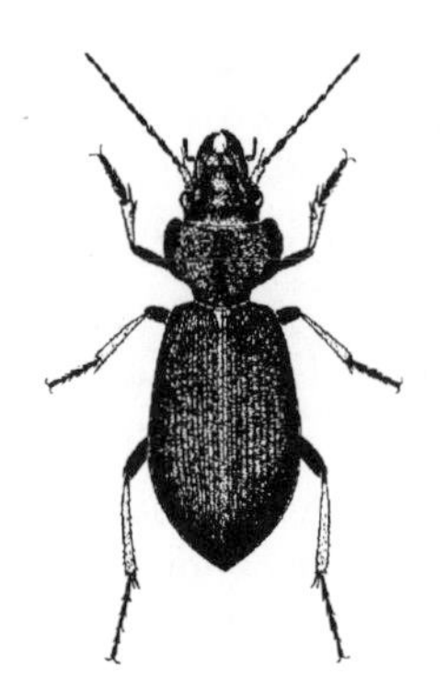

Figure 3.91 A ground beetle, *Craspedonotus tibialis* Schaum., adult. (USDA)

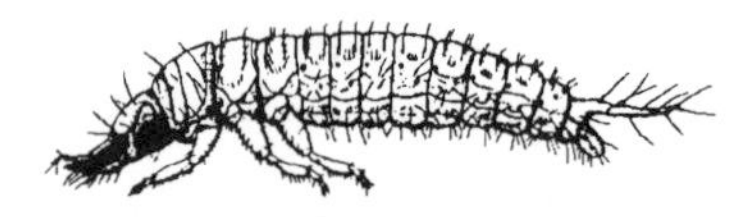

Figure 3.92 A ground beetle larva, *Harpalus pennsylvanicus* Dej. (Ill. Nat. Hist. Surv.)

Description of Stages: These elongate beetles have a complete life cycle with eggs, larval, pupal, and adult stages. This is a very large group with over 2500 species described from North America.

Eggs: The eggs are generally oval and cream colored. Each species has a different size egg.

Larvae: The larvae may be of various sizes and color patterns but they are usually light brown with darker spots. They have an elongate fleshy abdomen, well defined running legs, and projecting mouthparts. Most ground beetle larvae have a pair of pointed projections extending from the tip of the abdomen.

Pupae: The pupae are formed in cells in the soil and are generally cream colored.

Adults: The adults are usually brown or black, though some have iridescent colors or sheens. They may be 1/8–1.0 inch [3–25 mm] long, elongate oval in shape, have longitudinal ridges on the wing covers, projecting mandibles, threadlike antennae, and long running legs. An expert is needed to identify species.

Life Cycle and Habits: Most ground beetles have an annual life cycle with adults and larvae overwintering. The larvae are predaceous or scavengers, feeding on a wide variety of animal and plant food.

Control Options: Ground beetles do not need to be controlled for turfgrass protection. However, they often enter houses, and spot treatments or crushing are satisfactory. In fact, ground beetles have been shown to be important predators of eggs and young of sod webworms, cutworms, and chinch bugs.

ROVE BEETLES

Species: Many species of various genera may be found in turfgrass habitats. [Phylum Arthropoda: Class Insecta: Order Coleoptera: Family Staphylinidae] (Figure 3.93).

Distribution: Staphylinids are found worldwide.

Hosts: Rove beetles are general predators and/or scavengers.

Damage Symptoms: These beetles do not damage turf and are considered beneficial because they attack other insects or help remove organic material. Occasionally, they may dig into the soil of golf course greens and tees, leaving small mounds of soil on the surface.

Description of Stages: These beetles have a complete life cycle with egg, larval, pupal and adult stages.

Eggs: The eggs are generally oval, yellowish-white, and often have fine striations on the surface.

Larvae: The larvae are slender with tapering light colored bodies and darker head capsules. They may be $^{1}/_{16}$–$1^{3}/_{8}$ inch [2–35 mm] long and often have a fleshy cylindrical organ at the posterior of the body, the pusher or pseudopode.

Pupae: The pupae are light colored and resemble the adults in form.

Adults: The adults are usually $^{1}/_{8}$–$^{3}/_{4}$ inch [3–20 mm] long, slender and elongate, brown to black in color, and have beadlike antennae. The head projects forward and the forewings are reduced to short leathery pads which rarely reach halfway down the abdomen. They often look like earwigs, but lack the forceps on the abdomen.

Life Cycle and Habits: Little is known about rove beetle biology. Most are scavengers of molds, fungi, and decaying organic material. Some are predators and feed on the eggs, larvae, or pupae of sod webworms, grubs, and cutworms. Eggs are laid next to a food source and these hatch in 7 to 15 days. The larvae move and feed rapidly. Development may take 2 to 4 weeks. The

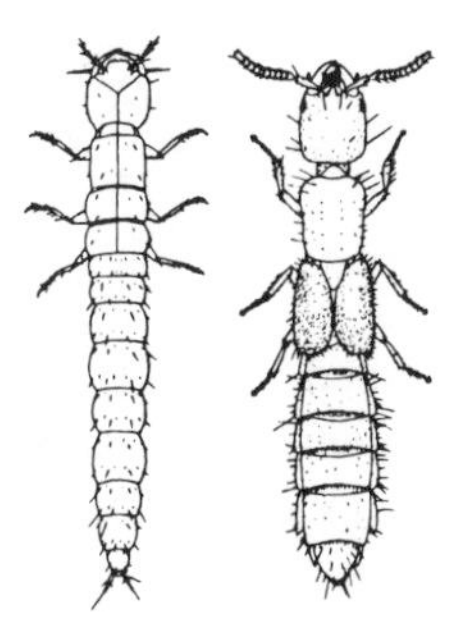

Figure 3.93 A rove beetle, *Nudobius pugetanus* Casey, larva (left) and adult (right). (redrawn from Struble)

larvae form a pupal cell, usually in the soil, and the pupa may take 6 to 14 days to mature. Rove beetles overwinter in the adult or larval stage.

Control Options: No controls are needed for these beneficial insects. Adults may be attracted to lights in great numbers in the spring or summer. Some of these produce a foul odor if handled or crushed. Spot treatments or crushing is adequate for control. Other species which dig into greens and tees may need periodic sprays of surface insecticides to reduce their activity.

FLEAS

Species: Usually the cat flea, *Ctenocephalides felis* (Bouche), is the one found in lawns. Occasionally the dog flea, *C. canis* (Curtis), or the human flea, *Pulex irritans* Linnaeus, may be present. [Phylum Arthropoda: Class Insecta: Order Siphonaptera] (Figure 3.94).

Distribution: These fleas have been distributed around the world by commerce and human activities.

Hosts: All three species may live on domestic animals and can bite humans. The larvae may feed on organic material in the turf but do feed on living plants.

Damage Symptoms: No damage is done to turf but the adult fleas can bite humans and pets, causing considerable skin irritation.

Description of Stages: Though highly modified for a parasitic existence, fleas have a complete life cycle with egg, larval, pupal, and adult stages.

Eggs: The eggs are shiny white, oval, and about 0.02 inch [0.5 mm] long.

Larvae: The larvae are slender and resemble primitive fly larvae. They are cream-colored, have long sparse hair, and a distinct head capsule.

Pupae: The pupae are formed inside a dirty white, oval, silken cocoon which is about 5/32 inch [4 mm] long.

Adults: The adults are wingless, laterally compressed insects with brown to black bodies and long hind legs used for jumping. They are 3/64–1/8 inch [1–3 mm] long and are covered with short spines which hook onto animal fir, making them difficult to pull out.

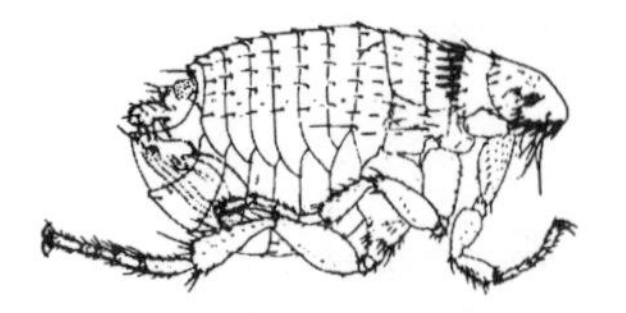

Figure 3.94 Cat flea, *Ctenocephalides felis* (Bouche), adult male. (redrawn from USDA)

Life Cycle and Habits: Domestic flea adults spend most of their time on mammal hosts found around the house. Mating and egg laying occurs on the animal. The large eggs eventually drop off the animal. This may be inside a house, in a pet pen, or out in the lawn. The eggs cannot survive freezing temperatures. In warmer weather the eggs hatch in 2 to 14 days and the rapidly moving larvae feed on organic debris, usually excrement from their host and blood pellets from adult fleas. Though the larvae prefer to be in the bedding of their host, they do well by feeding on general organic debris found in the lawns, carpeting, or upholstered furniture. The cat flea is best adapted for feeding in these nonpreferred areas. The larvae usually take two weeks to develop in moderate temperatures and with ample food supplies. In unfavorable conditions, the larvae may take a year to develop. Because the larvae often feed on the excrement of animals, they pick up the eggs of internal parasites, especially tapeworms. The parasites encyst inside the flea bodies and remain there until eaten by a cat or dog. After maturing, the larvae spin a tough cocoon in which they pupate. The pupa matures in 1 to 3 weeks during the summer months. In cool weather or when no animal activity is present, the pupae may enter dormancy for 6 to 12 months. The entire life cycle can be completed in as little as 2 to 3 weeks, but usually several months are required. This is why flea populations tend to become problems at the end of the summer after a couple of summer generations have been completed. The adult fleas must have a blood meal before eggs can be laid. Fleas do best in humidities of about 70% and temperatures between 65 and 80°F [18.3–26.7°C].

Control Options: Control of fleas takes a well planned and executed program. In essence, the adult fleas have to be killed on their host and the larvae have to be killed in their breeding areas. This usually means that the pets must be taken to a veterinarian for thorough treatment and holding while the house (if the pet was allowed inside) and lawn are treated for adult and larval fleas. This routine may have to be repeated in a few weeks to get any fleas which hatched from eggs or emerged from dormant pupae. Once a lawn and home have been treated and declared free of fleas, a constant monitoring of fleas on the pets and quick control of any new adults is necessary to reduce the chances of a new population buildup.

Option 1: *Biological Control—Parasitic Nematodes*—Recent tests have shown that products containing the insect parasitic nematode, *Steinernema carpocapsae* (= Exhibit®, Biosafe® Vector®), can control larval and pupal fleas in lawns.

Option 2: *Chemical Control—General Insecticide Applications*—General contact insecticide applications can reduce flea populations though treatments are usually necessary. Bendiocarb (=Dycarb®, Ficam®, Turcam®), carbaryl (=Sevin®), chlorpyrifos (=Dursban®), diazinon and fluvalinate (=Marvik®, Yardex®) are registered for flea control in turf.

MARCH FLIES (BIBIONIDS)

Species: Several species in the genera *Bibio* and *Dilophus* have larvae which inhabit soils around grass roots. [Phylum Arthropoda: Class Insecta: Order Diptera: Family Bibionidae] (Plate 3-19).

Distribution: Bibionids are found around the world, but they are known to damage roots of crops and grasses in Europe, Asia, and North America. Bibionids most commonly attack roots of cool-season turf, but several species are known from southern states. In fact, the famous "love bugs" are bibionids common to the Gulf states which fly while mating.

Hosts: The larvae generally feed on decaying organic matter but are known to feed on the roots of grasses. Larvae may be found in Kentucky bluegrass, annual bluegrass, bentgrass, and fescue lawns.

Damage Symptoms: Dead patches of turf, 2–10 inches [5–25 cm] in diameter, often appear in the spring after the snow cover has disappeared. Damage often resembles snow mold attack or early sod webworm damage. The dead turf easily lifts out to reveal the buff colored, wormlike larvae.

Description of Stages: This pest has a complete life cycle (egg, larva, pupa, and adult stages) with forms typical of more primitive flies.

Eggs: The eggs are cylindrical with rounded ends, slightly curved, and 1/32 × 1/128 inch [0.5 × 0.2 mm]. They are laid in clusters of several hundred and are cream colored when first deposited. The eggs often darken to a light reddish-brown before hatching.

Larvae: The first instar is a little over 1/16 inch [1.5 mm] long and has a large, light brown head and white, tapered body covered with long bristles. As the larva grows and molts, the bristles are lost and a row of short fleshy projections develop on each segment. The head becomes almost black and the body turns light yellow-brown when the larvae mature. Mature larvae may be 1/4–3/4 inch [6–20 mm] long, depending on the species.

Pupae: The pupa has wing pads, legs, and antennae easily observable. The abdomen is capable of twisting if the pupa is disturbed. Pupae are yellowish-white and 5/16–19/32 inch [8–15 mm] long.

Adult: The adults are dark colored, being black or reddish with brown or yellow legs. They are rather clumsy when walking. The wings are folded flat over the abdomen and each wing has a dark spot midway down the front margin. Some species have black wings. They have short beadlike antennae, and the males have large eyes which practically cover the head. Females have much smaller eyes.

Life Cycle and Habits: Adult March flies usually emerge in late March to May, depending on the species, location, and weather. Considerable numbers of adults may be found in suitable habitats such as moist lands with high organic debris present. The adults are quite active on sunny days, flying short distances close to the ground. They are frequent visitors to early spring flowers and flowering fruit trees. The males with small flat abdomens readily copulate with the females which have abdomens distended with eggs. Shortly after mating, the females seek out moist, organic soil. The females then dig a burrow the size of their body by folding back their front tarsi and using the enlarged spurs on the tibia. Within a day, the females may dig 3/4–1 5/8 inch [2 to 4 cm] into the soil where a slightly larger chamber is made. Within this chamber, 100 to 400 eggs are laid. The female then backs up the tunnel or tunnels a short distance from the egg cluster to die. The eggs hatch in 20–40 days depending on soil temperatures. The newly hatched larvae have no legs but long spines and bristles aid in moving through the soil. Apparently, the larvae eat decaying plant material or plants recently damaged by other insects. After a few weeks, the young larvae molt and shed their long hairs. By this time they have moved to areas with considerable food, usually under leaf litter, piles of grass clippings, or manure. Development is rather slow and the larvae may remain dormant in the soil during periods of drought. The larvae resume active feeding and growth in the fall wet periods and they may be active under snow cover. By February and March, the larvae have molted several times and are fully grown. During the winter they may have concentrated under patches of snow mold or white grub damaged turf. The mature larvae burrow 3/8–3/4 inch [1 to 2 cm] into the soil and pupate in an oblong cell. After a few weeks, the adults dig to the soil surface to begin the cycle again.

Control Options: March flies are generally considered nuisance pests and rarely attack healthy turf. In fact, no pesticides are registered for control of March fly larvae.

Option 1: *Habitat Modification*—March flies require fairly moist turf with high organic content. By allowing the turf to periodically dry, many of the larvae can be killed. Applications of manure should be avoided when March flies are known to be active. Follow turf management practices which reduce thatch buildup.

Option 2: *Reduce Turf Damage*—Many March flies seem to be opportunists. The larvae will often feed in areas damaged by other soil insects or diseases. Proper management of white grubs, billbugs, and diseases should reduce the chances of March fly damage.

ANTS—GENERAL

Species: There are over 600 species of ants in North America and many of these may nest in turf areas. Noxious ants are: harvester ants, *Pogonomyrmex* spp., which clear areas in turf in south and western states and commonly bite; and fire ants, *Solenopsis* spp., which make large soil mounds and often sting in mass (see below). The more common nuisance ants are: acrobatic ants, *Crematogaster* spp.; pavement ant, *Tetramorium caespitum* (Linnaeus); pyramid ant, *Conomyra insana* (Burkley); cornfield ant, *Lasius alienus* (Foerster); larger yellow ant, *Acanthomyops interjectus* (Mayr); smaller yellow ant, *Acanthomyops claviger* (Roger); the crazy ant, *Paratrechia longicornis* (Latreille); and, *Lasius neoniger* Emery.

Distribution: The harvester ants are most common from Arkansas through Texas and west. The native fire ant is also found in western states, while the imported fire ant species are present in the Gulf States. The rest of the nuisance species are found over much of North America. *L. neoniger* commonly builds mounds in northern golf course turf.

Hosts: Most ants do not directly attack turf, though the harvester ants may sting and cut down plants near the burrow opening. The fire ants may kill turf when the large mounds cover an area. Certain ants may be more common where weeds or grass species support honeydew producing insects such as aphids or mealybugs.

Damage Symptoms: Except as mentioned above, most ants do not attack turf directly. However, many build small mounds of soil as they excavate their burrows. When the mounds appear on golf course putting surfaces, the quality of ball roll is greatly reduced. Many people simply do not appreciate having small ants running across their lawn furniture, sidewalks, or porches.

Description of Stages: Ants have complete life cycles but with additional forms associated with social insects—workers, soldiers, queens, and drones.

Eggs: The translucent white eggs are usually elongate or bean shaped. They are commonly kept in a brood chamber and are stacked together until hatching.

Larvae: The larvae are generally white or cream colored, slightly C-shaped, and lack functional legs. These are commonly sorted and kept by workers, according to size.

Pupae: The pupae may spin tough, tan to yellow cocoons in which they transform into adults. Some species of ants do not form cocoons. These cocoons are often mistaken for eggs when an ant colony is dug up.

Adults: Ants are social insects with wingless worker, wingless soldier, and winged reproductive casts. Some species may have various sizes of workers and soldiers, each doing different tasks. The queens are often much larger than any other individual while the males may not be much larger than a normal worker. Some ants have multiple queens in the colony while others have only one. Ant identification is usually done using a normal worker, and an expert should be consulted if species determination is necessary. Unfortunately, most ant colonies are most noticeable when the new reproductives are swarming. These winged ants are much more difficult to identify.

Life Cycle and Habits: Most ants live in colonies of several hundred to thousands of individuals. They generally fall into three categories: plant or fungus feeders, sweet feeders, and fat or meat feeders. Harvester ants collect plant seeds and parts which are chewed up and used for food. Most of the sweet feeders tend aphids, scales, and mealybugs which produce a sugar-rich excrement called honeydew. The fat and meat feeders are usually general scavengers or predators that make use of other insects or animals living in the turf habitat. The acrobatic ants prefer to tend aphids, while the cornfield ant and yellow ants tend root aphids or mealybugs. The pavement ants, pyramid ants, and crazy ant are scavengers and predators. Most of these nuisance ants build small cone-shaped mounds of soil around the opening of their nests. They build the colonies during the warm months when food is plentiful. During this time the queen, or queens, are actively laying eggs which are tended by workers. The larvae are also fed by the workers. Most of these smaller ants take 30 to 50 days to complete the cycle from egg to new adult. Most ants produce new queens and drones in mid-summer to late fall. Often, just before or after a summer rain, these new winged reproductives will emerge in mass. Hundreds of workers will spill out of the nest opening and appear to be milling about while the winged forms take flight as soon as they reach the surface. After mating, the new queen seeks out a suitable site to dig a tiny burrow. She will close herself in this burrow and rear a small brood solely on food reserves stored in her body. When the first brood of workers emerge from the pupa, the initial chamber is opened and foraging for food begins. Many ants that swarm in the fall do not open the new nests until the following spring. During the winter months, most ants must survive on food stored in chambers or in their bodies. Some of the aphid and mealybug tending ants take these insects into their chambers and maintain the insects on plant roots.

Control Options: The common nuisance ants can be very difficult to manage because they are constantly reestablishing during the swarming period or they avoid pesticide applications by living in the soil. If ants are a real nuisance, attempt to get the species identified so that you can work with the knowledge of its food habits.

Option 1: *Cultural Control—Habitat Modification—* Many mound building ants will not tolerate constant disturbing from mowing, especially lower cutting heights. On the other hand, many ants need to have access to direct sunlight in order to help heat the brood. In these cases, taller cut, thick turf will discourage these ants. The aphid and mealybug tending ants must have access to these insects. Check ornamental trees and shrubs and flowers for aphid infestations tended by ants. Many of the commonly used surface insecticides eliminate the various insects which serve as ant foods.

Option 2: *Chemical Control—Surface Insecticide Applications—* Treatment of individual mounds of nuisance ants is usually not successful. Larger area applications are usually needed in order to kill the foraging workers and reduce the vigor of the colony. Treatments made in the early spring, when the first signs of ants are noticed, seem to be the most effective. This is probably due to the weakened condition of the colony from overwintering. If colonies need to be treated after the spring emergence, repeat applications may be needed. Bendiocarb (=Dycarb®, Ficam®, Turcam®), carbaryl (=Sevin®), chlorpyrifos (=Dursban®), cyfluthrin (=Tempo®), diazinon (not on golf courses or sod farms), fluvalinate (=Mavrik®, Yardex®), isazophos (=Triumph®) and malathion are registered for ant control in turf. In some of the southern states, baits may be available for harvester ant management.

FIRE ANTS

Species: Four economically important species of fire ants are found in the United States. They are: The native fire ant, *Solenopsis geminata* (Fabricius); the red imported fire ant, *S. invicta* Buren; the black imported fire ant, *S. richteri* Forell; and the southern fire ant, *S. xyloni* McCook. [Phylum Arthropoda: Class Insecta: Order Hymenoptera: Family Formicidae] (Figure 3.95).

Distribution: The red imported fire ant, the most important species, is a native of Brazil but was introduced into Alabama between 1933 and 1945. This pest now inhabits the area south of a line running from mid-North Carolina across to the Dallas-Ft. Worth, Texas area and down to Corpus Christi. The black imported fire ant is a native of Argentina and Uruguay which was imported into the Mobile, Alabama area as early as 1918. This ant is still localized in the area where Mississippi and Alabama join. The native fire ant was originally located across the southern states but has been displaced by the imported species. Colonies are still scattered from South Carolina across Texas. The southern fire ant is a native to Arkansas, Oklahoma, and Texas.

Hosts: Mounds may be built in any turf situation.

Damage Symptoms: Fire ants do not attack turf, but cause problems when they build earthen mounds for warming their eggs, larvae, and pupae. They also have a sting which may cause a burning and itching sensation at its minimum effect and serious welts or even allergic shock at its worst. Unsuspecting people and pets playing on lawns have been severely injured when they accidentally stand or sit on the ant mounds.

Description of Stages: Fire ants are social insects with a complete metamorphosis. Though they have egg, larval, pupal, and adult stages, the adults belong to several castes—-wingless workers of several sizes, winged queens,

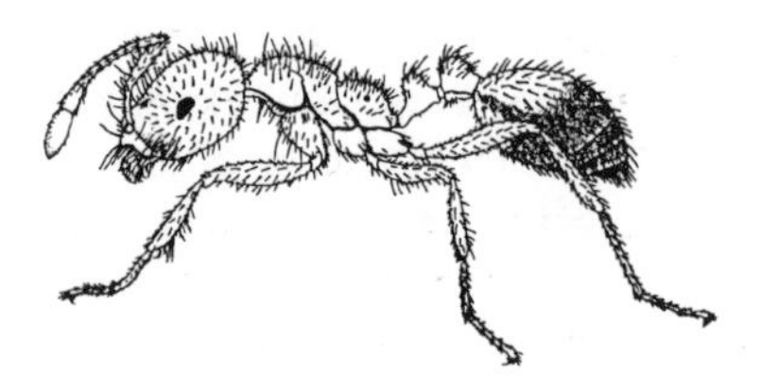

Figure 3.95 Southern fire ant, *Solenopsis xyloni* McCook, worker. (Calif. Agr. Exp. Sta.)

and winged males. The following descriptions are based on the red imported fire ant.

Eggs: The translucent white eggs are laid in clusters which are picked up and stored by workers in the upper part of the mounds. They are slightly bean-shaped and about 1/32 inch [0.75 mm] long.

Larvae: The larvae are grublike with no legs and a small head capsule. The larvae are usually white in color. Worker larvae get to be up to 1/4 inch [6 mm] long, while the reproductives may be twice as long.

Pupae: Fire ant pupae do not form a silk cocoon like some other ant species. The pupa looks like the adult but has the antennae, legs, and wing pads (if it is a reproductive) held close to the body. The pupa is first white but becomes darker before the adult emerges.

Adults: The workers are 1/16–1/4 inch [1.6 to 6 mm] long, and have the typical elbowed antennae of ants and a thin waist. The constriction at the waist, the petiole, has 2 nodes; the antennae are 10-segmented, with the last two segments forming a distinct club. The reproductives are winged and the queens may be twice the size of the workers. Depending on the species and locality, fire ants may be yellow to reddish brown to black in color. An expert should be consulted if species determination is necessary.

Life Cycle and Habits: Established colonies produce new queens and winged males during the warmer spring and summer months. These winged reproductives swarm periodically, usually 5 to 9 times, often after a rain. Mated queens attempt to establish a new colony by digging a small hole in the soil and closing up the entrance. Inside this chamber, the queen begins to lay 15 to 20 eggs in 2–3 days. Eggs are added to the pile over the next week, by which time the first eggs have hatched. The first instar larvae feed on several of the surrounding eggs. At this time the queen picks up the young larvae and sorts them into another pile. The larvae are then feed a liquid regurgitated by the queen. Apparently the queen uses body stores and degenerated wing muscles to provide larval food. A queen may lose over half of her body weight rearing the first brood. After 20 to 25 days the larvae pupate and the resulting tiny workers emerge 4 to 7 days later. These diminutive first workers are about 1/5 the size of the smallest workers found in an older colony. These workers, called minimis, break open the nesting chamber and begin foraging for insect food and start to enlarge the nest. The queen is now fed and she begins to lay more eggs which are properly cared for by the workers. If adequate food and water is in the areas, the colony steadily grows over the next few months. If a colony is established in June, it may contain 6,000 to 7,000 individuals by the following December. As the soil temperatures drop, the colony growth slows. By the following June, a one year old colony may have 10,000 to 15,000 workers and is producing new winged forms. Colonies 2 to 3 years old may have 20,000 to 200,000 workers. Established mounds will have a central pile of granular soil with openings and often smaller mounds around the parameter. The mound is established as a solar collector to store and incubate the larvae and pupae.

Fully mature colonies may have mounds 40 to 60 cm in diameter and 25 to 30 cm high. In clay soils, the mounds may be very hard and have been known to damage plows or mowing equipment. The mounds are generally not as tall in sandy soils. Considerable evidence exists that colonies may move their mounds and nests in search of food or when pesticides are applied. This is why mounds seem to appear overnight. The foraging workers rarely harm people, but when the mound is disturbed, the ants emerge in mass to swarm the intruder. Newborn livestock have been killed by attacks of disturbed ants.

Control Options: Since this is an important agricultural as well as urban pest, considerable effort has been expended in developing control techniques. Unfortunately, this group of pests have extremely great powers of reproduction and the ability to move considerable distances.

Option 1: *Chemical Control—Spot Treatments of Mounds*—Individual mounds can be treated with contact and stomach pesticides, and as the workers pick up the material they pass it through the colony. This is an effective method for a small number of mounds, especially in turfgrasses. However, often the ants detect the poison before the queen is killed and they may simply move the colony to another location.

Option 2: *Chemical Control—Use Poison Baits*—Several ant baits have been developed which are attractive to the foraging ants. By picking up the food bait, the workers distribute the poison throughout the colony. Take care to use the bait in such a manner that birds and other animals do not feed on it.

Option 3: *Chemical Control—Use Insect Growth Regulators (IGRs)*—Some new pesticides which are actually IGRs have been developed for use on fire ants. These act by sterilizing the queen or disrupting the development of the larvae. The IGRs generally have low toxicity, but may take 1–2 months to be effective.

Option 4: *Chemical Control—Area Wide Treatments*—Aerial applications of insecticides have been used to reduce fire ant populations in agricultural and recreational lands or to limit spread. Unfortunately, most of the pesticides developed for this technique have caused undesirable side effects to the environment. Areawide applications of insecticides with standard ground equipment have been effective and less detrimental to the environment.

CICADA KILLER

Species: *Sphecius speciosus* (Drury) [Phylum Arthropoda: Class Insecta: Order Hymenoptera: Family Sphecidae] (Figure 3.96; Plate 3–20).

Distribution: Found in all states east of the Rocky Mountains. Most common in southern states where the annual cicadas are prevalent.

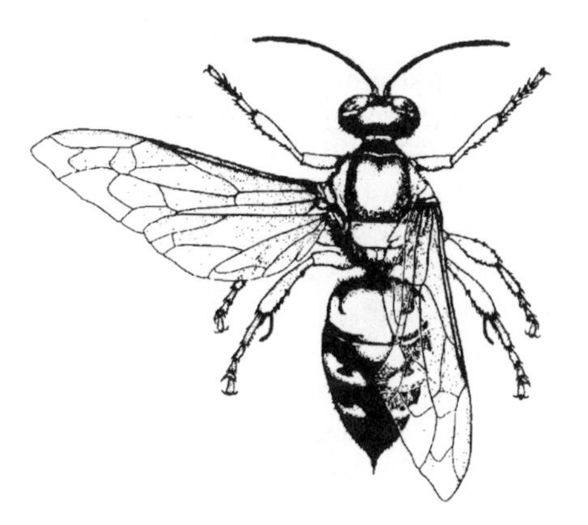

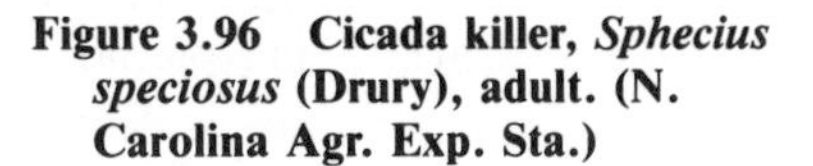

Figure 3.96 Cicada killer, *Sphecius speciosus* (Drury), adult. (N. Carolina Agr. Exp. Sta.)

Hosts: Does not attack turf. Selects light textured or sandy soils which are well drained. Prefers to dig burrow in sparsely vegetated areas such as golf course sand traps.

Damage Symptoms: Mounds of soil piled up around a hole approximately $^{1}/_{2}$ inch [12 mm] in diameter. Males are very protective of territory and often buzz or dive bomb people who enter the territory being protected. Females can sting but will not do so unless severely provoked.

Description of Stages: This is a wasp with a typical annual complete life cycle.

Eggs: The greenish-white eggs are $^{1}/_{8} \times ^{3}/_{64}$ inch [3-6 × 1-1.5 mm] and are attached to a cicada buried underground.

Larvae: The larva is grublike, but has no legs or well defined head capsule. Mature larvae are $1^{1}/_{8}$-$1^{3}/_{8}$ inch [28–32 mm] long.

Pupae: Mature larvae spin a tough spindle-shaped case in which the pupae reside. The pupae are at first white, but take on the adult colors before emergence.

Adults: The large adults are impressive with rusty red heads and thoracic areas. The wings are infused with orange and the abdomen is banded with black and yellow. The adults may be $1^{3}/_{16}$-$1^{3}/_{4}$ inch [30–45 mm] long, and have a wingspan of $2^{3}/_{8}$-4 inches [60–100 mm].

Life Cycle and Habits: This insect overwinters as a prepupa in a cocoon located 7-20 inches [17–50 cm] under the surface. When spring temperatures warm the soil, the pupal stage is formed and the adults dig to the surface in June and July. Males tend to emerge before the females and they establish flight territories, usually where females are to emerge and build new nests. The males fight off any other infringing males and are known to attack any other moving object in the area. Fortunately these males have no sting, but unsuspecting intruders are often shocked by the loud buzzing and strafing activities. Males have been known to dive bomb into people's heads and backs. The females are quite docile and after mating set out to construct a new tunnel. The tunnels are dug straight into the soil or are angled slightly. At the end of each tunnel are secondary tunnels which end in a chamber. The females search tree trunks and limbs for cicadas. When a cicada is located, the female swoops down, grabs the screaming prey and drops to the ground where a paralyzing sting is applied. The paralyzed cicada is then grasped under the wasp's body and flown back to the burrow. Each chamber is provided with 1 to 3 cicadas. An egg is laid on a cicada in the chamber and the chamber is sealed. Additional chambers are provided for, until the cicada population no longer supplies victims. The wasp egg hatches in 2 to 3 days and the voracious larva quickly devours the food. Only the cicada shells remain. By the fall, the mature larvae spin a cocoon, shrink, and prepare to overwinter. A single generation occurs per year.

Control Options: Though this is considered a beneficial insect, people often get upset with the activities of the territorial males, and the mounds thrown up by the females can be unsightly. Therefore, no controls are recommended unless this pest greatly interferes with human activities.

Option 1: *Cultural Control—Habitat Modification*—Since this insect prefers open sandy soil to build its burrow, good turf culture which promotes the growth of thick turf will often discourage burrow building.

Option 2: *Mechanical Control—Male Destruction*—Since the males are the offending aggressors, capture them with an insect net or swat them down and crush. This usually ends complaints by golfers and allows the females to capture the cicadas.

Option 3: *Chemical Control—Insecticide Application*—Dusting of the burrow openings with insecticides is very effective in killing the females as they go about their daily activities. Bendiocarb (=Turcam®, Ficam®), carbaryl (=Sevin®), and diazinon (not on golf courses or sod farms) are registered for this purpose.

NUISANCE VERTEBRATE PESTS

The vertebrates belong to the phylum Chordata which have bilateral symmetry, like insects, but the nerve cord is dorsal and the circulatory system is closed. Next to the nerve cord is a stiff rod of cartilage, the notocord. The notocord in vertebrates is replaced by a series of bony segments, the vertebrae or backbones. All chordates with these vertebrae are in the subphylum Vertebrata. The vertebrates are again subdivided into the fish, superclass Pices, and the four-legged animals, superclass Tetrapoda. The tetrapods are divided into four classes: the Amphibia (frogs, toads, salamanders), the Reptilia (lizards, snakes, turtles), the Aves (birds), and the Mammalia (mammals). Of these groups, some birds and mammals are occasional pests of turf, usually during their search for insect or weed food.

The **Class Aves** is a large and diversified group. About 1,700 species in over 90 families are found in North America. Fortunately, the vast majority of these birds inhabit turfgrass areas only on rare occasions. Some of the pest birds, usually imported species or blackbirds, damage turf only in their search for insect food. All other birds are considered beneficial and are protected by international, federal, and state laws. The most common damaging birds are the starling (Family Sturnidae) and the blackbirds (Family Icteridae). The blackbirds which may damage turf are the grackles, cowbirds, and red-winged blackbird.

Crows, robins, and various water birds may also be nuisances on turf. The plant feeding water fowl, especially ducks and geese, cause problems when their flocks forage on the high quality turf of golf courses, commercial facilities, and home lawns. Their copious feces, often laden with weed seeds can become a real nuisance.

The **Class Mammalia** also is a diverse group with more than 350 species found in North America. Mammals have fur on the body, warm blood, and mammary glands for feeding the young. Orders of mammals which commonly damage turf are the Marsupialia (opossum), Insectivora (shrews and moles), Carnivora (raccoons, skunks), Rodentia (pocket gophers, voles, and ground squirrels), and the Xenarthra (armadillos). The opossums are true marsupials, with a pouch for carrying the premature young for development. The insectivores are small mammals with pointed noses and tiny beadlike eyes. The carnivores are medium to large animals with prominent canine teeth. Carnivores are general meat eaters but they may feed on fruits, nuts, and berries. The rodents are small to medium animals with two pairs of prominent front teeth which continually grow and must be worn down through chewing. The armadillo is an unusual mammal with degenerated teeth and a protective armor of hairy plates.

COMMON GRACKLE

Species: *Quiscalus quiscula* (Linnaeus) is sometimes confused with the boat-tailed grackle, *Q. mexicanus* (Gmelin), and the great-tailed grackle, *Q. major*. [Phylum Chordata: Class Aves: Family Icteridae].

Distribution: The common grackle can be found in the summer east of the Rocky Mountains from southern Canada to the Gulf states. They usually migrate south for the winter and are most common in the Mississippi valley states and coastal states.

Damage Symptoms: These species commonly congregate with other blackbird species during the fall and winter. Large flocks are a nuisance and health hazard in urban areas. These birds rarely attack turf but will actively forage for various insects. Flocks may scrape away grub-damaged turf in late summer and fall. Grackles are generally considered beneficial when small groups feed on insects.

Description of Stages: An iridescent blackbird which is about 12 inches [30 cm] long, the common grackle has a shorter tail than its two relatives. Grackles generally have bright yellow eyes and long tails. These noisy birds make a series of clucks and rising screeches.

Life Cycle and Habits: These common birds are numerous in the urban habitat where they prefer to forage in lawns, parks, and open fields. They are often gregarious and prefer to breed in tall conifers. However, the large stick nest lined with fine grass may also be found in lower shrubs. Usually five pale blue eggs with irregular black markings are laid in the nest. Usually a single generation per year is reared and the birds form large migratory flocks with other blackbird species in the fall. This bird has an extremely diverse diet and will eat most any animal that is small enough to swallow. They also occasionally raid other birds' nests. They will feed on grains, wild fruits, and cultivated small fruits. Urban grackles readily take food at parks and can be a nuisance around trash baskets.

Control Options: See Starling. This species is generally not a pest until the gregarious migratory flocks are formed in late summer and fall.

STARLING

Species: *Sturnus vulgaris* Linnaeus. [Phylum Chordata: Class Aves: Family Sturnidae].

Distribution: The starling is an introduced pest from Eurasia. Since 1890, it has spread across North America from Quebec to southern Alaska down to northern Mexico.

Damage Symptoms: This anthropophilic (prefers to live around humans) species is well adapted to urban life and often roosts in large numbers on buildings and in trees. Large migratory flocks are a noise nuisance and a health hazard from their feces. Flocks in the fall may attack grain crops and can damage turf which is infested with grubs. The birds insert their beak into the thatch, spread it, and make a hole. This is the common searching activity used when looking for caterpillars or other surface and thatch dwelling insects. Grubs are located by scraping away the turf, like chickens scratching. This species is considered beneficial when they nondestructively feed on insects.

Description of Stages: The adults are slightly smaller than a robin, about 7.5–8.5 inches [19–22 cm] long. They have a short tail, generally black iridescent plumage, and are often flecked with white spots. They have a long pointed bill which is yellow-white in the summer and brownish in the winter. They have a large repertoire of calls consisting of clicks, squeaks, and whistles. They sometimes mimic the calls of other birds.

Life Cycle and Habits: Starlings prefer urban and suburban habitats. They build a rough nest consisting of twigs, trash, and grass. The inner cavity is lined with fine grasses and feathers. The nests are usually built in spaces on buildings, especially small confined cavities, and in larger trees. Nesting occurs at the first hint of spring and sometimes a second brood is reared in the summer. Large roosts may be located on buildings after the nesting season and thousands of birds may be present. Sometimes starling roosts are mixed with other fall gregarious black birds. These roosts, once established, are very difficult to move or drive away. Starlings are generally more aggressive than most native birds and outcompete them for food and nesting sites.

Control Options: See Birds—Introduction. These highly urban-adapted birds are best controlled by reducing nesting sites and food, or by moving roosts.

Option 1 : *Cultural Control—Restrict Roosting and Nesting Sites*—Buildings with fancy facades with many cavities should be covered with fine mesh or sticky repellents to discourage roosting.

Option 2: *Cultural Control—Move Roosts*—This is a very difficult task, especially if a roost has been established for several nights. Generally, specialists armed with loud speakers to play tapes of bird distress calls and exploding cartridges are needed to relocate a roost.

Option 3: *Chemical Control—Spray Wetting Agents*—This chemical technique relies on a soap, the wetting agent, and cold weather. Roosts are sprayed by plane or helicopter with a wetting agent just before a cold rainstorm in the fall. The birds become wet from the rain and the cold causes an outbreak of bird pneumonia. This technique is used by cities with major bird roosts which are creating health hazards.

REDWINGED BLACKBIRD

Species: *Agelaius phoeniceus* (Linnaeus). [Phylum Chordata: Class Aves: Family Icteridae].

Distribution: This migratory bird breeds during the summer from southern Alaska across to the state of Maine and south into central Mexico. It generally migrates southward for winter and can be found from New Jersey across to northern California and southward into Mexico.

Damage Symptoms: This pest is usually a major part of the blackbird flocks which gather in the fall and winter. These flocks may damage turf when they search for grubs by scraping away the surface vegetation. The large flocks are also considered a health hazard in urban areas because of the accumulations of wastes.

Description of Stages: These birds are about the same size as the starling but more slender. The males have a bright red-orange shoulder patch which fades slightly during the fall and winter. The females are streaked with dusky brown.

Life Cycle and Habits: At the very first hint of spring, the males begin to stake out territory along marshes, swamps, meadows, and pastures. They tend to prefer moist areas but may nest some distance from water. The territorial males fight off any intruder males and are constantly calling for the females. The nests are usually constructed of marsh grass or reeds woven into growing brush or reeds. The nest may contain 3 to 5 light blue eggs which are spotted or streaked with dark brown and purple. Depending upon the locality and season, the nesting pair may raise 2 to 3 broods in a summer. A new nest is constructed for each brood. These birds collect insects and some vegetable matter for food during the summer. In the fall, the males lose their territorial instincts and the birds begin to gather into large flocks. These flocks are usually mixed with cow birds, starlings, and grackles. These flocks often attack fall crops such as corn but they generally utilize any available food source.

Control Options: See Starling.

MOLES

Species: The eastern mole, *Scalopus aquaticus* (Linnaeus), and star-nosed mole, *Condylura cristata* (Linnaeus), are common turf pests in the eastern half of the United States. Several species of *Scalopus* occur in turf in the West Coast states. [Phylum Chordata: Class Mammalia: Family Talpidae].

Distribution: The eastern mole is common from Massachusetts west to lower Michigan, into Nebraska and south to all the Gulf States. The star-nosed mole is usually found in damp areas in the northeast from Virginia to Minnesota and north into Canada.

Damage Symptoms: Moles tunnel under the soil surface in search of insects and earthworms. This tunneling throws up ridges in turf. Deeper tunnels are often marked with cone-shaped earthen mounds. Turf often dies along the surface tunnels, especially in dry weather. Moles may become especially active in grub infested turf or areas where other soil invertebrates are numerous.

Description of Stages: Moles are plump, elongated, smooth-furred mammals with elongate enlarged skulls which end in a pointed snout. There are no external ears, the ear holes being hidden by thick fur, and the tiny eyes apparently can only detect light and dark. The front feet are greatly enlarged, clawed, and face outward. The eastern mole has a slightly pink, pointed nose with a short tail, while the star-nosed mole has a pink fleshy nose armed with 22 flesh projections and a long hairy tail. Adult moles are 3.5 to 8.75 inches [9–22 cm] long.

Life Cycle and Habits: Moles are true insectivores like their smaller cousins, the shrews. The eastern mole prefers to forage in well drained, loose soils of fields, lawns, and gardens but it may nest and set up base burrows along the edge of wooded areas. The star-nosed mole is semiaquatic and is commonly found in wet woodlands, fields, or swamps. It also digs in turf located next to streams, rivers, and ponds. Both species use earthworms as their major food item but they quickly forage for white grubs when available. Moles make longitudinal tunnels at the soil surface which are their major thoroughfares. Deeper tunnels are constructed for hiding and resting. These deeper tunnels are usually marked by soil thrown out of a hole, the cone-shaped "molehills." Most moles are active at dusk and dawn but tunneling may be observed on overcast days. A mole can travel up to a foot per minute in loose soils. Moles are generally solitary, though considerable numbers can be found in an area. Males seek out females in the early spring and the females produce a single litter of 2 to 6 young. The young are kept in an underground nest until old enough to begin to forage on their own, usually within a month. The star-nosed mole often builds its nest in a raised area that does not get swamped. Moles are considered beneficial because of their insect feeding behavior and ability to aerate the soil. However, they are a disaster in well maintained turf.

Control Options: Moles are generally not a problem in soils with few earthworms and no insect larvae. Since moles do not feed on plant material like pocket gophers, elimination of invertebrate food will effectively control mole activity. However, this type of management, where all the invertebrates are eliminated, is not recommended for ecological reasons. Flooding the tunnels through a water hose, using automobile exhaust, and placing chewing gum in the tunnels are not appropriate or effective management options.

Option 1: *Chemical Control—Fumigation*—Fumigation with toxic gases is moderately effective if currently active mole tunnels can be located. If considerable activity is noticed in an area, roll down the tunnels with a lawn roller and watch for fresh activity. In home lawns, moles do not respect property lines and moles may have to be controlled in several adjacent lawns. Fumigation often fails if the mole detects the gas before toxic levels are reached. The pests simply plug the tunnel with soil when the gas is detected.

Option 2: *Cultural Control—Trapping*—Several types of traps (e.g., harpoon, strangle, or clasping) can be effective to eliminate individual moles. Again, currently active tunnels must be located. Select one of the major long tunnels, not the short branching side tunnels, for trap placement. Harpoon traps work by pressing down a tunnel under the trap and setting the trap trigger over the spot. When the mole reopens the tunnel, it must press against the harpoon trigger. Strangle and clasping traps usually require that the burrow be opened carefully and the trap placed in the runway. Care must be taken to not overly disturb the tunnel or to leave behind human scents. Since moles patrol currently active tunnels several times a day, if a trap has not captured the pest within two days, relocate the trap.

Option 3: *Chemical Control—Poison Baits*—Baits marketed for gophers are not effective against moles. Moles usually will not eat the bran or peanut based baits.

POCKET GOPHERS

Species: Several species in the genus *Thomomys* are found from the Rocky Mountains to the West Coast. East of the Rocky Mountains, most pocket gophers are species of *Geomys*. The plains pocket gopher, *G. bursarius* and the southeastern pocket gopher, *G. pinetis*, are common turf pests.

Distribution: The plains pocket gopher is common in the prairie areas from Wisconsin to eastern North Dakota south to eastern New Mexico and eastern Texas. The southeastern pocket gopher is common in southern South Carolina, Georgia, and Alabama into the north half of Florida.

Damage Symptoms: Pocket gophers tunnel near the soil surface in search of plant roots and tubers. This tunneling often throws up ridges in turf areas. Periodically, pocket gophers dig deeper tunnels for food storage and shelter. These deeper tunnels are often accompanied with a mound of soil deposited at the soil surface. Since gophers are especially fond of tubers, such as spring flowering bulbs, they may be very active around flower beds or in turf containing "naturalized" bulb plantings. Pocket gophers occasionally eat insects encountered in the soil but they do not seek out insects like moles.

Description of Stages: Pocket gophers are stout-bodied rodents with the front incisor teeth always exposed. This is different than the teeth found in moles. Gophers have enlarged front feet armed with long claws used for digging. They also have short fur, small eyes and ears, and a short tail. Their common name comes from the presence of large fur-lined cheek pouches which open outside the mouth. These pouches extend back to the shoulders. These pouches can be emptied by squeezing with the front legs. The *Thomomys* spp. have smooth front teeth with no grooves, while the *Geomys* spp. have two conspicuous grooves running down the middle of the teeth. Most gophers are 7.75–14 inches [20–36 cm] long.

Life Cycle and Habits: Pocket gophers are solitary animals and they usually fight over disputed territory. Males are especially fierce and may kill other male intruders. In the spring, males search for females but separate after mating. Females give birth to 2 to 11 young in a deep chamber. Within several weeks the young may follow their mother in the tunnels and after two months, the young begin to dig their own tunnels and disperse. At three months the young are sexually mature. A second litter may be produced by some individuals. A gopher working in a suitable area will dig numerous surface tunnels, collecting roots and tubers and throwing up mounds of excess soil. Periodic deeper tunnels are dug and extra food, tucked into the cheek pockets, are deposited here. Activity is reduced in the winter months and stored food is eaten when fresh roots cannot be obtained.

Control Options: Pocket gophers, like moles, are extremely difficult to control because of their subterranean habits. Gophers are even more troublesome because they feed on plants rather than insects and worms. Special care should be taken to locate currently active burrows. These can be located by looking for fresh soil mounds. Attempts to drowned the animals, use automobile exhaust fumes, or chewing gum are not effecting for managing this group of pests.

Option 1: *Cultural Control—Use Killing Traps*—Several traps are available which spear or choke active gophers. Unfortunately, gophers often shy away from disturbed tunnels. They occasionally shove soil ahead of them in a tunnel and this soil is "speared" by the traps. If trapping fails the first time, repeat setting the traps in new locations.

Option 2: *Chemical Control—Use Fumigants*—Several gas producing canisters are available for gopher fumigation. These occasionally work if the gopher is trapped between the gas and a new section of the tunnel. If the gopher is in the old tunnels, it can often outrun the gas, or simply plug the tunnel with soil. Fumigants are also rarely effective in loose sandy soils.

Option 3: *Chemical Control—Use Poison Baits*—Gophers are rather fond of seeds and grains. Carefully drop poisoned grain into tunnels which have been cautiously opened using a sharp pointed stick. Better success is obtained if these baits are placed in more recent tunnels. Do not handle the baits with bare hands, as the human scent will often repel the animals.

SKUNKS AND CIVET CATS

Species: The striped skunk, *Mephitis mephitis* (Schreber), and spotted skunk or civet cat, *Spilogale putorius* (Linnaeus) are the two North American pests found foraging in turf. [Phylum Chordata: Class Mammalia: Family Mustelidae].

Description: The striped skunks have a head and body length of 13–18 inches [33–45 cm] with a 7–10 inch [17–25 cm] tail. They may be 6 to 14 lb [2.7–6.5 kg]. The spotted skunks are smaller with a head and body length of 9 to 13 inches [23–33 cm] and tail 5 to 9 inches [13–23 cm]. They weigh 1–2.5 lb [0.45–1.1 kg]. The striped skunks are usually solid, shiny, black with a broad, white stripe running from the head to mid-body where it divides into two lines. The spotted skunk has several irregular broken white stripes along the side and top of the body. The spotted skunk usually has a white tip to the tail.

Description of Stages: Both of these mammals prefer open country or sparse woods and brushland. They commonly inhabit suburban areas where food and cover abound. The young are born in May to June and 5 to 6 are usually in the litter. The young are kept in the burrow for several weeks before they emerge with their mother to forage. Striped skunk young often follow their mother in single file. By July and August, the young have learned to feed and care for themselves. Burrows are built in the ground, under old buildings or in rock and wood piles. These omnivores feed on fruits, vegetables, and nuts as well as small rodents, birds, and insects. Most activity begins shortly after sundown and ceases at sunrise. Apparently, skunks and civets are active most of the year but several females may den together in the winter. Because they may feed on rodents and insects, skunks are considered beneficial. However, they commonly catch rabies, which quickly spreads to most of the individuals in an area. In the fall, when white grubs reach their third instar stage, skunks often forage as single individuals or in groups. Once a suitable location has been found, the skunks will return night after night until the free meal is depleted.

Control Options: Skunks and civets are most easily disposed of by shooting at night. They are difficult to handle when trapped because of the scent glands. Unfortunately, many states have fur-bearing animal laws which protect skunks and civets. Other municipalities do not allow for the discharging of firearms within city limits.

Option 1: *Cultural Control—Reduce Den Habitat and Food*—Skunks and civets commonly nest in disorganized wood piles, hollow trees, or logs, under crawl spaces of porches or houses, or in partially closed culvert drains. Survey the area and attempt to eliminate these nesting sites. Keep wood piles neatly stacked with space between the rows, and cut down or remove hollow trees and logs. Thoroughly close in access to crawl spaces and keep drain culverts open.

Option 2: *Cultural Control—Manage Food Sources*—Skunks and civets usually become pests where garbage cans are not covered, dog or cat food is left outside and uncovered, or white grub populations are allowed to reach large numbers. Where trash is kept outside, make sure that lids or doors seal tightly and cannot be pried open. Remove open dog or cat food containers at dusk. Monitor for grub populations and control them before the third instar stage is reached.

Option 3: *Cultural Control—Trapping*—Various live traps are made which can capture skunks and civets. Unfortunately, captured animals usually release their scent and handling can be difficult. If a live trap is used, release the animals in a location recommended by the local animal control agency. Many municipal agencies will remove trapped animals, either for free or for a small fee. Remember that skunks and civets often carry rabies and special care should be used to avoid scratches or bites.

RACCOON

Species: *Procyon lotor* (Linnaeus). [Phylum Chordata: Class Mammalia: Family Pracyonidae].

Distribution: Common in North America south of the Canadian border. Not common in the higher areas of the Rocky Mountains.

Damage Symptoms: Raccoons usually dig up irregular patches of turf which is infested with grubs. This damage occurs at night and several raccoons may be active in an area. The raccoons often return night after night if grubs are easy to find. These pests are common carriers of rabies.

Description of Stages: This well known "masked bandit" usually has gray and black body hair, a length of 18 to 28 inches [46–71 cm], and a ringed tail, 8 to 12 inches [20–30 cm] long. They may weight 12 to 35 lb [5.4–15.9 kg]. The head has a black colored mask around the eyes.

Life Cycle and Habits: Raccoons do not hibernate but live in dens located in hollow logs, rock crevices, or ground burrows. Apparently urban raccoons will establish residence in storm sewers and under porches or house crawl spaces. Females give birth to 2 to 7 young, usually four, in April or May. The young are nursed in the den and open their eyes in about three weeks. After two months, the young may leave the den to forage with their mother. The mother teaches the young how to catch insects, frogs, crayfish, and other small animals. They will also feed on fruits and nuts. They often attack small grain crops such as corn. By the fall, the young are weaned and they have to locate their own territory. Raccoons are nocturnal but occasionally forage during the day.

Control Strategies: See Skunks and Civet Cats. Raccoons are considered game and fur-bearing animals in most states. Be careful in handling live animals, as they may carry rabies.

NINEBANDED TEXAS ARMADILLO

Species: *Dasypus novemcinctus texanus* Bailey. [Phylum Chordata: Class Mammalia: Family Dasypodidae].

Distribution: The armadillo is currently found from mid-Georgia across to southwestern Missouri and south across Oklahoma, Texas, and Florida. This interesting animal used to be rare in Oklahoma and Arkansas, but it has been steadily expanding its range northward.

Damage Symptoms: The armadillo eats any animal food that it can catch during its foraging in the soil. They prefer worms, insects, spiders, millipedes, and centipedes. However, they have been known to eat crayfish, amphibians, and bird eggs. They dig and root around in soils much like raccoons or skunks. The damage is often mistaken for damage by other animals.

Description of Stages: The armadillo is the only mammal found in the United States with heavy, leathery, ringlike armored plates over the body. Adults are 24 to 31 inches [61–79 cm] long with a 9.5 to 14.5 inch [24–37 cm] long tail. The head is covered with small scaly plates and has a pointed snout. The shoulders and pelvis are covered with wide scaly plates and these are joined together by nine narrow, jointed plates. This lets the armadillo roll into a ball if sufficiently disturbed. The tail consists of a series of ringed plates.

Life Cycle and Habits: Armadillos are truly unique and unusual animals. They generally hide in the daytime in burrows dug into the ground, often along the banks of streams or rivers. After dark, armadillos emerge to forage for food by rooting through plants and digging in the soil. They will eat most any small arthropod or animal which can be picked up. They often travel in bands of several individuals which grunt and snort constantly. Mating occurs in the fall and each female gives birth to exactly four identical quadruplets. Apparently the fertilized egg divides twice before being implanted in the uterine wall. The young are born in a breeding burrow lined with leaves and grass. The young are able to walk within hours after birth, and they have their eyes open. This litter follows their mother around, often in single file in search of food. Armadillos can run quite rapidly when disturbed but are often startled or dazed by bright lights at night. They may swim by gulping air or they may cross small streams by walking across the bottom. These animals cannot survive freezing temperatures over extended periods and are active year-round.

Control Options: Armadillos digging in turf are generally an indication that insects or worms are present in abundance. They will feed for several nights in an area if ants, grubs, or worms are easy to find. They usually move on after a night of feeding, unlike skunks or raccoons. See Skunks and Civet Cats.

REFERENCES

Baker, J.R. (Ed.) Insects and Other Pests Associated with Turf, Some Important, Common, and Potential Pests in the Southeastern United States. North Carolina Agr. Ext. Ser. Ag-268: (1982) 108 pp.

Bohart, R.M. 1947. Sod Webworms and Other Lawn Pests in California. *Hilgardia* 17: 267–308 (1947).

Bottrell, D.R. Integrated Pest Management. U.S. Council on Environmental Quality. U.S. Government Printing Office: 041-011-00049-1. (1979) 120 pp.

Leslie, A.R. and R.L. Metcalf (Eds.) Integrated Pest Management for Turfgrass and Ornamentals. U.S. EPA, Office of Pesticide Programs, Washington, DC: U.S. Government Printing Office: 1989-625-030. (1989) 337 pp.

Principles of Plant and Animal Pest Control, Vol. 3. Insect-Pest Management and Control. National Academy of Science, Washington, DC. Publ. 1695: (1969) 508 pp.

Niemczyk, H.D. *Destructive Turf Insects*. HDN Book Sales, Wooster, OH, 1981, 48 pp.

Niemczyk, H.D., and B.G. Joyner (Eds.) *Advances in Turfgrass Entomology*. ChemLawn Corporation, Columbus, OH, 1982, 149 pp.

Shartleff, M., R. Randell, and T. Fermanian. *Controlling Turfgrass Pests*. Prentice-Hall, Inc., NY, 1987, 305 pp.

Shetlar, D.J., P.R. Heller, and P.D. Irish. 1990. *Turfgrass Insect and Mite Manual*, 3rd ed. The Pennsylvania Turfgrass Council, Bellefonte, PA, 1990, 67 pp.

Streu, H.T., and R.T. Bangs (Eds.) *Proceedings of Scotts Turfgrass Research Conference*. Vol. 1—Entomology. May 19–20, 1969. O.M. Scotts and Sons Co., Marysville, OH. 89 pp.

Tashiro, H. Turfgrass Insects of the United States and Canada. Cornell Univ. Press, Ithaca, NY, 1987, 391 pp.

Ware, G.W. 1978. *The Pesticide Book*. W.H. Freeman & Co., San Francisco, CA. 197pp.

Index

Abamectin. *See* Avernectin-B
Acanthomyops
  claviger. *See* Smaller yellow ants
  interjectus. *See* Larger yellow ants
Achillea millefolium. *See* Yarrow
Acrobatic ants, 325
Adalia bipunctata. *See* Twospotted lady beetles
Agallia spp. *See* Leafhoppers
Agelaius phoeniceus. *See* Redwinged blackbirds
Agricultural wastes, 97. *See also* specific types
Agriolimax reticulatus. *See* Gray garden slugs
Agropyron repens. *See* Quackgrass
Agrostis stolonifera. *See* Creeping bentgrass
Agrotis ipsilon. *See* Black cutworms
Algae, 93, 150–151. *See also* specific types
  blackish blue-green, Plate 2–38, Plate 2–62, Plate 2–71
  fungicides for control of, 168
  as indicator of correctable conditions, 2
Aliette. *See* Fosetyl-aluminum
Allium vineale. *See* Wild garlic
Amaranthus blitoides. *See* Prostrate pigweed
Ambrosia artemissifolia. *See* Ragweed
Ambylomma americanium. *See* Lone star ticks
American dog ticks, 175, 308
Anilazine, 160
Annedids, 174. *See also* specific types
Annual bluegrass, 13–15, 84–85
  anthracnose in, Plate 2–31
  bacterial wilt in, Plate 2–68
  diseases of, 88, 153
  Leptosphaerulina blight in, Plate 2–35
  summer patch in, Plate 2–51, Plate 2–52, Plate 2–53
  tolerance of to herbicides, 82–83
Annual bluegrass weevils, Plate 3–11, 262–264
Annual grasses, 2, 3. *See also* specific types
  summer. *See* Summer annual grasses
  winter. *See* Winter annual grasses
Anthemis cotula. *See* Dog fennel
Anthracnose basal rot, Plate 2–31, Plate 2–32, Plate 2–34, 106, 120–121, 155, 158
  control of, 165
  season of occurrence of, 157
Antonina graminis. *See* Rhodesgrass mealybugs
Ants, 325–330. *See also* specific types
Aphelinus mali. *See* Chalcid wasps
Aphidius testaceipes. *See* Aphid wasps
Aphid wasps, 194
Arachnids, 174. *See also* specific types
Armadillium vulgare. *See* Pillbugs
Armadillos, 341–342
Armyworms, Plate 3–18, 227–231, 233–238. *See also* specific types
Arsenicals, 5. *See also* specific types
Arthropods, 174, 175–177, 206. *See also* specific types
Ascochyta leaf blight, 121–122
Ascochyta spp. *See* Ascochyta leaf blight
Asiatic garden beetles, Plate 3–10, 282, 288–289
Asulam, 77, 82, 84
Ataenius spretulus. *See* Black turfgrass ataenius
Atrazine, 77, 80, 82, 84
Attractants, 199–200. *See also* specific types
Augmentation, 190

Avermectin-B, 198
Azadirachtin, 198

Bacteria, 2, 93, 97, 145–146, 194–195. *See also* specific types
Bacterial wilt, Plate 2–68, Plate 2–69, 145–146, 156
Bahiagrass, 82–83, 84–85
Banks grass mites, 210–212
Banner. *See* Propiconazole
Banol. *See* Propamocarb
Barbarea verna. *See* Yellow ricket
Barnyardgrass, 4, 9, 84–85
Basal rot, anthracnose. *See* Anthracnose basal rot
Basamid. *See* Dazomet
Basidiomycetes, 133, 138. *See also* Fairy rings; Localized dry spot
Bayleton. *See* Triadimefon
Bedstraw, 42, 80–81
Beetles. *See also* Scarabs; specific types
  Asiatic garden, Plate 3–10, 282, 288–289
  green June, 282, 291–292
  ground, 191, 192, 204, 319–320
  Japanese, Plate 3–10, 271, 280, 282, 293–295
  June. *See* June beetles
  lady, 191
  May and June, 300–302
  Oriental, Plate 3–10, 282, 299–300
  rove, 191, 204, 321–322
  twospotted lady, 191
Beggarweed, 72
Bellflower, 68
Bellis perennis. *See* English daisy
Beneficial insects, 204. *See also* specific types
Benefin, 77, 80, 82, 84
Benomyl, 160, 164
Bensulide, 77, 79, 82, 84
Bentazon, 77, 80, 82, 84
Bentgrass, 14. *See also* specific types
  algae in, Plate 2–62
  anthracnose basal rot in, Plate 2–32
  black cutworms in, Plate 3–17
  blackish blue-green algae in, Plate 2–62
  black stem of, Plate 2–34
  brown patch in, Plate 2–38, Plate 2–39, Plate 2–62
  colonial, Plate 2–40
  creeping. *See* Creeping bentgrass
  diseases of, 88, 89, 94, 102, 153. *See also* specific types
  dollar spot in, Plate 2–62
  fairy rings in, Plate 2–60
  pink patch in, Plate 2–10
  Pythium blight in, Plate 2–44, Plate 2–45
  spring dead spot in, Plate 2–36
  superficial fairy rings in, Plate 2–62
  take-all in, Plate 2–15, Plate 2–16, Plate 2–17, Plate 2–18
  Toronto creeping, Plate 2–69
  velvet, Plate 2–56
  yellow patch in, Plate 2–5, Plate 2–6
  yellow tuft in, Plate 2–37
Benzimidazoles, 159, 160, 164, 168. *See also* specific types
Bermudagrass, Plate 2–12, Plate 2–21, Plate 2–22, Plate 2–23, 5, 22
  diseases of, 89, 94, 154
  rhodesgrass mealybugs on, Plate 3–7
  scales, 252–254
  tolerance of to herbicides, 82–83
  witchesbrooming of, Plate 3–1
Bermudagrass decline disease, 123–124, 155, 157
Bermudagrass mites, Plate 3–1, 206–208
Bermudagrass scales, Plate 3–8, 252–254
Betony, 71
Bibionids, 324–325. *See also* specific types
Bibio spp. *See* March flies
Biennial weeds, 44–47. *See also* specific types
Bigeyed bugs, 191–192, 316–317. *See also* specific types
Billbugs, Plate 3–12, Plate 3–13, 254–262. *See also* specific types
Bindweed, 67
Bipolaris spp.. *See* Damping-off disease; Helminthosporium leaf spot
Bird toxicity, 204
Black armyworms, 229
Blackbirds, 335–336
Black cutworms, Plate 3–17, 231–233
Blackish blue-green algae, Plate 2–38, Plate 2–62, Plate 2–71
Black-layer, 93, 150–152
Blacklegged ticks, 309
Black medic, 47
Black stem, Plate 2–34
Black turfgrass ataenius, Plate 3–10, 282, 286–288
Blight. *See also* specific types
  Ascochyta leaf, 121–122
  Cercospora leaf, 155

cultural practices and, 92
Curvularia, 131–132
Fusarium, 130, 155
gray snow mold. *See* Gray snow mold
Leptosphaerulina, Plate 2–35, 121–122, 155, 157
Nigrospora, 93, 132–133, 155
Pythium. *See* Pythium blight
Rhizoctonia. *See* Brown patch
season of occurrence of, 157
snow, 102
southern, Plate 2–55, 130–131
Typhula, 104–105
white, 140–141, 155
Blissus
insularis. *See* Southern chinch bugs
leucopterus. *See* Hairy chinch bugs
Bluegrass, Plate 2–14. *See also* specific types
annual. *See* Annual bluegrass
diseases of, 88, 89, 91, 92, 93, 153. *See also* specific types
Kentucky. *See* Kentucky bluegrass
Bluegrass billbugs, Plate 3–12, Plate 3–13, 256–260
Bluegrass sod webworms, 219–220
Blue-green algae, Plate 2–38, Plate 2–62, Plate 2–71
Boat-tailed grackles, 333
Boric acid, 197
Botanicals, 198–199. *See also* specific types
Bracbonid wasps, 193
Brass buttons, 37
Broadleaf weeds, 3. *See also* specific types
hard-to-control, 80–81. *See also* specific types
perennial, 47–75. *See also* specific types
postemergence control of, 80–81
preemergence control of, 80
summer annual, 26–37
winter annual, 37–44
Brome, 19
Bromus
catharticus. *See* Rescuegrass
inermus. *See* Smooth brome
Bronzed armyworms, 229
Bronzed cutworms, 228
Brown dog ticks, 308
Brown garden snails, 305
Brown patch, 91, 93, 94, 97, 98, 124–126, 155, 158
in bentgrass, Plate 2–29, Plate 2–38, Plate 2–39, Plate 2–40, Plate 2–62
in colonial bentgrass, Plate 2–40
control of, 165
cool temperature. *See* Yellow patch
in creeping bentgrass, Plate 2–29
cultural practices and, 92
diagnosis of, 88
management of, 125–126
season of occurrence of, 157
in tall fescue, Plate 2–41
threshold of, 89
Bryobia praetiosa. *See* Clover mites
Buckhorn, 52
Buffalograss, 82–83
Bull thistle, 59–60
Bur clover, 73
Burrowing sod webworms, 265–267

Cacodylic acid, 77, 79, 85
Cadmium-based fungicides, 159. *See also* specific types
Calendar date applications, 182–183
Callipus spp. *See* Millipedes
Calo Clor. *See* Mercury
Campanula rapunculoides. *See* Bellflower
Canada thistle, 57
Capsella bursa-pastoris. *See* Shepherdspurse
Carbamates, 199, 204. *See also* specific types
Carpetweed, 33–34
Carrot, 80–81
Caterpillars, 198, 238–242. *See also* specific types
Cat fleas, 322
Catnip, 80–81
Cenchrus pauciflorus. *See* Sandbur
Centipedegrass, Plate 3–9, 82–83, 154
Centipede mosaic virus, 144–145, 156
Centipedes, 174, 175, 176, 177, 312–313
Centuroides vitatus. *See* Scorpions
Cerastium vulgatum. *See* Mouse-ear chickweed
Cercospora leaf blight, 155
Cercospora leaf spot, 140, 155
Cercospora spp. *See* Cercospora leaf spot
Chafers. *See also* specific types
European, Plate 3–10, 282, 289–291
masked, 282, 295–299
northern masked, Plate 3–10, 295–297

southern masked, Plate 3-10, 297-299
Chalcid wasps, 193
Chemical hydrolysis, 202-203
Chenopodium album. *See* Lambsquarters
Chickweed, 38-39, 53
Chicory, 58
Chiggers, 307-308
Chilopods, 174. *See also* specific types
Chinch bugs, 185-186, 192, 242-246. *See also* specific types
hairy, 178, 242-246
southern, Plate 3-4, 246-248
Chipco 26019. *See* Iprodione
Chitin, 198
Chlamydomonas spp., 150
Chloroneb, 105, 127, 144, 161, 163, 164, 166, 167
Chlorosis, Plate 2-67
Chlorothalonil, 105, 108, 110, 121, 122, 126, 130, 133, 140, 144
as contact fungicide, 161
encouragement of disease by, 160
incompatibilities of, 169
modes of action of, 161
properties of, 164
scheduling of, 165, 166, 168
Chlorsulfuron, 77, 80, 82, 84
Chrysanthemum leucanthemum. *See* Ox-eye daisy
Chrysoperla oculata. *See* Green lacewings
Chrysoteuchia topiaria. *See* Cranberry girdlers
Cicada killers, Plate 3-20, 330-332
Cichorium intybus. *See* Chicory
Cinquefoil, 68-69
Cirsium
arvense. *See* Canada thistle
vulgare. *See* Bull thistle
Cirtrus oils, 197
Civet cats, 339-340
CL 3336. *See* Thiophanate
Clandosan. *See* Chitin
Cleary's 3336. *See* Thiophanate-ethyl
Clopyralid, 79, 81, 83
Clover, 50-51, 73. *See also* specific types
Clover mites, Plate 3-2, 208-210
CMA, 77, 82, 84
Cold weather stress damage, 102
Colletotrichum graminicola. *See* Anthracnose
Colonial bentgrass, Plate 2-40
Common armyworms, 233-235
Common chickweed, 38-39
Common earthworms, 303
Common grackles, 333-334
Common plantain, 51-52
Composting, 97
Condylura cristata. *See* Star-nosed moles
Conomyra insana. *See* Pyramid ants
Conservation, 190
Contact fungicides, 158, 161, 162, 163. *See also* specific types
Convolvulus arvenis. *See* Bindweed
Cool temperature brown patch. *See* Yellow patch
Copper spot, Plate 2-56, 133, 165
Coprinus kubickae. *See* Superficial fairy ring
Cornfield ants, 326
Corn speedwell, 41
Cotinis nitida. *See* Green June beetles
Cotula australis. *See* Brass buttons
Crabgrass, 4, 5-6, 84. *See also* specific types
Crabs, 175
Crambus spp. *See* Webworms
Cranberry girdlers, 264-265
Crane flies, 267-269. *See also* specific types
Crayfish, 175
Crazy ants, 326
Creeping bentgrass, Plate 2-1, Plate 2-25, Plate 2-29, Plate 2-52, 15-16, 82-83
diseases of, 88. *See also* specific types
Toronto, Plate 2-69
Creeping oxalis, 74-75
Creeping red fescue, Plate 2-52, Plate 2-54
Creeping speedwell, 64
Crematogaster spp. *See* Acrobatic ants
Crickets. *See also* specific types
mole, 181, 271-280
short-winged mole, 278-279
southern mole, 277-278
tawny mole, Plate 3-3, 275-277
Crop rotation, 188-189
Crown hydration, 102
Crown rust, 160
Cruspedonotus tibialis. *See* Ground beetles
Crustaceans, 174, 175-176. *See also* specific types
Ctenocephalides
canis. *See* Dog fleas

felis. *See* Cat fleas
Cultural practices. *See also* specific types
diseases and, 90–95
pests and, 188–189
weeds and, 2, 4
Cup changer samples, 185
Curalan. *See* Vinclozolin
Curly dock, 59, 80–81
Curvularia
geniculata. *See* Curvularia blight
lunata. *See* Curvularia blight
spp. *See* Damping-off disease
Curvularia blight, 131–132
Cutworms, 186, 227-231. *See also* specific types
black, Plate 3–17, 231-233
bronzed, 228
granulate, 229
life cycles, 229
variegated, 228
Cyclocephala
borealis. *See* Northern masked chafers
lurida. *See* Southern masked chafers
Cyclocephala spp. *See* Masked chafers
Cynodon dactylon. *See* Bermudagrass
Cyperus
esculentus. *See* Yellow nutsedge
rotundus. *See* Purple nutsedge
Cyproconazole, 163, 164

2,4-D, 1, 77, 78, 80, 81, 84, 169
Daconil 2787. *See* Chlorothalonil
Dactylis glomerata. *See* Orchardgrass
Daddy-long-legs, 175, 176
Daisies, 61, 73–74
Dallisgrass, 4, 11, 84–85
Damping-off disease, 142–144, 156
Dandelion, 49–50
Dasypus novemcinctus. *See* Ninebanded Texas armadillos
Daucus carota. *See* Wild carrot
Dazomet, 136
DCPA, 78, 80, 81, 84
Dead spot, spring. *See* Spring dead spot
Deer ticks. *See* Blacklegged ticks
Denver billbugs, 255
Dermacentor variabilis. *See* American dog ticks
Desiccants, 102, 200. *See also* specific types
Desmodium canum. *See* Beggarweed
Diagnosis of disease, 88, 98, 159, 161
Diatomaceous earth, 197, 200
Dicamba, 77, 78, 80, 81, 82, 84, 169
Dichlorprop, 77, 78, 80, 81, 85
Diclofop, 78, 82, 85
Digitaria
ischaemum. *See* Smooth crabgrass
sanguinalis. *See* Large or hairy crabgrass
Dilophus spp. *See* March flies
Dinitroanalines, 3. *See also* specific types
Diplopods, 174–175, 176. *See also* specific types
Diquat, 78, 82, 85
Disclosing solution, 184–185
Diseases, 87–170. *See also* specific types
agents causing. *See* Pathogens
bacterial, 145–146
biological control of, 95–97
cultural practices and, 90–95
diagnosis of, 88, 98, 159, 161
environmental conditions and, 90–92
fall-initiated, 106–123. *See also* specific types
fertility and, 93–94
fungicide control of. *See* Fungicides
fungicide encouragement of, 160
histories of, 90
irrigation and, 93
listing of major, 153–156
mechanical control of, 96
monitoring of, 88–90
mowing and, 92
nitrogen and, 93–94
in pest control, 190
resistance to, 91, 92
root, 155, 163. *See also* specific types
sample collection for analysis of, 98–102
season of occurrence of, 157
seedling, 142–144, 156. *See also* specific types
severity of, 91
signs of, 87, 88, 98
soil compaction and, 95
soil pH and, 95
spring-initiated, 106–123. *See also* specific types
stem, 155. *See also* specific types
summer-initiated, 123–144. *See also* specific types
symptoms of, 87, 88, 98
thresholds of, 88–90
traffic and, 95
viral, 144–145, 156, 195. *See also* specific types

winter, 102–106. *See also* specific types
Disuron, 79
Dithane M-45. *See* Mancozeb
Dithiopyr, 5, 78, 80, 82, 84, 85
Dock, 59, 80–81
Dog fennel, 43
Dog fleas, 322
Dollar spot, 93, 94, 97, 98, 117–118, 155, 158
in bentgrass, Plate 2–29, Plate 2–38, Plate 2–62
control of, 165
in creeping bentgrass, Plate 2–29
cultural practices and, 92
fungicide encouragement of, 160
in Kentucky bluegrass, Plate 2–26, Plate 2–28
in perennial ryegrass, Plate 2–27
season of occurrence of, 157
Downy mildew, 122–123. *See also* Yellow tuft disease
Draeculacephala spp. *See* Leafhoppers
Drechslera
poae. *See* Helminthosporium leaf spot
spp. *See* Damping-off disease; Helminthosporium leaf spot
Dry spot, localized, Plate 2–60, Plate 2–63, 138–139, 155
DSMA, 78, 81, 82, 85
Dyrene. *See* Anilazine

Eagle. *See* Myclobutanil
Earthworms, 174, 204, 303–304
Earwigs, 314–316. *See also* specific types
Eastern moles, 336
Echinochloa crusgalli. *See* Barnyardgrass
Elemental sulfur, 197
Eleusine indica. *See* Goosegrass
Endria inimica. *See* Painted leafhoppers
English daisies, 73–74
Entomopathogenic nematodes, 195
Environmental conditions and diseases, 90–92
Eriophyes cynodoniensis. *See* Bermudagrass mites
Erysiphe graminis. *See* Powdery mildew
Ethazole, 127, 144, 161, 164, 166
Ethofumesate, 78, 82, 84, 85
Ethylenbis-dithiocarbamate, 168
Etridiazole, 164
Euborellia annulipes. *See* Ringlegged earwigs
Eugenol, 200
Euphobia supina. *See* Spotted spurge
European chafers, Plate 3–10, 282, 289–291
European crane flies, 267–269
European earwigs, 314
Excavation, 136
Exitianus exitiosus. *See* Gray lawn leafhoppers
Exomala orientalis. *See* Oriental beetles

Fairy rings, Plate 2–57, Plate 2–58, Plate 2–60, 93, 133–137, 155
control of, 165
season of occurrence of, 157
superficial, Plate 2–60, Plate 2–62, 137–138, 155
Fall armyworms, 235–237
Fall-initiated diseases, 106–123. *See also* specific types
Fall panicum, 4, 10, 84–85
Fatty acid salts, 197
Fenarimol, 112, 113, 114–115, 121, 129, 160
modes of action of, 163
phytotoxicity of, 160
properties of, 164
scheduling of, 165, 166, 167, 168
Fennel, 43
Fenoxaprop, 5, 78, 82, 85
Fertility, 93–94
Fescue. *See also* specific types
creeping red, Plate 2–52, Plate 2–54
diseases of, 88, 89, 153–154. *See also* specific types
fine, 82–83, 88, 153–154
tall. *See* Tall fescue
Festuca arundinacea. *See* Tall fescue
Field sandbur, 84–85
Fiery skippers, 240–242
Fine fescue, 82–83, 88, 153–154
Fire ants, 325, 328–330
Fish toxicity, 204
Flag smut, 118–119, 155, 157, 166
Fleas, 322–323
Flies. *See also* specific types
European crane, 267–269
frit, 269–270
march, 324–325
syrphid, 179
tachiniid, 193, 194
Floral scents, 200
Florida pusley, 35

Fluazifop, 78, 85
Flutolanil, 105, 108, 116, 126, 131
  encouragement of disease by, 160
  modes of action of, 163
  properties of, 164
  scheduling of, 165, 166, 167
Foliar diseases, 155. *See also* specific types
Foliar/stem zone, 172
Forcipata spp. *See* Leafhoppers
Fore. *See* Mancozeb
Forficula auricularia. *See* European earwigs
Fosetyl-aluminum, 106, 116, 123, 127, 144
  incompatibilities of, 169
  modes of action of, 161, 163
  properties of, 164
  scheduling of, 166, 167
  as systemic, 161
Foxtail, 4, 8, 84–85. *See also* specific types
Freezing, 102
Frit flies, 269–270
Fuligo spp. *See* Slime mold
Fumigants, 136. *See also* specific types
Fungi, 87, 97, 98. *See also* specific types
  in pest control, 195
  red thread. *See* Red thread
  in weed control, 2
  yellow ring, Plate 2-59, 139, 155
Fungicides, 156–169. *See also* Pesticides; specific types
  applications of, 162–169
  cadmium-based, 159
  contact, 158, 161, 162, 163
  disease encouragement by, 160
  on golf courses, 159–161
  granular, 162
  incompatibilities of, 168–169
  in lawn care, 158–159
  mixing of, 168–169, 203
  modes of action of, 161, 163
  nontarget animal effects of, 204–205
  phytotoxicity of, 160
  preventive, 105, 156, 158
  properties of, 164
  resistance to, 159, 160
  site absorption, 161, 163
  sterol-inhibiting, 160, 160–161
  sulfur-containing, 160
  systemic, 161, 163
  types of, 161–162
Fungo 50. *See* Thiophanate
Fusarium
  culmorum. *See* Fusarium blight
  poae. *See* Fusarium blight
  spp. *See* Damping-off disease
Fusarium blight, 130, 155
Fusarium patch. *See* Pink snow mold

Gaeumannomyces graminis. *See* Bermudagrass decline disease; Take-all patch
Galium spp. *See* Bedstraw
Garden millipedes, 313
Garlic, 48–49, 80–81
Geocoris spp. *See* Bigeyed bugs
Geomys spp. *See* Pocket gophers
Geraniol, 200
Girdlers, 264–265. *See also* specific types
Glechoma hederacea. *See* Gound ivy
Gloeocercospora sorghi. *See* Copper spot
Glyphosate, 78, 81, 85
Good turf management, 189
Goosegrass, 2, 4, 5, 6–7, 84–85
Gophers, 338–339
Grackles, 333–334
Granular fungicides, 162. *See also* specific types
Granulate cutworms, 229
Grass loopers, 239–240
Grass webworms, 226–229
Grassworms, 239–240
Gray garden slugs, 305
Gray lawn leafhoppers, 318
Gray leaf spot, 139–140, 155, 157, 165
Gray snow mold, Plate 2-1, Plate 2-4, 97, 102, 104–105, 155
Great-tailed grackles, 333
Greenbugs, Plate 3-6, 214–216
Green foxtail, 4, 8, 84–85
Green June beetles, Plate 3-10, 282, 291–292
Green lacewings, 191, 192
Ground beetles, 191, 192, 204, 319–320. *See also* specific types
Ground ivy, 55, 80–81
Ground pearls, Plate 3-9, 279–280
Growth regulators. *See also* specific types
  insect, 199
  plant, 14, 110, 168–169
Grubs, 181, 186, 271, 280–302. *See also* specific types
  insecticides in control of, 284–286
  North American, Plate 3-10

Grubworms. *See* Grubs

Hairy chinch bugs, 178, 242–246
Hairy crabgrass, 5
Hantzschia spp., 150
Harvester ants, 325
Hawkweed, 62, 80–81
Healall, 60–61, 80–81
Heavy metal salts, 197
Helix aspersa. *See* Brown garden snails
Helminthosporium leaf spot, 93, 102, 108–110, 131, 155, 157, 166
Helminthosporium melting-out, 93, 108–110, 131, 155, 166
Henbit, 39
Herbicides, 1, 2, 3, 168. *See also* Pesticides; specific types
  alternatives to, 2
  characteristics of, 77–79
  mixing of, 203
  nontarget animal effects of, 204–205
  postemergence, 3, 5, 26, 80–81
  preemergence, 3, 4, 5, 26, 80
  tolerance of, 82–83
  trade names for, 77–79
Herpetogramma
  licarsisalis. *See* Grass webworms
  phaeopteralis. *See* Tropical sod webworms
Hieracium pratense. *See* Hawkweed
Hippodamia convergens. *See* Lady beetles
Human fleas, 322
Hunting billbugs, 260–262
Hydrolysis, 202–203
Hylephila phyleus. *See* Fiery skippers

Ice damage, 102
Ichneumonid wasps, 193
Imazaquin, 78, 80, 81, 82, 84, 85
Incompatibilities of fungicides, 168–169
Inorganic pesticides, 197. *See also* specific types
Insect attractants, 199–200. *See also* specific types
Insect growth regulators (UGRs), 199. *See also* specific types
Insecticides, 168, 169. *See also* Pesticides; specific types
  equipment for application of, 205–206
  in grub control, 284–286
  mixing of, 203
  nontarget animal effects of, 204–205
  volatilization of, 201
Insect pheromones, 184, 199–200
Insect repellents, 199. *See also* specific types
Insects, 171–342. *See also* specific types
  beneficial, 204. *See also* specific types
  biological control of, 189–195
  chemical control of, 195–200
  cultural practices and, 188–189
  goals of management of, 171–206
  identification of, 171, 173–177
  Integrated Pest Management in control of, 171, 172, 186–189, 195, 196
  invertebrate, 302–303
  leaf infesting, 206–238. *See also* specific types
  life cycles of, 171–172, 177–180
  mapping of, 185–186
  metamorphosis of, 178–179
  monitoring of, 181–186
  resistance of, 203–204
  stem infesting, 206–238, 242–270. *See also* specific types
  thatch infesting, 242–271. *See also* specific types
  turf as unique environment for, 172–173, 180–181
Integrated Pest Management (IPM)
  in insect and mite control, 171, 172, 186–189, 195, 196
  in weed control, 75–76
Invertebrate pests, 302–303. *See also* specific types
IPM. *See* Integrated Pest Management
Iprodione, 104, 105, 108, 110, 116, 122, 126, 130, 133, 144, 161
  encouragement of disease by, 160
  modes of action of, 163
  properties of, 164
  scheduling of, 165, 166, 167
Iron sulfate, 168
Irrigation, 93
Isoxaben, 78, 80, 82, 84
Ivy, 55, 80–81
Ixodes scapularis. *See* Blacklegged ticks

Japanese beetles, Plate 3–10, 271, 280, 282, 293–295
June beetles, 179, 180
  green, Plate 3–10, 282, 291–292
  May and, Plate 3–10, 300–302

Kentucky bluegrass, 13–14
  diseases of, 88, 89, 91, 92, 93, 153. *See also* specific types

dollar spot in, Plate 2-26, Plate 2-28
fairy rings in, Plate 2-57
leaf spot in, Plate 2-13
melting-out in, Plate 2-11
necrotic ring spot in, Plate 2-19, Plate 2-20
rust in, Plate 2-65
stripe smut in, Plate 2-30
summer patch in, Plate 2-48, Plate 2-49, Plate 2-50
tolerance of to herbicides, 82-83
Knawel, 80-81
Knotweed, 27-28, 80-81
Koban. *See* Ethazole
Kochia, 34
Kochia scoparia. *See* Kochia

Labidura riparia. *See* Shore earwigs
Lacewings, 191, 192
Lady beetles, 191
Lambsquarters, 31
Lamium amplexicaule. *See* Henbit
Large crabgrass, 4, 5, 84
Large patch of zoysiagrass, Plate 2-24, Plate 2-42, Plate 2-43, 97, 115-116, 155
Larger sod webworms, Plate 3-14, 221-222
Larger yellow ants, 326
Lasium alienus. *See* Cornfield ants
Latalus spp. *See* Leafhoppers
Lawn armyworms, 237-238
Lawn moths, 217-219, 223-224
Leaf blight. *See also* specific types
Ascochyta, 121-122
Cercospora, 155
Leptosphaerulina, Plate 2-35, 121-122, 155, 157
Leafhoppers, 317-319. *See also* specific types
Leaf infesting pests, 206-238. *See also* specific types
Leaf smut, 158
Leaf spot, Plate 2-13, Plate 2-14. *See also* specific types
Cercospora, 140
control of, 165, 166
gray, 139-140, 155, 157, 165
Helminthosporium, 93, 102, 108-110, 131, 155, 157, 166
red, 131
season of occurrence of, 157
Leatherjackets, 267-269
Legumes, 2
Lepdium virginicum. *See* Peppergrass
Leptosphaeria
herpotricha. *See* Spring dead spot
korrae. *See* Spring dead spot
narmari. *See* Spring dead spot
Leptosphaerulina leaf blight, Plate 2-35, 121-122, 155, 157
Leptosphaerulina trifolii. *See* Leptosphaerulina leaf blight
Lesco 4. *See* Mancozeb
Life cycles, 171-172, 177-180
of ants, 326-327, 329-330
of armadillos, 342
of armyworms, 229
of bigeyed bugs, 317
of centipedes, 313
of cicada killers, 331-332
of cutworms, 229
of earthworms, 304
of earwigs, 315
of fire ants, 329-330
of fleas, 323
of grackles, 334
of ground beetles, 320
of leafhoppers, 319
of millipedes, 313-314
of mole crickets, 272
of moles, 336-337
of pillbugs, 311
of pocket gophers, 338-339
of raccoons, 341
of redwinged blackbirds, 336
of rove beetles, 321-322
of slugs, 305
of snails, 305
of sowbugs, 311
of spiders, 307
of starlings, 334-335
of ticks, 310
Light traps, 184
Limax
flavus. *See* Tawny garden slugs
maximus. *See* Spotted garden slugs
Listronotus anthracinus. *See* Annual bluegrass weevils
Lithobius spp. *See* Centipedes
Localized dry spot, Plate 2-60, Plate 2-63, 138-139, 155
Lone star ticks, 308-309
Loopers, 239-240. *See also* specific types
Low temperature Pythium blight, 106
Lumbricus
rubellus. *See* Red earthworms
terrestris. *See* Common earthworms
Lynx. *See* Terbuconazole

Magnaporthe poae. *See* Summer patch
Maladera castanea. *See* Asiatic garden beetles
Mallow, 54, 80–81
Malva neglecta. *See* Mallow
Mancozeb, 110, 116, 126, 127, 144
  as contact fungicide, 161
  encouragement of disease by, 160
  modes of action of, 161
  properties of, 164
  scheduling of, 165, 166, 168
Manures, 97
March flies, Plate 3-19, 324–325
Margarodes meridionalis. *See* Ground pearls
Masked chafers, Plate 3-10, 282, 295–299
May and June beetles, Plate 3-10, 300–302
MCPA, 78, 81, 85
MCPP. *See* Mecoprop
Mealybugs, Plate 3-7, 250–252
Mecoprop (MCPP), 77, 78, 80, 81, 82, 84, 85, 169
Medicago
  lupulina. *See* Black medic
  minima. *See* Bur clover
Melanotus phillipsii. *See* White blight
Melting-out disease, Plate 2-11, Plate 2-12, 94, 108–110, 131, 155, 166
Mephitis mephitis. *See* Striped skunks
Mercury, 160
Metalaxyl, 123, 127, 144, 163, 164, 166, 167
Metam. *See* Metam-sodium
Metam-sodium, 136
Methyl bromide, 136
Metolachlor, 78, 82, 84
Metribuzin, 5, 78, 81, 82, 85
Metsulfuron methyl, 79, 81, 82, 85
Microbial degradation, 202
Microbial toxins, 197–198
Microdochium nivale, 102. *See also* Pink snow mold
Mildew, 93, 119, 122–123, 155, 157, 166. *See also* specific types
Millipedes, 174–175, 176, 177, 313–314
Mites, 171–342. *See also* specific types
  banks grass, 210–212
  bermudagrass, Plate 3-1, 206–208
  biological control of, 189–195
  chemical control of, 195–200
  clover, Plate 3-2, 208–210
  cultural practices and, 188–189
  goals of management of, 171–206
  identification of, 171, 173–177
  Integrated Pest Management in control of, 171, 172, 186–189, 195, 196
  invertebrate, 302–303
  leaf infesting, 206–238. *See also* specific types
  life cycles of, 171–172, 179–180
  mapping of, 185–186
  monitoring of, 181–186
  resistance of, 203–204
  spider. *See* Spider mites
  stem infesting, 206–238, 242–270. *See also* specific types
  thatch infesting, 242–271. *See also* specific types
  turf as unique environment for, 172–173, 180–181
  twospotted spider, 180
  winter grain, Plate 3-2, 212–214
Miticides, 201, 204–206. *See also* specific types
Mocap, 168
Molds. *See also* specific types
  control of, 166
  gray snow, Plate 2-1, Plate 2-4, 97, 102, 104–105, 155
  pink snow, Plate 2-1, Plate 2-2, Plate 2-3, 94, 102–104, 155
  slime, Plate 2-64, 141, 166
  snow. *See* Snow molds
Mole crickets, 181, 271–280. *See also* specific types
  short-winged, 278–279
  southern, 277–278
  tawny, Plate 3-3, 275–277
Moles, 336–338
Mollugo verticillata. *See* Carpetweed
Mollusks, 174. *See also* specific types
Monitoring
  active, 183–185
  of diseases, 88–90
  in Integrated Pest Management, 187–188
  of pests, 173, 181–186
Moss, Plate 2-72, 70, 93
Moths, 217–219, 223–224. *See also* specific types
Moth sex pheromones, 200
Mouse-ear chickweed, 53
Mowing, 92
MSMA, 78, 79, 81, 82, 84, 85
Muhlenbergia scherberi. *See* Nimblewill
Mushrooms, 134
Mycelium, 87

Myclobutanil, 163, 164

Nancozeb, 106
Napropamide, 79, 83, 84
Necrotic ring spot (NRS), Plate 2-20, Plate 2-21, 93, 94, 97, 113, 155, 157, 158, 168
Nemacur, 168
Nematodes, Plate 2-70, 87, 98, 146-150, 155, 173-174. *See also* specific types
  assays of, 99-101
  control of, 168
  entomopathogenic, 195
  management of, 149-150
  in pest control, 195
  symptoms of injury from, 147
Neocoprop, 77
Nephelodes minians. *See* Bronzed cutworms
Net-blotch disease, 109, 166
Nicotine, 198
Night crawlers. *See* Common earthworms
Nigrospora blight, 93, 132-133, 155
Nigrospora sphaerica. *See* Nigrospora blight
Nimblewill, 21
Ninebanded Texas armadillos, 341-342
Nitrogen and disease, 93-94
North American grubs, Plate 3-10
Northern masked chafers, Plate 3-10, 295-297
Nostoc spp., 150
NPV. *See* Nuclearpolyhedrosis virus
NRS. *See* Necrotic ring spot
Nuclearpolyhedrosis virus (NPV), 195
Nutsedge, 24-25, 84-85

Odonaspis ruthae. *See* Bermudagrass scales
Oils, 197
Oligonychus pratensis. *See* Banks grass mites
Onion, 48, 80-81
Oniscus spp. *See* Sowbugs
Ophiobolus patch. *See* Take-all patch
Orchardgrass, 17-18
Organic arsenicals, 5. *See also* specific types
Organic feeders, 181. *See also* specific types
Organic pesticides, 199. *See also* specific types
Organochlorines, 199. *See also* specific types
Organophosphates, 199. *See also* specific types
Oriental beetles, Plate 3-10, 282, 299-300
Oryzalin, 77, 79, 80, 82, 83, 84
Oscillatoria spp., 150
Oscinella frit. *See* Frit flies
Oxadiazon, 79, 80, 83, 84
Oxalis, 26-27, 74-75, 80-81
Oxalis
  corniculata. *See* Creeping oxalis
  stricta. *See* Oxalis
Ox-eye daisy, 61
Oxidus gracilis. *See* Garden millipedes

Painted leafhoppers, 318
Panicum, 4, 10, 84-85
Panicum dichotomiflorum. *See* Fall panicum
Parapediasia teterrella. *See* Bluegrass sod webworms
Parasites. *See also* specific types
  nematode. *See* Nematodes
  in pest control, 189, 190, 192-193
Paratrechia longicornis. *See* Crazy ants
Paspalum dilatatum. *See* Dallisgrass
Patches, 155. *See also* specific types
  brown. *See* Brown patch
  control of, 166
  cool temperature brown. *See* Yellow patch
  fungicide encouragement of, 160
  Fusarium. *See* Pink snow mold
  large, Plate 2-24, Plate 2-42, Plate 2-43, 97, 115-116, 155
  ophiobolus. *See* Take-all patch
  pink, Plate 2-9, Plate 2-10, 107-108, 155, 157, 166
  season of occurrence of, 157
  summer. *See* Summer patch
  take-all, Plate 2-15, Plate 2-16, Plate 2-17, Plate 2-18, 93, 94, 110-112, 155, 157
  yellow, Plate 2-5, Plate 2-6, 102, 105, 155, 157, 165
  zoysia. *See* Large patch of zoysiagrass
Pathogens, 87, 189, 194-195. *See also* Diseases; specific types
Pavement ants, 325
PBO. *See* Piperonyl butoxide
PCNB. *See* Pentachloronitrobenzene

Pediasia trisecta. *See* Larger sod webworm
Pendimethalin, 79, 80, 83, 84
Pennywort, 80–81
Pentachloronitrobenzene, 104, 110, 160, 165, 166, 167
Penthaleus major. *See* Winter grain mites
Peppergrass, 44
Perennial ryegrass
  diseases of, 88, 154
  dollar spot in, Plate 2–27
  fairy rings in, Plate 2–58
  large patch in, Plate 2–42, Plate 2–43
  Pythium blight in, Plate 2–46, Plate 2–47
  red thread in, Plate 2–7
  tolerance of to herbicides, 82–83
Perennial weeds, 3, 15–25. *See also* specific types
  broadleaf, 47–75. *See also* specific types
Peridroma saucia. *See* Variegated cutworms
Pesticides, 197–200. *See also* specific types
  botanical, 198–199
  calendar date applications of, 182–183
  equipment for application of, 205–206
  inorganic, 197
  nontarget animal effects of, 204–205
  organic, 199
  preventive applications of, 105, 156, 158, 181–182
  reactive applications of, 183
  synthetic organic, 199
  tank mixing of, 203
  timing of, 183–186
  toxicity of, 204–205
  use of, 200–206
  volatilization of, 201
Pest management vs. pest eradication, 186
Pests. *See also* specific types
  biological control of, 187, 189–195
  chemical control of, 195–200
  cultural practices and, 188–189
  eradication of, 186
  identification of, 171, 173–177
  insect. *See* Insects
  invertebrate, 302–303
  leaf infesting, 206–238. *See also* specific types
  life cycles of, 171–172, 177–180
  mapping of, 185–186
  mechanical control of, 189
  mite. *See* Mites
  monitoring of, 173, 181–186
  physical control of, 189
  resistance of, 203–204
  resistance to, 189
  stem infesting, 206–238, 242–270. *See also* specific types
  thatch infesting, 242–271
  turf as unique environment for, 172–173, 180–181
  vertebrate, 332–342
Petroleum oils, 197
Pheromones, 184, 199–200
Phleum pratense. *See* Timothy
Phoenician billbugs, 255
pH of soil, 95
Phyllophaga spp. *See* June beetles
Physarum spp. *See* Slime mold
Phytotoxicity of fungicides, 160
Pigweed, 30
Pillbugs, 176, 311–312. *See also* specific types
Pineappleweed, 80–81
Pine oils, 200
Pink patch, Plate 2–9, Plate 2–10, 107–108, 155, 157, 166
Pink snow mold, Plate 2–1, Plate 2–2, Plate 2–3, 94, 102–104, 155
Piperonyl butoxide (PBO), 198
Pitfall traps, 183–184
Plantago
  lanceolata. *See* Buckhorn
  major. *See* Plantain
Plantain, 2, 51–52
Plant growth regulators, 14, 110, 168–169. *See also* specific types
Poa annua. *See* Annual bluegrass
Pocket gophers, 338–339
Pogonomyrmex spp. *See* Harvester ants
Polyamia spp. *See* Leafhoppers
Polydesmus spp. *See* Millipedes
Polygonum aviculare. *See* Knotweed
Popillia japonica. *See* Japanese beetles
Porcelleo spp. *See* Pillbugs
Portulaca oleracea. *See* Purslane
Postemergence control, 3, 5, 26, 80–81, 84–85. *See also* specific types
Potassium salts, 79
Potentilla norvegica. *See* Cinquefoil
Powdery mildew, 93, 119, 155, 157, 166
Predators in pest control, 189, 190, 191–192

Preemergence control, 3, 4, 5, 26, 80, 84–85. *See also* specific types
Preventive applications of pesticides, 105, 156, 158, 181–182
Procellio spp. *See* Sowbugs
Procyon lotor. *See* Raccoons
Prodiamine, 79, 83, 84
Pronamide, 79, 83, 84, 85
Propamocarb, 127, 144, 163, 164, 166
Propiconazole, 104, 105, 112, 116, 119, 121, 129, 142
  modes of action of, 163
  phytotoxicity of, 160
  properties of, 164
  scheduling of, 165, 166, 167
Prosapia bicincta. *See* Twolined spittlebugs
ProStar. *See* Flutolanil
Prostrate pigweed, 30
Prunella vulgaris. *See* Healall
Pseudaletia unipuncta. *See* Common armyworms
Puccinia spp. *See* Rust
Pulex irritans. *See* Human fleas
Puncturevine, 32
Purple nutsedge, 24, 84–85
Purslane, 29–30
Pusley, 35
Pyramid ants, 325–326
Pyrethrin, 198
Pyrethroids, 199. *See also* specific types
Pyricularia grisea. *See* Gray leaf spot
Pythium
  aphanidermatum. *See* Pythium blight
  aristosporum. *See* Pythium root rot
  graminicola. *See* Pythium root rot
  myriotylum. *See* Pythium blight
  ultimum. *See* Pythium blight
  vanterpooli. *See* Pythium root rot
Pythium blight, 93, 98, 106, 126–128, 155, 158
  in bentgrass, Plate 2-44, Plate 2-45
  control of, 166
  cultural practices and, 92
  diagnosis of, 88
  management of, 127–128
  in perennial ryegrass, Plate 2-46, Plate 2-47
  season of occurrence of, 157
Pythium root rot, Plate 2-25, 116, 155, 157
Pythium snow blight, 102, 106, 157
Pythium spp. *See* Damping-off disease

Quackgrass, 2, 20
Quintozene, 164
Quiscalus
  major. *See* Great-tailed grackles
  mexicanus. *See* Boat-tailed grackles
  quiscula. *See* Common grackles

Raccoons, 341
Ragweed, 33
Reactive applications of pesticides, 183
Red earthworms, 303
Red fescue, Plate 2-52, Plate 2-54
Red leaf spot, 131
Red sorrel, 80–81
Red thread, Plate 2-8, 94, 98, 106, 107–108, 155, 158
  control of, 166
  in perennial ryegrass, Plate 2-7
  season of occurrence of, 157
Redwinged blackbirds, 335–336
Repellents, 199. *See also* specific types
Rescuegrass, 4, 25
Resistance
  to disease, 91, 92
  to fungicides, 159, 160
  of pests, 203–204
  to pests, 189
Rhizoctonia
  cerealis. *See* Yellow patch
  oryzae. *See* Brown patch
  solani. *See* Brown patch; Damping-off disease; Large patch
  zeae. *See* Brown patch
Rhizoctonia blight. *See* Brown patch
Rhizoctonia diseases, 88. *See also* specific types
Rhizotrogus majalis. *See* European chafers
Rhodesgrass mealybugs, Plate 3-7, 250–252
Rhodesgrass scales, 250–252
Richardia scabra. *See* Florida pusley
Ringlegged earwigs, 314
Ring spot, necrotic, Plate 2-20, 93, 94, 97, 113, 155, 157, 158, 168
Ripicephalus sanguineus. *See* Brown dog ticks
Root diseases, 155, 163. *See also* specific types
Root rot, Plate 2-25, 116, 155, 157. *See also* specific types
Rotenone, 198
Roundworms. *See* Nematodes
Rove beetles, 191, 204, 321–322. *See also* specific types
Rubigan. *See* Fenarimol

Rumex
acetosella. *See* Sheep sorrel
crispus. *See* Curly dock
Rust, Plate 2-65, Plate 2-66, 141-142, 155, 157, 160, 166. *See also* specific types
Ryania, 199
Ryegrass
diseases of, 88, 154. *See also* specific types
perennial. *See* Perennial ryegrass

Sabadilla, 198
Sample collection, 98-102
Sampling, 183-185
Sandbur, 4, 12, 84-85
Sanitation, 188
Scales, Plate 3-8, 250-252. *See also* specific types
Scalopus aquaticus. *See* Eastern moles
Scapteriscus
abbreviatus. *See* Short-winged mole crickets
borellii. *See* Southern mole crickets
vicinus. *See* Tawny mole crickets
Scarabs, 280. *See also* Beetles; Grubs; specific types
Scarab sex pheromones, 200
Schizaphis graminum. *See* Greenbugs
Sclerophthora macrospora. *See* Yellow tuft disease
Sclerotia, Plate 2-55, 87, 88
Sclerotinia homoeocarpa. *See* Dollar spot
Sclerotium rolfsii. *See* Southern blight
Scoliid wasps, 193
Scolla dubia. *See* Scoliid wasps
Scolopendra obscura. *See* Centipedes
Scorpions, 175, 176
Scutigerella immaculata. *See* Symphylans
Sedges, 3, 15-25. *See also* specific types
control of, 84-85
as indicator of correctable conditions, 2
Seedling diseases, 142-144, 156. *See also* specific types
Segmented worms, 174. *See also* specific types
Selagimella spp. *See* Moss
Sentinel. *See* Cyproconazole
Setaria
glauca. *See* Yellow foxtail
viridis. *See* Green foxtail
Sethoxydim, 79, 83, 85
Sex pheromones, 200
Sheep sorrel, 2, 56, 80-81
Shepherdspurse, 40
Shore earwigs, 314
Short-winged mole crickets, 278-279
Shrimp, 175
Siduron, 83, 84
Signs of disease, 87, 88, 98
Silica gel, 200
Simazine, 79, 80, 83, 84, 85
Site absorption fungicides, 161, 163
Skunks, 339-340
Slime molds, Plate 2-64, 141, 166
Sludges, 97
Slugs, 174, 305-306
Smaller yellow ants, 326
Smooth brome, 19
Smooth crabgrass, 4, 5, 84
Smut. *See also* specific types
control of, 166
flag, 118-119, 155, 157, 166
fungicide encouragement of, 160
leaf, 158
stripe, Plate 2-30, 118-119, 155, 157, 160, 166
Snails, 174, 305-306
Snow blight, 102, 106, 157
Snow molds, 91, 93, 102, 158. *See also* specific types
gray, Plate 2-1, Plate 2-4, 97, 102, 104-105, 155
pink, Plate 2-1, Plate 2-2, Plate 2-3, 94, 102-104, 155
season of occurrence of, 157
Sod webworms, Plate 3-15, 185, 217-219. *See also* specific types
bluegrass, 219-220
burrowing, 265-267
larger, Plate 3-14, 221-222
tropical, Plate 3-16, 224-228
Soil compaction, 95
Soil pH, 95
Solenopsis spp. *See* Fire ants
Sonchus oleraceus. *See* Sow thistle
Sorrel, 56, 80-81
Southern blight, Plate 2-55, 130-131
Southern chinch bugs, Plate 3-4, 246-248
Southern masked chafers, Plate 3-10, 297-299
Southern mole crickets, 277-278
Sowbugs, 311-312. *See also* specific types
Sow thistle, 36

Speedwell weeds, 3, 41, 63, 64, 80–81. *See also* specific types
Sphecius speciosus. *See* Cicada killers
Sphenophorus
  cicatristriatus. *See* Denver billbugs
  parvulus. *See* Bluegrass billbugs
  phoeniciensis. *See* Phoenician billbugs
  spp. *See* Billbugs
  venatus. *See* Hunting billbugs
Spider mites, 175, 180. *See also* specific types
Spiders, 174, 175, 306–307. *See also* specific types
Spilogale putorius. *See* Spotted skunks (civet cats)
Spittlebugs, Plate 3–5, 248–250
Spodoptera
  frugiperda. *See* Fall armyworms
  mauritia. *See* Lawn arymworms
  ornithogalli. *See* Yellowstriped armyworms
Spore production, 93
Spotrete. *See* Thiram
Spotted garden slugs, 305
Spotted skunks (civet cats), 339–340
Spotted spurge, 28–29
Spring dead spot, Plate 2–21, Plate 2–22, Plate 2–23, Plate 2–36, 94, 114–115, 155, 157
Spring-initiated diseases, 106–123. *See also* specific types
Spurge, 28–29, 80–81
St. Augustine decline, Plate 2–67, 144–145, 156
St. Augustinegrass, 5, 154
Stachys floridana. *See* Betony
Starlings, 334–335
Star-nosed moles, 336
Stellaria
  graminea. *See* Stitchwort
  media. *See* Common chickweed
Stem diseases, 155. *See also* specific types
Stem infesting pests, 206–238, 242–270. *See also* specific types
Stem/thatch zone, 172
Sterol-inhibiting fungicides, 160–161. *See also* specific types
Stitchwort, 66
Stress damage, 102. *See also* specific types
Striped grassworms, 239–240
Striped skunks, 339
Stripe smut, Plate 2–30, 93, 118–119, 155, 157, 160, 166
Sturnus vulgaris. *See* Starlings
Subdue. *See* Metalaxyl
Sulfur-containing fungicides, 160. *See also* specific types
Summer annual grasses, 4–12. *See also* specific types
  broadleaf, 26–37
Summer-initiated diseases, 123–144. *See also* specific types
Summer patch, 88, 94, 97, 128–129, 155, 158, 163
  in annual bluegrass, Plate 2–51, Plate 2–52, Plate 2–53
  in creeping red fescue, Plate 2–54
  cultural practices and, 93
  fungicide encouragement of, 160
  in Kentucky bluegrass, Plate 2–48, Plate 2–49, Plate 2–50
  management of, 129
  season of occurrence of, 157
Superficial fairy rings, Plate 2–60, Plate 2–62, 137–138, 155
Symphylans, 175, 177. *See also* specific types
Symptoms of disease, 87, 88, 98
Synthetic organic pesticides, 199. *See also* specific types
Syrphid flies, 179
Systemic fungicides, 161, 163. *See also* specific types

Tachinid flies, 193, 194
Take-all patch, Plate 2–15, Plate 2–16, Plate 2–17, Plate 2–18, 93, 94, 110–112, 155, 157
Tall fescue, 16–17, 84–85
  brown patch in, Plate 2–41
  diseases of, 88, 89, 154
  gray snow mold in, Plate 2–4
  large patch in, Plate 2–42
Tank mixing, 203
Tarantulas, 306–307
Taraxacum officinale. *See* Dandelion
Tawny garden slugs, 305
Tawny mole crickets, Plate 3–3, 275–277
Tehama bonifatella. *See* Western lawn moths
Terbuconazole, 163, 164
Teremec SP. *See* Chloroneb
Terraclor. *See* Pentachloronitrobenzene
Terraneb SP. *See* Chloroneb
Tersan 1991. *See* Benomyl

Tetramorium caespitum. *See* Pavement ants
Tetranychus urticae. *See* Twospotted spider mites
Texas armadillos, 341–342
Thatch, Plate 2-59, 94, 110, 155, 160, 201
Thatch infesting pests, 242–271. *See also* specific types
Thatch/soil zone, 172
Thiophanate, 113, 121, 126, 129, 133, 160, 163, 165, 167, 168
Thiophanate-ethyl, 164, 167
Thiophanate-methyl, 164, 165
Thiram, 160, 161, 164, 165
Thiramad. *See* Thiram
Thistle, 36, 57, 59–60. *See also* specific types
Thomomys spp. *See* Pocket gophers
Thresholds of diseases, 88–90
Thyme-leaf speedwell, 63
Ticks, 175, 308–311. *See also* specific types
Tillage, 189
Timing of pesticides, 183–186
Timothy, 18
Tiphia popillavora. *See* Tiphiid wasps
Tiphiid wasps, 193
Tipula
  paludosa. *See* European crane flies
  spp. *See* Crane flies
Tolerance of herbicides, 82–83
Toronto creeping bentgrass, Plate 2-69
Touche. *See* Vinclozolin
Traffic, 95
Trechispora
  alnicola. *See* Yellow ring
  spp. *See* Superficial fairy ring
Triadimefon, 104, 105, 108, 112, 116, 119, 121, 129, 131, 133, 142
  modes of action of, 163
  phytotoxicity of, 160
  properties of, 164
  scheduling of, 165, 166, 167
Triazoles, 161, 163, 168. *See also* specific types
Tribulus terrestris. *See* Puncturevine
Triclopyr, 78, 79, 81, 83
Trifluralin, 77, 82, 84
Trifolium repens. *See* White clover
Trombicula spp. *See* Chiggers
Tropical sod webworms, Plate 3-16, 224–228
Turf as unique environment, 172–173, 180–181
Twolined spittlebugs, Plate 3-5, 248–250
Twospotted lady beetles, 191
Twospotted spider mite, 175, 180
Typhula spp. *See* Gray snow mold

UGRs. *See* Insect growth regulator
Ultraviolet light degradation, 202

Vapam. *See* Metam-sodium
Variegated armyworms, 229
Variegated cutworms, 228
Velvet bentgrass, Plate 2-56
Veronica
  arvensis. *See* Corn speedwell
  filiformis. *See* Creeping speedwell
  serpyllifolia. *See* Thyme-leaf speedwell
Vertebrate pests, 332–342. *See also* specific types
Vervain, 80–81
Vinclozolin, 108, 110, 122, 126, 130, 144
  modes of action of, 161, 163
  properties of, 164
  scheduling of, 165, 166, 167
Viola spp. *See* Violet
Violet, 65, 68, 80–81
Viruses, 144–145, 156, 195. *See also* specific types
Voids, 2–3
Volatilization, 201
Vorlan. *See* Vinclozolin

Wasps, 193, 194. *See also* specific types
Wastes, 97. *See also* specific types
Weather-mediated predictive models, 185
Webworms. *See also* specific types
  bluegrass sod, 219–220
  burrowing sod, 265–267
  grass, 226–229
  larger sod, Plate 3-14, 221–222
  sod. *See* Sod webworms
  tropical sod, Plate 3-16, 224–228
Weeds, 1–85. *See also* specific types
  annual grasses. *See* Annual grasses
  biennial, 44–47
  broadleaf. *See* Broadleaf weeds
  categories of, 3
  chemical control of, 1, 3, 4
  cultural strategies in control of, 2, 4
  as indicators of correctable conditions, 2

Integrated Pest Management in control of, 75–76
perennial. *See* Perennial weeds
sedges. *See* Sedges
speedwell, 3, 41, 63, 64, 80–81
strategy in control of, 3–4
summer annual. *See* Summer annual grasses
winter annual. *See* Winter annual grasses
Weevils, Plate 3-11, 262–264. *See also* specific types
Western lawn moths, 223–224
Wetting agents, 203. *See also* specific types
White blight, 140–141, 155
White clover, 50–51
Wild carrot, 46, 80–81
Wild garlic, 48–49, 80–81
Wild onion, 48, 80–81
Wild violet, 68
Wilt. *See also* specific types
Wilt, bacterial, Plate 2-68, Plate 2-69, 145–146, 156
Winter annual grasses, 12–15. *See also* specific types
broadleaf, 37–44
Winter diseases, 102–106. *See also* specific types
Winter grain mites, Plate 3-2, 212–214
Winthemia quadripustulata. See Tachiniid flies
Witchesbrooming, Plate 3-1
Wolf spiders, 175
Woodsorrel, 74, 80–81

Xanthomona campestris. See Bacterial wilt

Yarrow, 69–70, 80–81
Yellow ants, 326
Yellow foxtail, 4, 8, 84–85
Yellow nutsedge, 24, 84–85
Yellow patch, Plate 2-5, Plate 2-6, 102, 105, 155, 157, 165
Yellow ring, Plate 2-59, 139, 155
Yellow rocket, 45
Yellowstriped armyworms, 228, 235
Yellow tuft, Plate 2-37, 122–123, 155, 157, 160. *See also* Downy mildew
Yellow woodsorrel, 74, 80–81

Zineb, 162
Zoysiagrass, Plate 2-66, 23
diseases of, 97, 115–116, 154. *See also* specific types
herbicide tolerance in, 82–83
large patch of, Plate 2-24, Plate 2-42, Plate 2-43, 97, 115–116, 155
Zoysia japonica. See Zoysiagrass
Zoysia patch. *See* Large patch of zoysiagrass